ALL ■ IN ■ ONE

CompTIA

Project+™ Certification

EXAM GUIDE

(Exam PK0-005)

ABOUT THE AUTHOR

Joseph Phillips, PMP, PMI-ACP, PSM, CompTIA Project+, Certified Technical Trainer+, is the Director of Education for Instructing.com, LLC. He has managed and consulted on projects for industries including technical, pharmaceutical, insurance, manufacturing, and architectural, among others. Phillips has served as a project management consultant for organizations creating project offices, maturity models, and best practice standardization.

As a leader in adult education, Phillips has taught on successful implementation of project management methodologies, information technology project management, risk management, and other topics. Phillips has taught for Columbia College, University of Chicago, and Ball State University, among others. He is a Certified Technical Trainer and has taught over 650,000 professionals on Udemy.com, Instructing.com, and in-person seminars. He has contributed as an author or editor to more than 35 books on technology, careers, and project management.

Phillips is a member of the Project Management Institute and is active in local project management chapters. He has spoken on project management, project management certifications, and project methodologies at numerous trade shows, PMI chapter meetings, and employee conferences in the United States and in Europe. When not writing, teaching, or consulting, he can be found behind a camera or on the golf course. You can contact Phillips through https://instructing.com.

About the Technical Editor

Mike Griffiths is a project management author, trainer, and consultant. He helps bridge the agile to the traditional chasm by providing examples, success stories, and cautionary tales. Mike helped create one of the first agile approaches, DSDM, in 1994 and has been using a variety of agile techniques ever since. He served on the board of the Agile Alliance and the Agile Leadership Network and helps organize agile conferences worldwide. Mike is also active in the PMI community and contributed to *PMBOK Guide, Third Edition* through *Seventh Edition.* He was chair of the group that wrote the *Agile Practice Guide* and co-chair of the *PMBOK Guide, Seventh Edition.*

ALL ■ IN ■ ONE

CompTIA

Project+™ Certification

EXAM GUIDE

(Exam PK0-005)

Joseph Phillips

New York Chicago San Francisco
Athens London Madrid Mexico City
Milan New Delhi Singapore Sydney Toronto

McGraw Hill books are available at special quantity discounts to use as premiums and sales promotions, or for use in corporate training programs. To contact a representative, please visit the Contact Us pages at www.mhprofessional.com.

CompTIA Project+™ Certification All-in-One Exam Guide (Exam PK0-005)

1 2 3 4 5 6 7 8 9 LCR 27 26 25 24 23

Library of Congress Control Number:

ISBN 978-1-264-85131-7
MHID 1-264-85131-6

Sponsoring Editor
Lisa McClain

Editorial Supervisor
Janet Walden

Project Manager
Nitesh Sharma, KnowledgeWorks Global Ltd.

Acquisitions Coordinator
Caitlin Cromley-Linn

Technical Editor
Mike Griffiths

Copy Editor
William McManus

Proofreader
Rachel Fogelberg

Indexer
Claire Splan

Production Supervisor
Thomas Somers

Composition
KnowledgeWorks Global Ltd.

Illustration
KnowledgeWorks Global Ltd.

Art Director, Cover
Jeff Weeks

I dedicate this book to my mom,
Virginia Ruth Phillips.
She's managed five boys
and all their projects.

CONTENTS AT A GLANCE

CONTENTS

ACKNOWLEDGMENTS

Books, like projects, are never done alone.

I'd like to thank Mike Griffiths for keeping me on track and focused on the CompTIA requirements, objectives, and test-centric ideas. A big thank you goes to my editor and friend, Lisa McClain, for her patience, guidance, conversation, and overall support for this book. Thank you to Caitlin Cromley-Linn for her keen organizational skills, attention to detail, and ever-present communication and support. I appreciate your coaching and patience. Bill McManus—thank you for tightening my writing, clarifying my thoughts, and helping me to be a better writer. Thanks also to Nitesh Sharma at KnowledgeWorks Global Ltd. and the talented people in the KGL production department for all of their hard work.

Thank you to my friends and colleagues for their encouragement as this book was created: Mark and Catherine Kudlacik, Don Kuhnle, Greg and Mary Huebner, Bill Tribble, Marvin Lloyd Hoffman, Fred and Carin McBroom, my golf coaches Pauly Sinklair Jr., Levi Root, and Nick Stine, and my best friends and brothers who taught me all the troublesome ways: Steve, Mark, Sam, and Ben. Thank you to my team: Amanda Chambers, Kristin Mitchell, and Angela Lemmons: you all keep me organized, on-task, and make my life so much better. A special thank you for listening and coaching to my dear friend Alison White-Mazza.

INTRODUCTION

Managing a project is not unlike directing a movie, coaching a major league baseball team, or flying the space shuttle around the moon. Of course, if you were directing a movie, you'd be working with superstars. If you were coaching a major league team, you might win the pennant. And if you were flying the space shuttle, you'd have a great view. But with project management, you get to experience some of the same thrills I'm sure directors, coaches, and astronauts experience.

Relax. This book will help you become the superior project manager you've dreamed of becoming. *CompTIA Project+ Certification All-in-One Exam Guide (Exam PK0-005)* shows you how to get started, get funding, and get the project done. You'll discover advanced project management techniques, the mechanics of project management, and inspiration to keep moving toward the end result of any technical project. I'll explain how you can direct your team to work together and independently. I'll show you how you can motivate team members, get management fired up about your project, and keep yourself from burning out. This book takes you from project management basics to advanced concepts on such topics as creating budgets, devising work breakdown structures, and sustaining an exciting environment that will guarantee your success over and over.

As a project manager, you'll be challenged and forced to think on your feet, and you'll learn how to lead people rather than just manage them. Project management is a wonderful life experience that will stretch your brain and abilities further than you ever thought possible. Some people love project management so much they've dedicated their careers to it. These professionals love the exhilaration of finding a solution to a seemingly impossible predicament. They love the nirvana of resolving disagreements between coworkers—and watching their team become tight as family. They thrill over each success en route to the victory of completing the project on track and on budget.

My hope is that you'll become one of these people and that I can help get you there. This book is written based on my experiences as a project manager. How I wish a book of this caliber were available when I started my career! Fortunately for you, it's here now. You can (and I hope you do) read the book from cover to cover. Or, if you really want to, you can skip from chapter to chapter. Heck, read it backward if it helps you! Regardless of your reading tactics, the best way to learn is by doing. Do yourself a big favor and complete the exercises at the end of each chapter—they'll help to reinforce what you've read. If you're new to project management, try to discuss some of the issues in this book with an accomplished project manager. Once you've finished reading the book, teach someone else what you've learned. After all, teaching is just learning twice.

The CompTIA Project+ PK0-005 Certification Exam Objectives are what you want to focus on as you prepare to pass the exam. That's right—*pass* the exam, not just take the exam. The CompTIA Project+ objectives are all covered throughout this book, but it's a good idea to read over the objectives and make certain that you have a good understanding of all the terms and activities you'll be tested on for the exam.

The CompTIA Project+ PK0-005 exam will test you on waterfall and adaptive project management approaches, so be familiar with both.

Finally, if you'd like to discuss any of the topics in this book, feel free to drop me an e-mail. I try to respond to as many as possible. You can reach me at cs@instructing.com. I wish you all the best in your career and your endeavors as a project manager.

About the Exam

The CompTIA Project+ PK0-005 exam is based both on what you will learn in this book and your experience as a project manager. Prior to taking the exam, CompTIA recommends that you have at least 6–12 months of project management experience and that you have a good grasp on the terms, concepts, and processes of project management. In this book, I've identified all these concepts, but now it's up to you to put in the final efforts toward studying these objectives. Be confident! You can do this!

The skills and knowledge measured by this exam are derived from an industry-wide job task analysis and validated through an industry-wide survey. The results of this survey were used in weighing the domains and ensuring that the weighting is representative of the relative importance of the content. The exam is in a conventional linear format. There are, as of this writing, up to 90 multiple-choice and performance-based questions on the exam, and candidates have 90 minutes to complete them. The passing score for the PK0-005 exam is 710 on a scale of 100–900.

The following table lists the domains measured by this examination and the extent to which they are represented:

Domain	Percent of Examination
1.0 Project Management Concepts	33 percent
2.0 Project Life Cycle Phases	30 percent
3.0 Tools and Documentation	19 percent
4.0 Basics of IT and Governance	18 percent

Types of Questions

The candidate selects the option(s) that best completes the statement or answers the question from four or more response options. Distracters, or wrong answers, are response options that a candidate with incomplete knowledge or skills is likely to choose, given these choices are generally plausible responses for the content area. Test item formats used in this exam are as follows:

- **Multiple choice** The candidate selects one option that best answers the question or completes a statement.
- **Multiple response** The candidate selects more than one option that best answers the question or completes a statement.
- **Performance based** The candidate demonstrates practical, hands-on knowledge in a simulated setting.

You can find additional exam-specific review information in Appendix B.

The Objective Map

The objective map included in Appendix A has been constructed to help you cross-reference the official exam objectives from CompTIA with the relevant coverage in the book. References have been provided for the exam objectives exactly as CompTIA has presented them, with the chapter that covers each objective. The objectives in Appendix A are current at the time of this publication, but you should always refer to the Computing Technology Industry Association website at https://www.comptia.org/certifications to obtain the most current version of the exam objectives, as they are subject to change.

About the Online Content

This book includes access to the TotalTester Online practice exam software that provides you with a simulation of the CompTIA Project+ PK0-005 exam. In addition to the TotalTester access, other online resources, created by the author of this book, include video clips providing detailed examples of key certification objectives, as well as supplemental worksheets and templates (as Microsoft Word, Excel, and PowerPoint files) that are related to specified chapter exercises. See Appendix C for more information and how to access the online resources.

CHAPTER 1

Launching a Technology Project

This chapter covers the following topics:

- Defining the project management framework
- Working with predictive projects
- Introducing agile project management
- Developing the project charter
- Gathering project information
- Defining the project requirements
- Establishing the completion date

You are a project manager and you want to earn the CompTIA Project+ certification. First, you can do this! Many others have done this and so can you. Second, you know that managing an IT project is different from managing any other type of project that you may have worked on in the past. In the world of information technology, we face attacks on all fronts: ever-changing business needs, hardware compatibility, software glitches, security holes, and network bandwidth, not to mention careers, attitudes, and office politics. To pass the CompTIA Project+ exam successfully, you can rely in part on your experience to answer many of the exam questions, but you'll also need to know some facts, processes, and activities that a project manager should do in a formal project management approach, which are covered in this book.

IT projects are overwhelmingly the most common project type for most organizations. IT projects are also the most challenging and exciting place to be in a company. The IT projects that you manage will affect the entire organization, will have an impact on its success, and can boost your career, confidence, and life to the next level.

IT project management can be as exciting as a whitewater rafting excursion or as painful as a root canal; the decision is yours. What makes the difference between excitement and a sore jaw? Many things do: leadership, know-how, motivation, and, among other things, a clear vision of what each project will produce, what it will cost, and when it will end.

This first chapter will help you build a strong foundation for managing successful IT projects. Like anything else in the world, project management requires adequate planning, determination, and vision for success. Ready to start this journey? Let's go!

VIDEO For a more detailed explanation, watch the *Launching a Project* video now.

Defining the Project Management Framework

A project management framework is a structure the project exists in. The project management framework is the skeleton of the project and how the project operates. It defines the approach and expectations for how the project will operate. While every organization can have different rules and standard operating procedures (SOPs) for how projects work in their environment, there are some generally accepted practices when it comes to project management. The CompTIA Project+ exam will test you on the common, widely accepted practices of IT project management, not the nuanced variations of project management.

Everyone talks about projects: data projects, hardware replacement projects, software development projects, and a host of other activities that have the project label smacked onto them. But are all these activities really projects? It's important to define and understand clearly what a project is and is not.

Projects are temporary; they do not, thankfully, last forever. Another term, *operations*, describes the ongoing core business of an organization. Operations are the day-to-day tasks, business focus, and purpose of an organization; they're what companies do. Projects are unique endeavors that don't fit into the day-to-day model and activities of an organization. Projects are special undertakings that create unique products, services, and conditions.

A *project*, technically, is a temporary endeavor to create a unique product or service. A project is an undertaking outside of the normal operations of an entity. For example, you might roll out a new application, install new monitors, create a new portion of a website, or establish a new call center for application support. In some organizations, such as those composed of application developers or consultants, or IT integration companies, everything they do is a project, because they complete projects for other organizations. Consider a company that creates custom applications for other organizations. Its operation is an ongoing series of projects. The organization that completes the project work is the *performing organization*.

It's not unusual in the IT world to encounter companies that perform projects for other organizations. Your company might even be one of those entities, or it might purchase goods and services from a company that completes projects for it. An organization whose main income is generated by completing projects for others might be referenced as a company that does management by projects. Even in these companies, however, a distinction still exists between operations and projects, although the distinction is becoming less clear in companies that are adopting the emerging trend of merging project teams with operations to avoid the risk of knowledge loss at project handover. This concept is incorporated into the *product life cycle*, which describes the creation, support, and eventual sunsetting of a product.

Exploring Project Characteristics

How do you know you're managing a project rather than an operations task? Projects always have a foreseeable conclusion. Operations, on the other hand, have no end in sight and will continue to go on as long as the organization exists. So, the task of maintaining hardware or keeping software current is not a project—it's an operational activity. Operational activities are ongoing and projects are not. Operational activities support the organization and should have management, a budget, and their own set of processes and controls that are totally different from those of a project.

In addition to projects being temporary, they are unique. Every project is different: different goals, different stakeholders, different risks, and so on. No two projects are exactly alike—just like snowflakes, thumbprints, and IP addresses. Although organizations may carry out similar projects, such as building houses or designing applications, each project will always be unique. Projects are not assembly lines; they are unique and special.

EXAM TIP Projects don't last forever or maintain things. Projects are temporary, but the end result of the project, the deliverable, will likely last beyond the project. For the CompTIA Project+ exam, know that projects are temporary while operations are ongoing. You'll likely have to identify the difference between projects and operations.

Projects are led by *project managers (PMs)*, individuals who are responsible for the coordination of the project events, the orchestration of team members and activities, and the management of all facets of the project. All these facets, such as scope and costs, are integrated with one another. What the project manager does in one portion of the project has a direct effect on the other portions of the project. For example, if the project manager does a poor job in quality control, that will directly affect the risk management activities of the project. So the project manager is a coordinator not just of the people, but also of the processes that help achieve the results of the project.

Working with Project Management Entities

In many organizations, the project is a special activity in which people are assigned to the project and led by the project manager; when the project is done, the team disbands and everyone goes back to their day-to-day work. Nothing is wrong with that approach at all; in fact, that's pretty common in many organizations. There are other ways to structure projects than the simple (and clean) stand-alone project approach, especially in larger organizations, where politics, goals, and outcomes can be complex. In these types of organizations, a project may not live in its own bubble, but may be part of a larger structure with its own rules and procedures that supersede and direct the project manager's activities.

Identifying Programs and Projects

A program is the most common structure that projects live within. A *program* is a collection of related projects all working together toward a common goal. A program is led by a *program manager*, who orchestrates the activities of the projects, through the

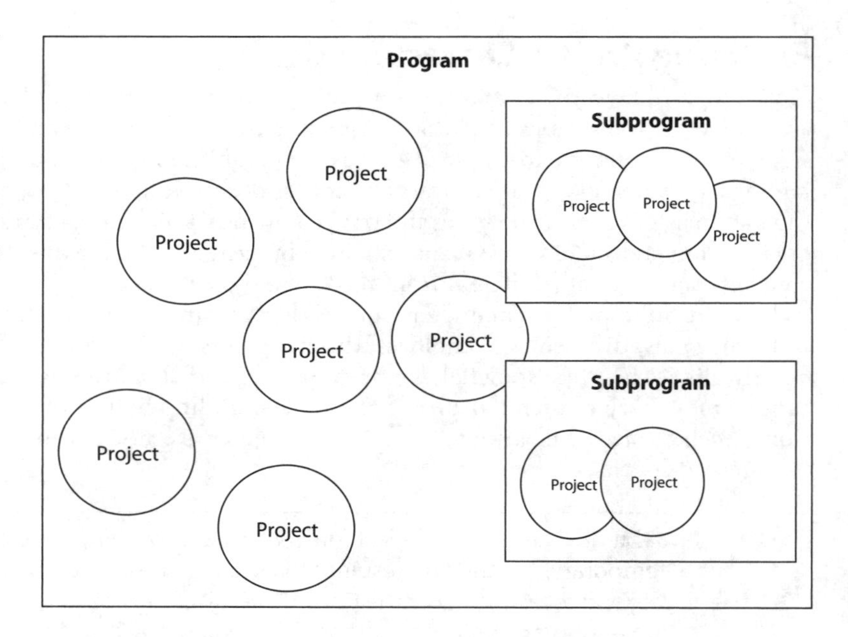

Figure 1-1 Programs are a collection of projects and subprograms working toward a common goal.

project managers, to achieve benefits that wouldn't be realized if all the projects acted independently of one another. The perfect example of a program is the construction of a skyscraper. Think of all the different types of projects that could happen in the construction of a skyscraper: concrete, framing and metalworking, plumbing, electrical, glass windows, and the list could go on. Each one of these types of major components could easily be a project. The program manager coordinates the projects so that they work together efficiently to save time, costs, and frustration.

Programs almost always are a collection of related projects, but a program could have a subprogram. In Figure 1-1, you can see how a program could be structured with subprograms and projects. Each program is led by a program manager, and each project is led by a project manager. Each project manager would report to her respective program manager, and each subprogram manager would in turn report to the program manager of the entire program. Note that it's possible to mix projects and subprograms within one program entity.

Reporting to a Project Management Office

A *project management office (PMO)* provides centralized management, support, and guidance for all projects within an organization. Rather than each department or line of business managing projects with its own approach and methodology, the PMO centralizes the project management approach for the organization. This ensures consistency of the practices, tools, reporting, and methodologies project managers use within the organization. It also helps to set expectations for stakeholders and provides consistency throughout all projects. The PMO can establish the formal project management framework for the organization that all project managers will be expected to perform within.

The PMO can also help the project managers within the organization better communicate with one another, share resources, provide training, and ensure that compliance is happening throughout the projects. Having a predetermined approach to project management that all project managers adhere to is ideal when many projects are in motion. The standardized approach helps everyone act the same way, provides consistency of practice, and ensures that the same processes, forms, software, tools, and techniques are being used throughout.

There are three types of PMOs:

- **Directive** The PMO directly manages the project for others. The PMO directs the entire project. You could say the directive PMO owns the project.
- **Controlling** The PMO controls the project through compliance with standards, a strict framework, methodologies, forms, templates, and the mechanics of project management.
- **Supportive** The PMO is more of a consultant to the project managers within the organization. The supportive PMO helps by providing templates, consultation, and lessons learned.

A PMO can be created for an entire organization, or, more likely, within each department or line of business. For example, the sales, marketing, IT, and human resources departments within the same company could each have a unique PMO that is implemented differently from the PMOs of other departments within the company.

Respecting the Organization Portfolio

A *portfolio* is a way to describe a company's collection of investments. Just like you might have a stock portfolio or an investment portfolio, when it comes to projects and programs, an organization's portfolio describes the collection of projects and programs it has invested in or will invest in. The items within the portfolio are seen as investments, so there's a financial aspect to the cost of each investment, the expected return on investment (ROI), and a business case for why it has been selected. The reason why a project is launched is the business value of doing the project. Projects aim to increase revenue or reduce cost.

Items within the portfolio are usually just projects and programs. Each project or program is reviewed by a portfolio committee or steering committee. This is a group of executives that reviews each proposed investment and determines the priority of the project or program to happen, its ROI, the risk of the investment, and other factors. Throughout the year, this committee will meet and review the status of the projects and programs. In some instances, the committee may determine not to invest in a project or program for financial reasons, technological advances that cause items to be no longer needed, or changes in the marketplace. The review is an opportunity to see that the items are doing well, or if they are not performing well, there may be a discussion on cancelling the investment.

Working with Predictive Projects

A predictive project management approach means that you and your project team predict what should happen in the project. Predictive projects involve deep planning before the work begins, and the project manager and the project team know, or at least predict, what should happen in the project. Before you hop into the launch of a predictive project, it's paramount that you understand the life cycle of project management.

EXAM TIP The CompTIA exam objectives use the term waterfall projects. Traditional project management flows from one phase of the project to the next, like a waterfall cascades from the top to the bottom. Frankly, the term is a little dated and is more commonly called predictive project management now. Predictive, like waterfall, tries to predict everything that'll happen in the project by defining all of the phases. When you see waterfall on your exam, and in this book, know that it's a predictive approach to project management.

Predictive projects move through a logical progression of activities to reach the project closing. You could examine a project in construction, health care, manufacturing, or technology, and you'd see the same set of project management processes that move the project forward. The framework that all projects share is called the *project management life cycle*—it's universal to all predictive projects. The project management life cycle describes the evolution of project process groups that will move a project from initiation to closure. Figure 1-2 captures the project management life cycle and shows how projects use different process groups to move the project toward its closing.

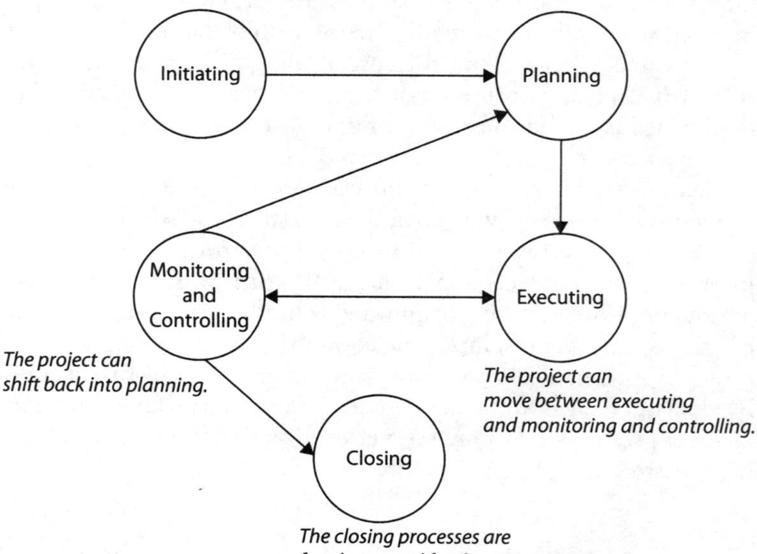

Figure 1-2 The project management life cycle uses process groups to move the project forward.

You might hear the terms "project life cycle" and "project management life cycle" used interchangeably. Technically, however, these are not the same thing. The project management life cycle is universal to all predictive projects and consists of the five process groups: initiating, planning, executing, monitoring and controlling, and closing. A project life cycle describes the unique phases of a project that are specific to the discipline and nature of the project. For example, you would not have the same phases in a construction project that you'd experience in a software development project. The phases of the project compose the project's life cycle, whereas all predictive projects use the project management life cycle that's composed of the process groups.

EXAM TIP The Project+ exam, however, calls process groups phases. And while this isn't what most organizations refer to them as, be aware of this nuance for your exam. Phases are unique to the type of work you're doing, such as construction, software development, or health care.

Initiating the Project

Project initiation is the official launch of the project. Initiation is based on identified business needs that justify the expense, risk, and allotment of resources for the project to exist. It's important for IT project managers to keep the idea of the business need in mind throughout the project. Companies don't launch projects because of cool technology or fast gadgets and gizmos or to be on the bleeding edge of technology—there must be a financial reason behind the project initiation. The business need is linked to the organization's strategies and tactics, goals and mission, and responsibility to its shareholders, owners, and customers. Your CompTIA exam will address the business need identification as part of the pre-project discovery phase.

I'll dive into project initiation more in this chapter, but for now, know that the initiating process group is responsible for creating the project charter and identifying the project stakeholders. The *project charter* is the official document that authorizes the project manager and the project to exist within the organization. The *project stakeholders* are all the people and organizations that are affected by the existence of the project and the project's outcome. If you're the project manager, you're a stakeholder. More on this in just a bit—I promise.

Planning the Project

Good projects need good plans. You, the project team, and many of your stakeholders will need to know where your project is going and how you plan on getting there. Planning is an iterative project process group that communicates the intent of the project manager. It shows which processes will be used in the project, how the project work will be executed, how you'll control the project work, and, finally, how you'll close down phases and the project at its end. Planning requires time, resources, and often a budget for testing, experimenting, and learning.

The primary result of the planning process group is the project management plan. This document is a collection of smaller plans that address different areas of the project. In Chapter 2, I'll go into the details of each of these project plans, but for now, here's a quick overview of what the planning processes help the project manager create:

- Scope management plan
- Scope baseline
- Change management plan
- Configuration management plan
- Requirements management plan
- Cost management plan
- Cost performance baseline
- Schedule management plan
- Schedule baseline
- Quality assurance plan
- Process improvement plan
- Resource management plan
- Communications management plan
- Risk management plan
- Procurement management plan
- Stakeholder engagement plan
- Transition plan/release plan

Some project documents, forms, and checklists can also go into this plan, but these are the headlines. Many of these plans don't have to be created from scratch each time—that'd be a pain. You can adapt previous, similar project plans as templates for your current projects to save time and effort and to use the benefit of historical information during planning. Planning, I want to stress, is an iterative activity. You'll come back to planning over and over throughout the project; planning is not a one-time activity.

TIP You don't have to create all of these plans and documents for every project. As a general project management rule, the bigger the project the more detail you'll need. You can also leverage older, similar project plans for your current project by adapting the older plans to fit the newer project. There's no need to work hard when you can work smart.

Executing the Project

Here's the heart of a predictive project: getting the work done. After being presented with your approved project, your project team goes about the business of getting the project work done and creating key results. Project execution is unique to each discipline and is led and directed by the project manager. This is also the process group where members of your project team will spend the bulk of their time and effort and where the project will cost the bulk of your budget. It's the heart of the project's mission: to create the product or service the stakeholders are expecting.

Project execution includes the quality assurance process, as the project team must create the project work correctly, ideally the first time. It's almost always more cost effective to do the work right the first time than to pay for it to be fixed later. In IT, simple mistakes can mushroom into costly wastes of time and materials. I'll talk all about quality and the IT projects in Chapter 11.

It is also in the project execution process group that you'll acquire, develop, and manage the project team. It's a fine line between managing your project team and leading the project team. Management is really all about key results: you want your project team to get their work done as planned, on time, and according to budget. You want your team to be as committed to the project work as you are. Good project management balances management with leadership. Leadership is about aligning, motivating, and directing your project team.

The final process in execution is linked to the costs of your project: *procurement*. You'll need to understand the procurement process, how contracts work, and the rules and policies your organization has established regarding the procurement process. Most IT projects need to purchase resources—that is, materials such as software and hardware—to satisfy the requirements of the stakeholders. Conducting the procurements according to the procurement management plan can be a time-consuming process, and when time is of the essence, that can cost your project.

Monitoring and Controlling the Project

In tandem with project execution is the monitoring and controlling process group. This set of processes ensures that the project work your team is doing is being completed accurately and according to plan. If there are problems, issues, or risks, then the project shifts back to project planning to figure the stuff out before moving back into execution. Monitoring and controlling the project is based on your project plans, the work of the project team, and shifting conditions within the project.

You'll manage scope, time, and cost changes with the monitoring and controlling processes. It's also in this process group that you'll work with the project stakeholders to verify that the project scope has met their requirements so that they'll accept the project deliverables the project team has created for them. Scope verification is an inspection-driven process that leads to acceptance decisions for the project.

Another inspection-driven process that's done without the stakeholders is quality control. Quality control involves you and the project team inspecting the project work to confirm that it's done correctly before the stakeholders look at what you've created. Quality control is all about your team keeping mistakes out of the customers' hands. This is actually a great example of how project execution and monitoring and controlling work together. Recall that quality assurance is about doing the project work correctly the first time. Quality control is about proving that the work was done correctly—and if it was not, the team takes corrective actions to fix the errors.

Monitoring and controlling also provides communication for reporting the overall performance of the project, the performance of key project deliverables, and information on project specifics, such as the time, cost, and risk portions of the project. Monitoring and controlling also requires that the project manager oversee and administer the procurement agreements with the project vendors.

Closing the Project

I'll address the project closure in detail in Chapter 12, but it's important to address project closing at the beginning of the project. Because projects are temporary, the project manager, project team, and other key stakeholders all need to be in agreement as to where the project is going. You'll need to define indicators that signal the project is complete. Because technology can change so quickly and frequently, it is vital that you define what constitutes the project closure. You don't want a project that drones on and on because of loosely defined requirements.

The closing process group allows project phases and the project as a whole to be closed. Some documentation, final reports, and communications happen in the final activities of the project. All of the project information should be archived for future usage—sometimes called *organizational process assets*. Basically, the work you've done in your project can be used for supporting the solution you've created, or other project managers can use your project files to help their projects.

The closing process group also includes the close procurement process. Contracts will define how the relationship between the buyer and the seller should end. This includes post-delivery support, warranties, inspections, and payments. When it comes to closing out the procurement, your company may require a procurement audit to determine how and where the project monies were spent, what was purchased, and that all the invoices and contracts are complete.

Introducing Agile Project Management

Agile project management focuses more on quick results and less on predicting and planning what should happen in a project. Agile project management acknowledges that predicting how everything will go in a project is really difficult, if not impossible, and instead embraces the idea of short bursts of work, changing requirements, and delivering the most important parts of a project first to the customers. Agile project management is very different than predictive project management and is most suited for software development projects.

If you're building a house, you really do have to predict what you're building. You'll work with architects and engineers to predict what the end result of the house should look like. You must do lots of planning for all the different parts of the house. And then the physical labor of building the house, the inspections for compliance, and you can see the progress of the construction. That's a great example of a predictive project: you plan, execute, control, and can see the work as the project moves toward closing.

Now think of a software development project. You have to plan for the big picture, but whereas you can easily see that an electrician is actively working on your house project, you can't simply look at a programmer and know whether they are thinking of a solution for your software development project or thinking of what they'll have for dinner.

Agile projects are typically suited for knowledge work projects, where the bulk of the work happens through mental efforts, creativity, and trial and error. Industrial work is what you'll often find in a predictive project—where you can see the effort taking place. One approach is not better than the other, it's just a different project management approach depending on the type of work the project is completing.

Exploring Agile Projects

There are several different frameworks for agile projects, but they all have one thing in common: change is welcome. Change in an agile project, which is almost always a knowledge-work project, is much easier to implement than in a predictive, industrial-work project. In other words, it's much easier to add a dialog box in a piece of software than it is to add a swimming pool inside a house. Agile projects welcome change at any time in the project.

Because agile projects welcome and expect change, they operate in short bursts of planning and work. Agile projects begin with a big list of features and functions that should be included in the project. That big ol' list is called the *product backlog* and it includes everything the customer could possibly want as a finished project. The product owner will examine the product backlog and prioritize the requirements from most important to least important. The stuff at the top, which is the most important, is what the development team will create first.

This is important: the prioritized backlog orders the requirements from most important to least important based on the customer's business value. Business value is why the project is being done. Business value is the profitability, the cost saving, the opportunity, the heart of the project's existence. In agile projects, we say that we eat our dessert first, meaning we get to create the most exciting things for the customer right away. For both the CompTIA Project+ exam and your agile projects, focus on the business value to see what's most important to the customer and why the project has been launched.

Let's go back to the predictive house project. The customer gets the business value only when the entire house is done. Typically, your customer can't move into the house until it's entirely done and they have a certificate of occupancy from the inspectors. Nobody wants to live in a house that's still under construction. In an agile project, however, you don't have to complete the entire piece of software for the customer to begin using what you've created. Business value can be realized while the software project continues.

You've probably visited websites or purchased early released software that has some significant business value to you. And then a bit later, the website and software is updated with more features. And then again later you'll see additional releases. That's a great example of an agile project. The whole project isn't done, but you, the customer, can start using the most important features of the software. You get to eat your dessert first.

EXAM TIP There are many different frameworks for agile project management, such as Scrum, Kanban, Lean, and eXtreme Programming. Right now, we're talking about agile at a conceptual level. The CompTIA Project+ exam will test you on predictive and agile project management. There are only so many exam questions and so many areas of project management. In other words, CompTIA will ask a wide variety of questions, but can't focus too intensely on any one area of project management and agile. Know the big concepts first and then focus on the smaller details later.

Comparing Iterative and Incremental Approaches in Agile

There's one additional big concept you need to know for agile project management: incremental and iterative approaches. It's really pretty straightforward, but these terms get confused by many in our industry:

- **Incremental** You and your project team create increments of a deliverable ready for the customer to use.
- **Iterative** You and your project team repeat the development of the product over and over until it's ready for use by the customer.

Imagine your customer wants a software solution that lets employees clock into work, track the different tasks they complete, reserve time for meetings, and even schedule their holidays and vacation days. You and the customer identify lots of features to include in this time management software. The customer insists that the most important feature of the software is the capability for employees to clock in and out of work; the other features are great, but not as important. Your team creates a software version that attacks the most important feature first—clocking in and out of work—and the customer can begin using the software in its limited initial version right away. The project team will continue to work on the other prioritized features and append them to the software in new versions and releases. Each release of the software with new features is an increment of the final product.

Now imagine this same project, but this time the customer wants everything complete and polished before any releases happen. So you and the team will go through rounds, or iterations, of creating, perfecting, and polishing the software until there is a wholly complete and approved project for the customer. It's like you're sketching a picture, then painting over the sketched lines, then adjusting some colors, and then doing the final touches, and then framing the artwork ready to be displayed. It's the same product, but there are lots of steps and iterations to achieve the final completed work.

Technically, a project can be both iterative and incremental. For example, the customer in our imaginary time-keeping software could have you move through iterations of product design, build, and perfections and then release a portion of the project for utilization. Then you'd repeat the work in other areas of the software. You're iteratively creating and perfect the software, but releasing it in big increments for the end users.

Developing the Project Charter

In order to have a project, you need a project charter. A project charter is a document that comes from someone in the organization that has authority over all the resources used and affected by the project. The project charter, which is the first process in project initiation, sets the high-level objectives of the project and gives the project manager authority to manage the project and its resources. Resources are people and physical things, like equipment, materials, and facilities the project will need. The project charter should be written and signed by the project sponsor, project steering committee, project management office, or any entity that has authority over all resources to be used in the project. The project manager might write or help write the charter, but it's signed by someone with organizational authority.

According to Moore's Law, a principle authored by Intel co-founder and Chairman Emeritus Gordon Moore, the processor chip speed of technology doubles every 18 months. Today that law is practically defunct, because processor speeds are no longer doubling at the same rate. The basic concept of Moore's Law, however, is often applied to practically all other areas of technology, which means that, in turn, the role of an IT project manager can be expected to change just as rapidly. IT project managers everywhere struggle with keeping teams, budgets, and goals focused. IT project management becomes even more tedious when you consider the economy, the instantaneous expectations of stockholders and management, the constant turmoil in the IT industry, and the flux of each team member's commitment to their own career.

Why do so many projects fail from the start? Projects fail for many different reasons: other projects take precedence, team members lose sight of the purpose of the project, and project managers try to do the work rather than lead the team, among other reasons. At the root is a fundamental problem: vision. Vision, in project management terms, is the ability to see the intangible clearly and recognize the actions required to get there. One of your jobs is to develop, nurture, and transfer the vision to everyone on your team. As a project manager, you cannot have a clear vision of the project if the project's needs are never clearly established. The project charter helps to define and communicate the project vision to keep everyone focused on the big picture of the future state the project aims to create.

Creating Reasonable Expectations

Once you've discovered your vision, create a goal. A goal should be a clearly stated fact, such as, "The new database will be installed and functional by December 6 of next year." A goal sums up the project plan in a positive, direct style. Every member of your team should know and pursue the goal. It's not all up to you. The goal establishes the direct need and purpose for undertaking the project. The goal can be expressed as part of the project charter or in an entirely separate document that comes after the project charter is created.

When creating a goal for your project, be reasonable. Just as it would be foolish to say, "I'm going to lose 60 pounds this month," it would be unreasonable for you to create an impossible goal.

A logical goal is not just an idea, a guesstimate, or some dreamy date to be determined. A goal is actually the end result of a lot of hard work. Each IT project will, of course, have different attributes that determine each goal.

Let's say, for example, that your company is going to be migrating your servers and desktops to the latest and greatest operating system. With this scenario, certain questions would have to be answered to determine the ultimate end date for the goal: Is the hardware adequate for the new OS? Will the applications work with the new OS? Will the team have adequate time to be trained and experiment with the new OS? These questions will help you create the end date for the goal.

Creating the Project Charter

The project charter is an official, detailed document in line with your organization's vision and goals. Obviously, a project can stem from a broad, general description of an IT implementation. A goal narrows the description and sets a deadline. A project charter formalizes the goal and serves as a map to the destination. Above all, however, a project charter formally authorizes the project.

Not only does a charter clearly define the project, its attributes, and its end results, but it also identifies the project authorities—usually the project sponsor, project manager, and team leaders (if necessary). The charter also specifies the role and contact information for each authority.

Why do you need a project charter? Why not just hop right in and get to work? In a small company, plowing right into the project may turn out just fine. However, in most companies a project charter is the foundation for success. Consider what the charter accomplishes:

- It authorizes the project.
- It defines the business need in full.
- It identifies the sponsor of the project.
- It identifies the project manager.
- It makes the project manager accountable for the project.
- It assigns authority to the project manager on behalf of the project sponsor.

Two items warrant additional discussion from the contents of the project charter: the project sponsor and the project manager. You know who the project manager is—that's you. The *project sponsor* is a person who has the authority to launch the project. The project sponsor is seen as the management champion of the project and has authority over all resources in the project. It's so important to ensure that the project sponsor for the project is the right person; choosing the wrong project sponsor can create risks, issues, and drama in a project.

Project Charter Elements

When the project charter is created, you can include just about any information pertaining to the project that you'd like to include. Generally, though, consider these elements:

- **Official name** Every project needs a name.
- **Project sponsor and contact information** The project sponsor should be someone in the organization who has the authority to assign the project manager power over the project resources.
- **Project manager and contact information** The project manager is officially named in the project charter.
- **Project purpose** The purpose defines the problem statement or opportunity the project will address.
- **Business case** The business case defines why the project needs to happen and synchronizes the project to the organization's strategic plan. The business case does not necessarily need to be included in the project charter, but it should be referenced.
- **Key deliverables** These are the primary products, services, or results the project should create.
- **Preliminary scope statement** The charter can include a preliminary scope statement that clearly defines the items that are in scope and out of scope. Identifying out-of-scope items is a way to clarify what the project won't accomplish and helps to address assumptions. For example, the project will create the new software, but it will not create the training and support documentation.
- **General statement about the team's approach to the work** You might reference a software model you'll use or an approach you've used on similar projects in the past. This is where you'll communicate if your project management approach will be predictive or agile.
- **Basic timeline for project milestones** A milestone is an event in the project that shows progress. Milestones typically come at the end of a project phase.
- **Resources, budget, staff, and vendors** Some of this information will be known at the launch of the project, given the nature of the work or the structure of the organization. Often, however, only the roles and responsibilities are known, and the project is organized after the project charter is created.
- **Summary budget** Depending on the organization's rules and project approaches, the budget is usually based on a rough order of magnitude cost estimate with a defined range of variance or a maximum dollar amount for the project.
- **High-level assumptions and constraints** An *assumption* is something that's believed to be true but hasn't necessarily been proven to be true, such as operating system and hardware compatibility. A *constraint* is anything that limits the project manager's options, such as a requirement to use the cinnamon roll software development model.

- **High-level risks** A *risk* is an uncertain event or condition that may have a positive or negative effect on the project. For example, data loss, network downtime, and the loss of key resources are all typical IT risks.

Every project needs a charter. It authorizes the project, creating a sense of responsibility for the project manager, a sense of ownership for the sponsor, and a sense of teamwork for the project team. The project charter will save you headaches, establish who's in charge, and move you to your goal more quickly and with more confidence.

Following is an example charter, based on a fictional company, Best Enterprises. While I've purposely kept the technology vague, a real project charter would call out the exact technology the project will install. The fictional company's network currently consists of 380 computers running Windows workstations. It has made a decision to move all the workstations to a current standard Microsoft operating system and all the servers to the latest version of Microsoft Server.

Sample Project Charter

Project: Systems Upgrade: Workstations and Servers
Project Sponsor: Sharon Brenley, Chief Information Officer (x. 233)
Project Manager: Michael Sheron, Network Administrator (x. 234)
Project Team: Edward Bass, Ann Beringer, Mike Tallent, Carol Fox, Charlotte Harving, Kyle Hardie, Casey Murray, Dustin Bossmeyer, Mark Turner, Frank Simmons

Project Purpose All desktops will be upgraded to the current Windows desktop platform by December 30. All servers will be upgraded and moved to five Windows servers by January 15 of the following year.

Business Case Windows Server and Windows desktop operating systems have served our company for the past several years. We've learned to love it, embrace it, and grow with it. However, it's time to let the past version go and standardized all of our workstations and servers. We'll be embracing a new technology from Microsoft, similar to the current Windows desktop, but far superior. This new Microsoft operating system will enable us all to be more productive, more mobile, more secure, and more at ease.

In addition, there are new technologies that work excellently with this platform, such as wireless networking for our manufacturing shop floors and new accounting software that will be implemented next year.

Of course, our company will continue to embrace our web presence and the business we've earned there. This project will enable us to follow that mindset and will create greater opportunities for us all.

As our company has experienced over the past year, our servers are growing old, slow, and outdated. We'll be replacing the servers with five new multiprocessor servers loaded with RAM, solid state drives, cloud technologies, and faster and reliable backup systems—which means faster, more reliable, more productive work for us all. The operating system we'll be implementing for all of our servers is the latest version of Microsoft Windows Server.

The latest version of Windows Server will enable our users to find resources faster, will help us keep our network up longer, and will provide ever-increasing security.

Project Results

- Windows upgraded on every desktop and portable computer.
- Windows Server installed on five new servers.
- All implementation complete by January 15.

Basic Milestone Timeline

- **September** Test deployment methods, capture user and application status, finalize deployment image, and create scripts.
- **October** Initial deployment of 100-user pilot group. Test, document, and resolve issues. Redeploy 100-user pilot group with updated images and scripts. Begin Windows Server testing and design.
- **November** Begin month-long four-hour training sessions. While participants are in class, the new operating system will be deployed to their desktops. Troubleshoot and provide floor support in coordination with Jamie Bryer, Help Desk Manager. Continue to test Windows servers. Three Windows servers will go live on November 15.
- **December** Finish deployment of the desktop OS. Install new Microsoft servers and create infrastructure. Move each existing server to Windows Server platform. Project completed by January 15.

Project Resources

- Budget: $775,000 (includes desktop OS, Microsoft Server, rack computers, client access licenses, consultants, training).
- Test lab reserved for four-month duration.
- On-site consultant from Donaldson IT.

Project Constraints, Assumptions, and Risks

- The project must be completed by January 15.
- The project must not exceed $775,000.
- Our preferred vendors should be able to deliver our new workstations and servers on time without affecting the project schedule.
- The learning curve for our employees to learn the new OS could affect productivity and efficiency.

Your project charter can include as much or as little information as you deem necessary. Project charters (with the exception of the budget) are often shared with the entire organization, so you may need to make a few revisions before the charter is complete.

Sharing a project charter with the entire organization, especially if it will affect all users as in the sample charter, can get everyone involved, excited, and aware of coming changes. A project charter also creates a sense of responsibility for all involved.

Your project team members will get distracted, be pulled in different directions, and lose interest. Vacations pop up, kids get sick, people quit. Realize at the onset that not everyone will be as dedicated to your project as you are. Do your best to inspire, motivate, and lead. Set aside politics, egos, and aspirations and work toward the goal.

Finally, keep in mind that a charter can be called different things in different organizations and that the level of detail can vary, depending on the organization and the project being created. Most charters, however, accomplish two primary things: authorizing the project work and defining the project work.

Gathering Project Information

Everybody talks about project management, but what is it exactly? In some organizations, any task or duty is considered a project that requires someone to manage it. *Puh-leeze!* Project management is the ability to administer a series of chronological tasks resulting in the desired goal. Some tasks can't be completed until others are finished, while other tasks can be done in parallel. Some tasks require the skill of a single individual; other jobs in the project require that everyone chip in and lighten the load.

IT project managers must be able to balance their love for and implementation of technology while leading and inspiring their team members. Of course, the goal of project management is not technology for technology's sake, but rather a movement toward things like improved customer service, enhanced product quality, and increased profitability. Add to that mix external factors such as market conditions, competition, demands for new technology, and even the changing pace of technology, and it's no wonder IT projects can become so frustrating. As you can see in Figure 1-3, project management is a high-wire balancing act.

The business value, or why a project has been created, really drives the implementation of a project. This is part of the project's discovery/concept phase. Business value can be to increase efficiency, to increase productivity, to respond to a customer request or a new regulation, or countless other reasons a project is initiated. The project manager must understand what's driving the project and how the project supports the business needs and mission of the organization, and how the deliverable of the project will be used by the stakeholders.

Establishing the Project Requirements

Before the actual project work can begin, the project manager must establish the project requirements with the project stakeholders. *Stakeholders* are any individuals, groups, or communities that have a vested interest in the outcome of the project. For some projects, the stakeholders may be just one department. For projects that affect every department, the stakeholders include people and departments throughout the entire organization. Identifying stakeholders is important, because their input to the project requirements early in the project can ensure the project's success.

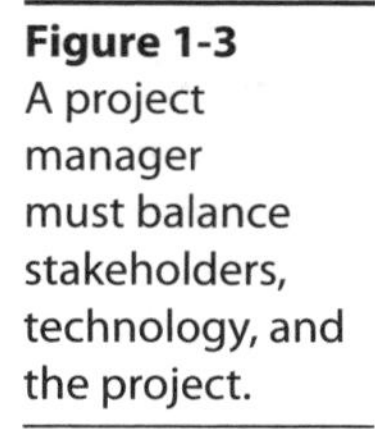

Figure 1-3 A project manager must balance stakeholders, technology, and the project.

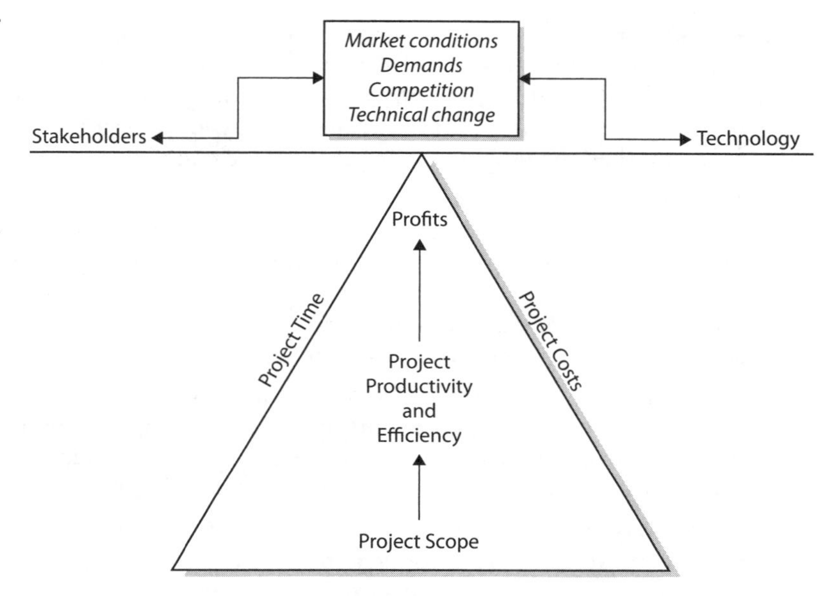

Of course, on most projects, there will be key stakeholders who influence the project's outcome: department managers, customers, directors, end users, and other folks who have direct power over the project work or results. With the input of these key stakeholders—specifically their requirements for the project, constraints on the project, and time and cost objectives for the project—the project manager will be able to gather the project requirements to begin building a project plan to create the project deliverables. Common stakeholders can include the following:

- **Customers and users** These are often called the end users, clients, or recipients of the project deliverables. These stakeholders could be internal to your organization or quite literally customers who purchase the deliverable your project creates.
- **Project sponsor** This is a person in the organization who has the authority to grant the project manager power over the project resources, assign a project budget, and support the existence of the project. This person also signs the project charter to launch the project officially and assigns the project manager to the project.
- **Portfolio review board** This group of stakeholders is responsible for determining which projects are worthy of the organization's capital. This board defines the governance of projects and programs within an organization and oversees the selection of the projects, while considering a number of factors such as return on investment, project value, risk to reward of proposed projects, and predicted financial outcomes of launching a new project.

- **Program managers** A program is a collection of projects working together to realize benefits that the organization could not realize if the projects were managed independently of one another. The program manager oversees all of these orchestrated projects in her program. If your project is operating within a program, then the program manager is a stakeholder.
- **Project management office (PMO)** Some organizations use a PMO to centralize and coordinate the management of projects within an organization, line of business, or department. PMO functions can vary by organization, though most offer project management support, guidance, and direction for projects within their business domain. It's not unusual for a PMO to direct the actual project management of a project.
- **Project team** These are the people who work on planning and executing the project plan. Depending on the organization, the project team members may work full-time or part-time on the project, and they can come and go as the project work warrants or stick around for the duration of the project.
- **Functional management** Functional management consists of managers of the administrative functions of an organization: consider finance, human resources, and accounting. Functional managers have their own staff and their own day-to-day duties to keep the operations stable.
- **Operations management** These are managers of the core business area such as design, manufacturing, and product development. Operations managers oversee and direct the salable goods and services of the organization.
- **Business partners** These are the sellers, vendors, and contractors that may be involved in a project through a contractual relationship. Business partners can provide goods and services such as hardware, software, and subject matter experts (SMEs) such as developers, technical writers, and software testers that you might need on your project.
- **Project manager** You are a stakeholder for your project. You're responsible for developing the project plans, keeping the project on track, monitoring and controlling the project, and communicating the project status and performance. As the saying goes in project management, if the project succeeds, it's because of everyone's efforts; if the project fails, blame the project manager.
- **Product owner** If you are working in an agile project management framework, you will have a stakeholder that represents the customers. This role is typically called the product owner, though you might also see it called the customer representative. The product owner is responsible for prioritizing the product backlog and explaining the requirements in detail to the project team.
- **Product manager** Some projects may include a product manager as a stakeholder. Product managers are responsible for identifying the features and functions of a product, how changes or improvements to the product may be allowed, and will monitor the project for quality in the product.

Clarity is paramount. When the decision has been handed down that your organization will be implementing some new technology and you'll be leading the way, you need a clear, thorough understanding of the project's purpose. Ambiguous projects are a waste of time, talent, and money. Before the project begins, you need to know what exact results signal the project's end. A project truly begins when you know exactly what the project will produce.

Once the predictive project is defined, you need clearly stated objectives, requirements, and boundaries for the project. Although management may have an ideal timeline for project completion, it'll take some planning and research to determine the exact duration of the project. The role of a project manager is not permanent, but temporary. As the project manager, you are responsible for setting the goal, developing the steps to get there, and then leading the way for your team to follow.

Agile projects operate with a fixed deadline and budget, but the scope of work can change. Recall that the agile project begins with a prioritized list of product features and requirements called the product backlog. The team works to complete as many items as possible from the top of the list, because those are most important, to the bottom of the list, as these are of least value. I'll talk more about the agile approach throughout this book. Stay tuned.

TIP Because agile projects deliver the highest business value items first, the items that won't fit within the schedule and cost constraints are of lesser value to the customer.

The end result of the project, the vision, is also called the Definition of Done, or the DoD. How will you know what the end result of the project is to be? Ask! Who do you ask? Stakeholders such as the project sponsor can answer these kinds of questions. You must have a clear vision of the end result, or the project will drone on and on forever and you'll never finish. Too often, IT projects can roll into project after project, stemming from an original, indecisive, half-baked wish list. Whether you are a full-time employee within an organization or a contract-based project manager, you must have a clear understanding of what the end results of the project will be.

Imagine your favorite archeologist maneuvering through a labyrinth of pitfalls, poison darts, and teetering bridges to retrieve a golden statue. In the movies, there's always some fool who charges past the hero straight for the booty and gets promptly beheaded. Don't be that person. Before you can rush off toward the goal of any given project, you've got to create a clear, concise path to get there.

To create this path, you'll have to interview the decision makers, the users the change will affect, and any principals involved in the development of the technology. These are the stakeholders—the people who will use the project deliverables on a daily basis or will manage the people who will use the project deliverables. You must have a clear vision of what the project takes to create it or you're doomed. Often projects start from a wish list and evolve into a catalog of complaints about the current technology. One of your jobs in the early stages of the project will be to discern valid input from useless gripes.

As you begin your project, consider the questions presented in the following sections.

Does the Predictive Project Have an Exact Result?

Predictive projects that are as indecisive as a six-year-old at an ice cream stand are rarely successful. As a project manager, you must ensure that the project has a definable, obtainable end result. At the creation of the project, every project manager, project sponsor (the initiator of the project), and team member should know and recognize the end result of the project. Beware of projects that begin without a clearly defined objective.

Although you should be looking for exact requirements that a project is to include, you must also look for requirements that are excluded from a project (for example, a project that requires all mail servers to be upgraded in the operating system, but not the physical hardware). As the project takes form, the requirements to be excluded will become obvious given management, the time allotted for the project's completion, and the given budget.

Are There Industry or Government Sanctions to Consider?

Within your industry, there may be governmental or self-regulating sanctions you will have to take into account for your project. For example, a banking environment will involve regulations dealing with the security of the technology, the backup and recovery procedures, and the fault tolerance for the hardware implemented. Government regulations vary by industry, and if your company is a government contractor, there are additional considerations for the project deliverables. These are compliance requirements you must plan for in your project.

You also need to be aware of other regulations and standards that apply to your industry. Regulations are "must-haves" that are required by law. For example, pharmaceutical companies, utility companies, and food packaging companies have regulations that dictate their practices. If companies break regulations, fines and lawsuits may follow. Standards, however, are generally accepted guidelines and practices within an industry. Standards are *heuristics*, sometimes called guidelines, which are not laws but are usually followed. The project manager must be aware of regulations and standards that affect the project's work and deliverables.

Does the Project Have a Reasonable Deadline?

Massive upgrades, software rollouts, application development, and system conversions take teamwork, dedication, and time. Projects that don't have a clearly stated, reasonable deadline need one. Projects should not last forever—they are temporary. Acknowledge the work. Do the work. Satisfy the stakeholders with deliverables of the project. Once you've accomplished this, the project is done.

We'll talk more about project scheduling in Chapter 7, but the project manager must be aware of the project calendar and the resource calendar. The *project calendar* defines the hours in which the project work can take place. For example, if your project is to rewire an entire building with new network cable, the project calendar may specify access to the building between the hours of 8:00 P.M. and 6:00 A.M. The *resource calendar* is specific to the project team members and takes into consideration the hours employees are available, their vacations, and company holidays.

In addition, the project manager must consider how many working hours project team members will be able to devote to the project in a given day. Six hours of productivity is typical of an eight-hour day, because of impromptu meetings, phone calls, and other interruptions. These factors directly influence the project schedule and whether the project can meet the project deadline with the given resources.

Does the Project Sponsor Have the Authority to Initiate the Project?

Most IT folks hate politics, but we all know politics, personal interests, and department leverage are a part of every organization. Make certain the project sponsor is the person who should be initiating the project—without stepping out of bounds. Make certain this individual has the resources to commit to the implementation and has the support of the people up the organizational chart. And do this with the full knowledge and support of management.

The project sponsor should be an individual within the organization who has the power to assign team members, allocate funds, and approve decisions on the project work. The project sponsor is typically above the functional managers of the project team members assigned to the project work.

Does the Project Have a Financial Commitment?

If you do not have a clear sense of a financial commitment to the completion of the project, put on your hard hat and don't stand under any fans. Technology costs money because it makes money. The goal of a project in the corporate world is the same goal of any company: to make or save money. A tech-centric project requires a financial investment for quality hardware, software, and talent. If the project you are managing has a budget to be determined somewhere down the road, you've got a wish list, not a project at all.

Is Someone Else Doing This Already?

In large companies, it's easy for two projects to be competing against each other for the same end result. This comes back to communication among departments, teams, and the chief information officer. In a perfect world, IT projects fall under one umbrella, information is openly shared among departments, and everyone works together for the common goal of a company (to make money). This process can be administered through a project or program management office where projects are tracked across the enterprise. Of course, that doesn't always happen. You should do some initial research to ensure that your project isn't being accomplished elsewhere in the company before you invest time, finances, and your career on it.

Do the Stakeholders Understand Agile?

If your organization is new to agile project management, there will be a learning curve. Stakeholders, including the project team members, may be used to working in a predictive environment, sometimes called a waterfall approach. Agile doesn't have all the top-heavy, front-loaded planning that you find in waterfall projects, but uses short bursts of planning and execution to move the project forward. If you and your stakeholders are new to agile, there will be a learning curve you'll need to consider to bring everyone up to speed on the agile framework.

Possessing Multiple Personas

Are you an optimist? A pessimist? A realist? A project manager has to be all of these when identifying project requirements. You have to be an optimist so that you can lead your people, manage the resources, and identify the technology needed for the project. You have to be a pessimist, secretly of course, because you need to look at the worst-case scenario for each piece of the technology implementation. You have to be a realist because you need to look at the facts of the projects completely, unattached, unemotionally, and unencumbered.

When your project is developing, you should play devil's advocate to each cornerstone of the project. You need to question the concepts, the technology, and the time it may take for each step of the implementation. As you can see in Figure 1-4, you should question everything before you begin.

The following sections present the questions to consider.

How Will This New Technology Affect Your Users?

Not all technology you implement has a direct effect on your users, but most of it does. Your life may be IT, but the accountant in the finance department doesn't like change. She likes everything the way it is now—that's everything from having to click OK on a redundant error message to installing her favorite screen saver. If your technology changes her world, you should let her know ahead of time; otherwise, she'll be certain to let you know afterward. Your primary objective must be to make her job easier.

As technology has become integrated in practically all areas of an organization, users have become more tech-sophisticated. They want to know why the change is happening, why the change is needed, and how it will help them. This brings us back to requirements gathering and communication. Ninety percent of a project manager's job is communication. If the project manager wants buy-in from the stakeholders, particularly the users, he must communicate the benefits and rationale behind the technology project.

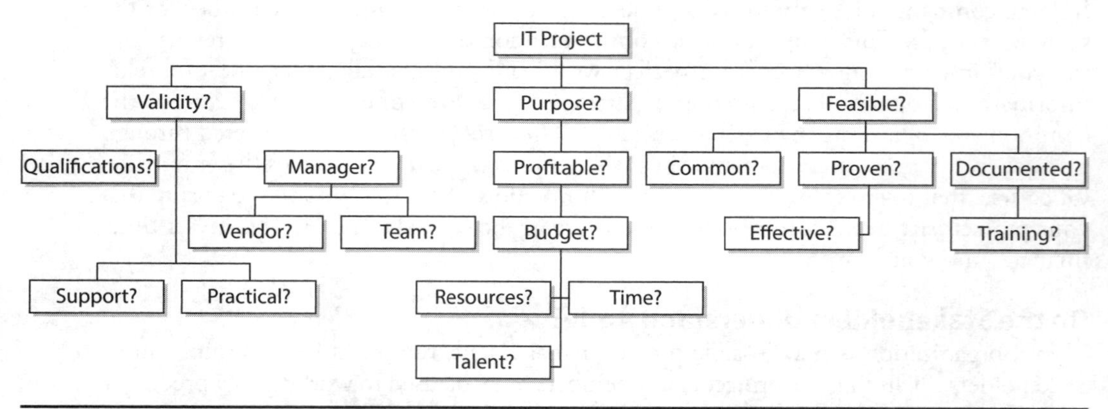

Figure 1-4 Project managers must question all aspects of a project.

Will This Technology Affect Other Solutions?

How many times have you installed software without testing it, only to discover that it disrupts something as unrelated as printing? I hope never, but it happens. You must question and test the ability of the new technology to work with your current systems. Of course, if you're considering a 100-percent change in technology, then there really isn't a software compatibility issue. You might see these issues called roadblocks, blockers, or impediments, especially in agile projects.

Will This Technology Work with Any Operating System?

How many operating systems are in your organization? Though the goal may be just one, I'd wager that two or three different OSs are floating around. Think about those graphic designers and their Macs. Remember those salespeople and their old Windows laptops? And what about those mainframe and server-based Linux users? If your company uses multiple operating systems, you've got to question the compatibility of the technology for each.

What Other Companies Are Using This Technology?

The assumption is that you are buying this solution rather than building it. Therefore, is it a bleeding-edge solution? Are you first in line? No one likes to be first, but someone has to be. When embracing and implementing a new technology, ask that question of the vendor's salesperson. Hopefully the salesperson will be happy to report about all the large companies that have successfully installed, tested, and implemented the vendor's product. That's a good sign. If someone else has done it, you can, too.

Does the Vendor of This Technology Have a Good Track Record in the Industry?

From whom are you buying this technology? Has the vendor been around for a while and implemented its product many times over? Does the vendor have a history of taking care of problems when they arise? This is not to say you should not buy from a startup—every major IT player was a startup at some time in its history. You should feel fairly confident, however, that the vendor selling the product today will be around to support it tomorrow.

What Is the Status of Your Network Now?

You may not always have to ask this question, but with so many network-intensive applications and new technologies today, it doesn't hurt. You don't want to install the latest bandwidth hog on a network that's already riding the crest of 90-percent utilization. You and your organization won't be happy. By asking this question, you may uncover a snake pit that needs to be dealt with before your project can begin.

What If...?

Finally, you need to dream up worst-case scenarios and see if there are ways to address each. Find out how the technology will react when your servers are bounced, lines go down, and processor utilization peaks. Ask these questions and have answers for them now rather than waiting for the crisis that hits during your four-week vacation to Alaska.

No Other Choices?

At the start of a project, at its very genesis, ensure that the proposed technology is the correct technology. Of course, sometimes you have no control over the solution that is to be implemented because some vice president and decision maker heard about the product from his golf buddy who is CIO at another large firm and is now having you install it everywhere. It happens.

Other times, hopefully most of the time, you have some input to the product implemented to solve a problem. You are the professional, the guru, so you should have a definite say regarding the technology that you'll be in charge of delivering. You'll need to create a list of questions and then find the appropriate technology that offers the needed solution, works with your current systems, and fits within your budget. Having the right technology to begin with ensures success at project's end.

Interviewing Management

To have a successful project, you need a clear vision of the delivered result. You need to know why the project is being implemented. You need a strong commitment to the project from management. You need to share management's vision of how the end results will benefit the company. How will you discover these facts? Ask!

When your boss comes to you, for instance, and reports that you are to manage a project to upgrade the mail servers, you need to find out why. It may not be that the manager really wants the mail servers upgraded; he could just be having trouble opening a cartoon his frat brother from Utah sent him and is blaming it all on the company's e-mail system.

When you approach management to find out why the project needs to happen, you aren't questioning their decision-making ability. You are, however, trying to understand their vision for the project. In your organization, your immediate manager may be the most technically savvy genius in the world, and her decisions are always right on target. In other organizations, if not most, managers know that a technology exists and can be implemented. However, they don't know exactly which technology they're after. Figures 1-5 and 1-6 show the difference between effective decision-making abilities and poor decision-making abilities.

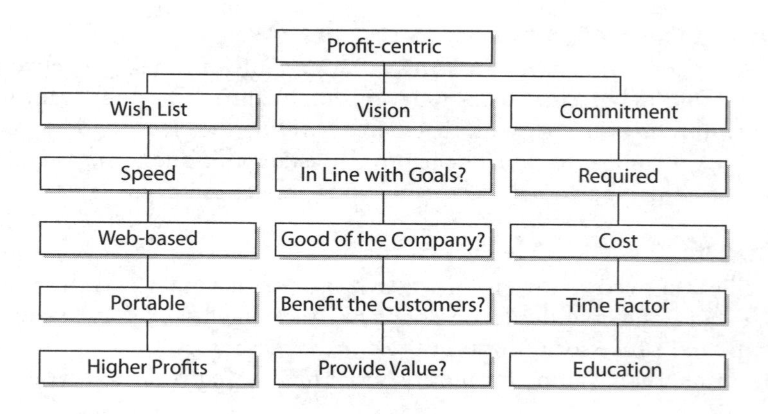

Figure 1-5 Well-informed decisions result in success for everyone, not just the project.

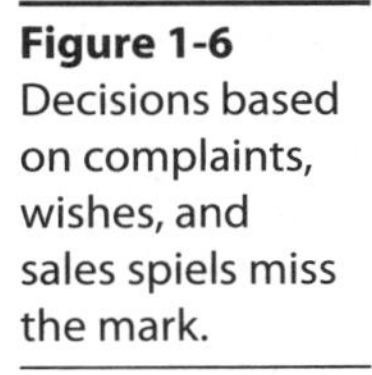
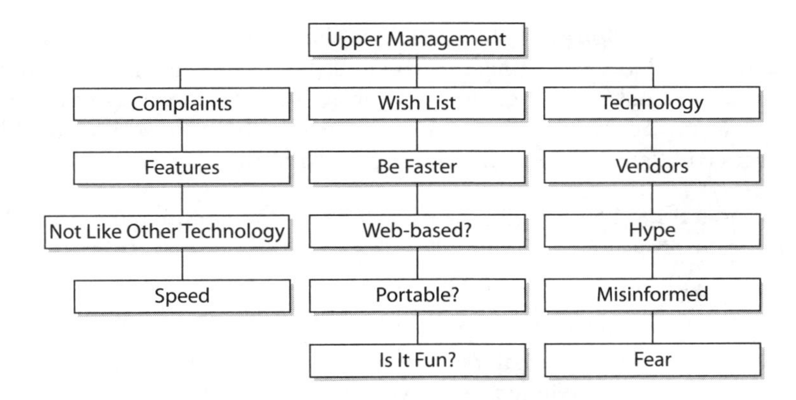

Figure 1-6 Decisions based on complaints, wishes, and sales spiels miss the mark.

As the project manager, your job is to ensure the success of your project and your career, and to ensure a successful impact on the bottom line. When you speak with management about the proposed project, you are on a fact-finding mission. Ask questions that can result in specific answers. For example,

- What do you want technology so-and-so to do?
- Why is this solution needed?
- How did you discover this product?
- What led you to the decision that this was the way for your organization to go?

Sometimes a manager may come to you with a specific problem for you to solve. In these instances, the project is wider and more open-ended, and you'll have to drill deeper into the problem presented. Let's say, for example, that a vice president is complaining about the length of time it takes her to retrieve customer information through the organization's database. She just wants it faster.

Your questions may be something like this:

- Can you show me how the process is slow?
- Is it slow all the time or just some of the time?
- How long have you experienced this lag?
- Have others reported this problem?

You could do several things to increase the speed of the process. Each may require a financial commitment initially but will result in faster responses for all of the database users. Do you want to investigate this route?

Notice how you're thinking like an executive. It's not technology for technology's sake. A new multiprocessor database server, gigabytes of memory, and faster switches are all cool stuff, but if they don't earn their keep, they are just toys. When you are inventing a project, think like an executive of a company and show how the investment in software, hardware, and talent can create more dollars by increasing productivity, safeguarding data, or streamlining business processes and ultimately making customers happy.

Your organization may shift much of these requirement-gathering duties to a business analyst (BA). That's fine, but you and the business analyst should work together to examine the goals, requirements, and objectives of the project that will eventually feed into your project scope. One approach that I've always liked is called SMART; for each project goal, determine whether it meets each of the following:

- **Specific** You must know what the specific requirements and deliverables are for your project.
- **Measurable** It's a good idea to avoid vague terms such as fast, good, and happy. You need measurable metrics for the project requirements.
- **Achievable** The goals of the project should be achievable considering the resources, cost, and time required versus what's available in the organization. Management and customers that ask for a long list of requirements without providing a balance of time and monies are setting themselves up for disappointment.
- **Relevant** The goal of the project shouldn't be based on someone's private agenda. The goals of the project should support the primary business need of the organization, provide an opportunity for the organization, or solve a problem. Basically, all projects should either increase revenue or cut costs.
- **Time-bound** Requirements that are dreamy, are open-ended, and don't provide an easy link to a conclusion aren't good requirements for accurate planning and goal creation.

Interviewing the Stakeholders

As you know, stakeholders are individuals, groups, or organizations that have a direct interest in the outcome of the project. Your project's success or failure will directly affect the way they complete their work, use their existing technology, or continue to buy from your company. Stakeholders can include

- Management
- Project manager
- Project team
- Project sponsors
- Customers
- End users
- Community

In a technical project, the largest group of stakeholders is typically the users. Any project that has an impact on users needs to be discussed with them. This can be done in several different ways. The most popular, and sometimes most disruptive, is by forming a focus group. Focus groups are often led by a professional, impartial moderator and are conversational in tone.

CAUTION If you don't have a good moderator who will direct the conversation to productive input, you might find that focus groups have a tendency to engage in gripe sessions about the problem rather than finding the solution. If you choose this route, take control of the discussion and keep the participants focused on the solution.

A focus group enables you to gather users from each affected department, present the project to them, and then listen to their input. You need to explain how the proposed technology will be better than the current one, how it will solve problems, and, if necessary, why the decision is being made to change. Input from focus groups can alter your entire project for better or worse.

Another way to interview users is through an intranet site. This method can be an effective form of communication because users have the opportunity to share their opinions and have some say about the project. Of course, with this route, it's best to create an intranet site that asks for responses to a survey so that the results can be tallied quickly. See Figure 1-7 for an example of an online survey.

Some project managers rely on the Delphi Technique, an approach that's often used in risk management, but can be applied to any consensus-gathering activity. The participants and their comments are anonymous. The participants are allowed to comment freely on the technology, including their concerns and their desires for requirements. All of the comments are then shared with all of the participants, and they can agree or disagree with the comments according to their opinions and experience. Because the process is anonymous, there is no fear of retribution or backlash or of offending other participants. After several rounds of discussion, a consensus is formed on what is needed. An intranet site can automate the method and keep users anonymous.

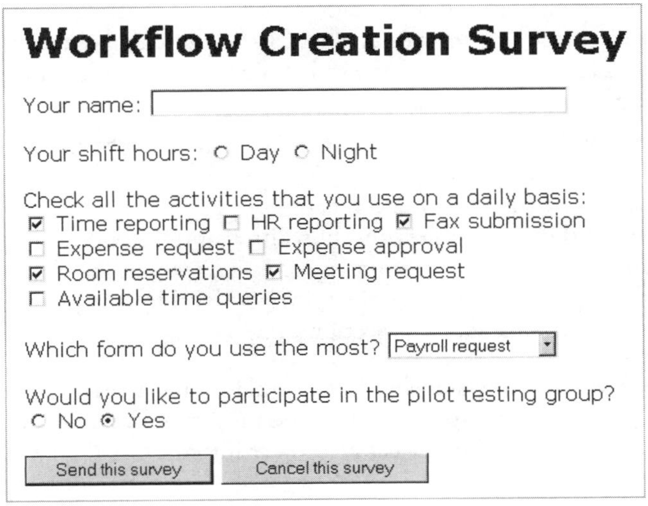

Figure 1-7 An online survey can quickly tally users' input regarding a new technology.

Workflow Creation Survey

Your name:

Your shift hours: Day Night

Check all the activities that you use on a daily basis:
Time reporting HR reporting Fax submission
Expense request Expense approval
Room reservations Meeting request
Available time queries

Which form do you use the most? Payroll request

Would you like to participate in the pilot testing group?
No Yes

Send this survey Cancel this survey

Finally, you should learn how the users do their work now. This is especially important for projects such as new software development, application upgrades, and new hardware technologies. This can be accomplished in a usability laboratory, where mock screens resembling the technology being implemented are made available. Feedback from users helps design the solution to be implemented. By working with a user one-on-one, you can experience how he is using the current technology, how the new technology will affect him, and what the ultimate goal of a technical change should be: increased productivity and increased profits. Don't lose sight of that fact.

This is really stakeholder observation, and it comes in two flavors:

- **Passive observation** The observer simply observes and documents the work and does not interact with stakeholders at all. It's sometimes called invisible observation.
- **Active observation** The observer interacts with the users, stops their work to ask questions, and can even get involved in the actual work to experience the users' processes. This approach is sometimes called visible observation.

As stakeholders are identified, they should be added to a stakeholder register. The stakeholder register helps with requirements gathering and also with project communication. The stakeholder register defines

- **Stakeholder identification information** This includes each stakeholder's contact information, role in the project, and organizational position.
- **Assessment information** This includes each stakeholder's specific requirements, project expectations, and project influence, along with the specific phases and deliverables each stakeholder is most interested in.
- **Stakeholder classification** Stakeholders who support your project are considered positive stakeholders. Stakeholders who oppose your project are considered negative stakeholders, or project resistors. Neutral stakeholders are indifferent to your project. This part of the stakeholder information may also include information on the stakeholder role in the organization, such as internal employee, customer, or vendor.
- **Stakeholder management strategy** This may be included in the stakeholder register, though it's often a separate document. The stakeholder management strategy defines how the project manager will increase support for the project among the stakeholders and how interruptions and objections to the project can be minimized. The strategy considers which stakeholders wield power and influence over the project, interest level for the project, and strategies to overcome stakeholder objections.

Understanding how stakeholders complete their work can help the project manager and the project team appreciate how the project deliverables will be used. Understanding the end result of the project at project initiation will enable accurate identification of project goals.

Working with a Business Case

A *business case* is a document that explains why the project needs to be launched. Business cases are usually written by a business analyst who has studied the project need, the likely cost to complete the project, the likelihood of project success, and all of the factors that support the project need. *Business objectives* are the specific goals defined in the business case. In some instances, you, the project manager, may be called upon to write the business case for your project. The business case helps upper management determine the need for the project, the cost of the project, and the return on investment for completing the project. The business case can describe the current state and the future state the project will create.

A typical business case has seven components that appear in the following order:

- **Executive summary** Although this high-level summary of the business is the first part of the final business case document, you'll probably write it last. The executive summary is an abstract of the actual business that provides all the quick and juicy information condensed into one page (usually); it's often the only part of the document that management and key stakeholders actually read.
- **Problem statement** This section defines the goal, problem, or purpose of completing the project. It defines the problem quickly and directly. The problem statement may discuss what the project will solve, the opportunity the project will realize, the benefits of doing the project, and any ramifications the organization may face by not doing the project.
- **Analysis** The analysis is the documented study of the problem or opportunity. You'll describe the problem, how the problem came into being, how the problem is affecting the organization and/or its customers, and how not solving the problem (or opportunity) will affect the organization in the future. Return on investment analysis clearly defines the likely cost of the project and the expected return for the investment.
- **Proposed solutions** The business case will offer at least two solutions regarding the problem, but often more. The two solutions are to do something about the problem or ignore the problem. More likely, the business case will include two or three specific actions to address the problem. Each solution will include a description of the actions, the expected outcomes, and any effects the solution will create.
- **Project overview** Next comes an overview of the project, should it get initiated, and the intended goals of the project. This section defines the likely costs, schedule, and resources needed to complete the project. Of course, this is a high-level overview, so you'll also need to document any assumptions you're making regarding the project. The business case doesn't get into all of the research that the project manager, team, and experts will do during the project planning activities, but provides only a high-level insight into the proposed project.

- **Cost-benefit analysis** The business case also requires the proposed cost of completing the project and what costs the organization may incur if it doesn't do the project. The costs of the project are usually described with a modifier, such as "+/– 50 percent," depending on how confident you are about the costs of the project you've described in the project overview section. Note that this section also includes the costs of not completing the project: fines, waste, frustration, and other negative costs that aren't purely financial.
- **Recommendations** Finally, you'll offer your recommendations for the solution, the project initiation, and what the key performance indicators (KPIs) of the project should be. Your recommendations shouldn't be too much of a sales pitch for the project, but should be factual and based on your analysis and study of the problem. It is entirely possible, and acceptable, for the proposed solution not to be used to implement a project, especially if the costs are too high or the benefits for completing the project are too few.

Business cases are not needed for every project, but larger projects almost always pass through the business case process before they begin. As mentioned, the process may be performed by a business analyst who interviews the customers, project managers, management, and other key stakeholders to formulate the description of the problem and the proposed solution for the organization.

TIP You might never write a business case. Business cases are often created by a business analyst and before the project manager is even involved. A well-written business case can provide immediate insight to the requirements and the business value of the project.

Establishing the Completion Date

There's a cartoon that's probably posted in every auto mechanic's garage. In the cartoon, a bunch of people are rolling around laughing uncontrollably. Above all this mayhem is the caption, "You want it when?" Of course, as an IT project manager, you can't take that same approach, but a reasonable deadline has to be enforced.

A firm end date accomplishes a few things:

- It creates a sense of responsibility toward the project.
- It gives the team something to work toward.
- It signifies a commitment from sponsors, team members, and the project manager.
- It confirms that this project will end.

How do you find the completion date for a project, and how do you know if it's reasonable? The magic end date is based on facts, research, and planning. In upcoming chapters, you'll get a more detailed look at project end dates for predictive and agile

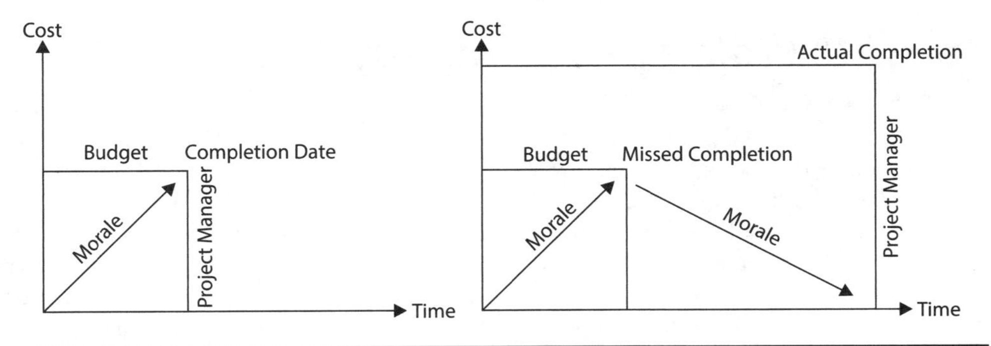

Figure 1-8 If a project schedule stays on track, so will the budget and the morale.

projects and how you establish them. For now, know that projects are a sequence of steps, and each step will take time. The completion of each step will predict when a project should end. And it's not just the project execution that you must account for. A project manager must also consider the time for planning, meetings, and responses to project risks, issues, and the consideration of change requests that are inevitable in technology.

Some project managers in predictive projects create a flexible deadline. Don't do this. If you allow yourself a deadline that is not firm, you'll take advantage of it. And so will your team, your sponsor, and your management. Set a deadline based on an informed opinion, and then stick with it. The charts in Figure 1-8 demonstrate how a missed completion date is bad for the project, the organization, and morale.

A rule of economics that affects scheduling is Parkinson's Law, which states that work will expand to fill the time allotted to it. In other words, if you give yourself extra time to complete a project, the project will "magically" fill the extra time. A firm deadline gives the project manager and the project team a definite date to work toward.

Other factors can have an impact on your projected deadline:

- **Business cycles** Does your project deadline coincide with busy times of the year? Think of a retail giant. How willing do you think it would be to overhaul the database that handles shipping and store management in December?
- **Financial situations** An organization may be more (or less) willing to invest in new hardware or software at a particular time of the year due to taxes, fiscal year ending, or the advent of a new budget. You've got to consider these factors when you request finances for your project.
- **Time commitments throughout the year** When will your team members take vacation? How will their vacation plans coincide with the project's deadline? What other internal time commitments do they have? Will they be traveling to other sites? These factors can delay a project for weeks and months—ultimately resulting in a missed deadline. Work with your team members to ensure that their availability coincides with their responsibilities within the project plan.

CompTIA Project+ Exam Highlight: Initiating Processes

If you're using this book as a study guide for the CompTIA Project+ exam, you're in luck. In each chapter you'll find a sidebar like this one. In this sidebar, I'll highlight the specifics that you absolutely must know in order to pass your CompTIA Project+ exam. All of these exam sidebars reflect the exam objectives as of this writing. I strongly encourage you to visit https://www.comptia.org/certifications/project and download the most recent exam objectives, because it's possible they've changed since this writing.

This first chapter, "Launching a Technology Project," maps to several portions of the CompTIA Project+ exam objectives. To help you study, I've called attention to the specific CompTIA exam objectives partially covered within this chapter. Following are the objectives and key points covered in this chapter.

1.1 Explain the basic characteristics of a project and various methodologies and frameworks used in IT projects Projects are short-term endeavors to create unique products, services, or conditions. Projects aren't part of operations, but they often create or deliver products and services that operations will use, such as software, workstations, servers, and even networks. Because projects do not last forever, they have a definite start and finish. This is evident in technology when you consider the temporary nature of technology—it's always changing and evolving.

You know the five process groups of the project management life cycle, right? As a reminder, here they are

- **Initiating** This is the preproject setup and project launch.
- **Planning** This is an iterative group of project processes that communicate the intent of the project.
- **Executing** This is the process group where the project manager directs project plan execution and the project team does the work to complete the project objectives. It's where the bulk of project funds and time is spent.
- **Monitoring and controlling** This process group happens in tandem with project execution. It oversees the project work to ensure that the work is being done according to plan.
- **Closing** These processes are used to close out a project phase and to close the project altogether. It's also where you'll find the close procurement process.

Although these process groups have distinct responsibilities and activities, there is plenty of overlap and movement between them. Consider how planning, executing, and monitoring and controlling all allow the project manager and the project team to shift between these groups of processes. These project management process groups are usually associated with a predictive project, but agile projects are also initiated, planned, executed, monitored and controlled, and closed, just in iterations rather than as a logical flow of work from the launch to completion of the project.

Projects are constrained by the available resources. Resources are people, equipment, and money. Recall that the portfolio review board can examine potential projects and determine their worth in investment. Organizations have only so much capital to invest in projects, and it's the project manager's responsibility to be accountable for the investment. Human resources constraints also affect projects; consider the availability, competence, and interest of the project team. Finally, resource constraints in technology projects can be older, slower equipment and software that the project must use or interact with.

1.2 Compare and contrast Agile vs. Waterfall concepts Predictive projects are also called waterfall projects. They are phases of project work that fall into the next stage, like a waterfall. For example, in construction you do the foundation phase, then the framing phase, and then the electrical phase, and so on. In an IT project, a waterfall project you have phases of work that are conception, initiation, analysis, design, construction, testing, production/implementation, and maintenance. Waterfall projects plan, or predict, everything up-front, so this is why they are also called predictive projects.

Agile projects have much less emphasis on planning and upfront definition for the project than a waterfall project. Agile projects create a prioritized product backlog and the team works in short iterations of planning and execution. There are many different variations of agile project management that I'll discuss later in this book. For now, know that agile projects welcome and expect change and work with a fixed deadline, a fixed budget, and a changing scope.

Predictive projects are plan-driven and are resistant to change, while agile projects are iterative, incremental, and welcome change.

2.2 Given a scenario, perform activities during the project initiation phase In this chapter, this objective centered on why your organization needs a project, who's demanding the project, and what opportunity the project will realize or what problem the project will solve. Remember that the business analyst may assist in needs assessment and requirements gathering and writing the business case, but you and the business analyst should work together to fully capture and understand the project purpose.

You'll also need to prepare a project charter in order for the project to be authorized officially in the organization. The person who signs the project charter, the project sponsor, should have the authority to grant the project manager power over the resources the project needs to be successful. Some project charters may rely on contractual agreements if the vendor is the performing organization.

Charters authorize projects to exist in an organization, and they officially name the project manager. Charters have

- Key project deliverables
- High-level milestones
- High-level cost estimates
- Stakeholder identification
- A general project approach
- A problem statement

- High-level constraints and assumptions
- High-level risks
- Overall project objectives

These components of a project charter are identified at a high level at project initiation. During project planning, the project manager and the project team will progressively elaborate on these components to fully create the project management plan.

The specific portion of this exam objective you learned about in this chapter is the justification of a project. All projects must be in alignment with the business case, the organizational strategic plan, and the mission of the organization. Projects that don't map to these requirements should not be chartered.

You also learned about the identification of the project stakeholders. Project stakeholders are any people or groups that are affected by your project's existence. Stakeholders invested in your project are considered positive stakeholders, while stakeholders who don't favor your project are considered negative stakeholders. Part of stakeholder identification is to create a stakeholder register with each stakeholder's contact information, project concerns, interests and threats, and information about how the stakeholder will be involved in the project.

Stakeholder observation is a strategy to help you identify stakeholder needs and requirements and to learn how the stakeholders will use the project's solution. Recall that passive observation simply records what the observer sees, while active observation allows the observer to interact with the stakeholders during their work.

Chapter Review

Have you ever seen one of those movies with the ace reporter scrambling into the newsroom with just minutes to go before his deadline? He writes a fantastic article on the mayor, the mob boss, or the sports team and submits it with just seconds to spare. Meanwhile, his cigar-chomping boss is ranting about this reporter's usual skin-of-the-teeth behavior.

That's how IT project management can be. The really awful part? Sometimes it's not even that close. Projects are consistently late, over budget, and half-cooked. IT project management is not about implementing a technology. It's about leadership, integrity, decision-making ability, planning, and time management.

To be a successful project manager, you have to start each project with a clear, concise vision of what the project will yield, when it will end, and how you can lead your team to that destination. Successful projects start with a solid business case, then the project charter, and into planning—all carrying forward the goals and vision of what the project will create and what success looks like for the project completion. Agile projects and predictive projects alike need a clear vision of the business value the project aims to create.

Project management is governed by business cycles, dedication, time, and sometimes weekends. Project management offices direct, support, and consult project managers in their projects. A clearly communicated framework of the project management approach is needed, whether it be agile or predictive, and the stakeholders all must agree to follow the defined structure. Successful project management takes more than implementing the latest whiz-bang technology. To succeed in project management is to succeed in leadership.

Exercises

These exercises allow you to apply the knowledge you have learned in this chapter and are followed by possible solutions.

Exercise 1-1: Analyze Objectives of Key Personnel

You are the project manager for Ogden Underwriters Insurance Company. This company has offices in Chicago, Des Moines, Seattle, Lincoln, and Atlanta.

You have been tasked with managing the rollout of a new web-based training program. You are to interview several members of your company to determine their goals for the project and work those into your plan as much as possible. As this is a simulated exercise, you'll find quotes from several key personnel in the following table:

Person	Title	Concerns	Objectives
Nancy Gordon	Chief Executive Officer	I am very excited about this project. All employees should have access to the website, no matter where they are located in the country. The training should supplement our existing classroom training and provide new information as needed. I would also like to see videos of common tasks for quick review. Finally, the web-based training must be searchable, user friendly, and easy for learners to stop and resume lessons with ease. Have fun!	
Cory Owens	Accountant	Will this thing really work? Our network seems pretty slow already. I don't have time to be waiting on images to load, downloads, and other stuff like that. My computer is so old, and so is everyone else's in this department, that we can't take another software upgrade if we have to. By the way, when are we going to get new computers? Mine at home is faster than the one here at work. Sigh.	
Sarah Sullivan	Claims Adjuster	This is a great idea; I just hope I have time to use it. I get interrupted a lot so I'd need to be able to pause and restart if necessary. Will this web-based training work with my computer? I'm using Linux here and Windows at home. I will be able to access it from home, won't I?	

(continued)

Person	Title	Concerns	Objectives
Michael Bogner	Chief Information Officer	Web-based training will allow for training on demand in modular pieces. The thing to remember though is that all users will need computers with the technology to support your solution and a standardized web browser to take advantage of this. In fact, there are 240 PCs that need to be replaced in six months. Go ahead and work that into your budget and your plan. They'll need to use the current Windows OS. I guess that'll mean these folks will need training, too. And what's the cost factor of the training?	
Jill Vaughn	Web Design Manager	I've wanted to do WBT (web-based training) forever. My team will be using screen recording software, web cams, and interactive software. Make certain all the users have the correct plug-ins for their browsers.	
Jackson Dahl	Web Integrations Team Leader	We can do it—if a few things come true. We'll need a fatter pipe to our ISP if we're going to host the pages here. Of course, if users are coming from all over the country to this thing, we're going to need to talk about security, authentication, and types of access. We'll probably need another server for Jill's database.	

To complete this exercise, analyze persons being interviewed and their concerns about the project, and then record the objectives of each.

Exercise 1-2: Write Project Charter

Now that you've gleaned the key pieces of information from each of the key staff members, you need to write a charter for the project. Your sponsor for this project is Nancy Gordon. Here are some facts that must be included in your project charter:

- Official project name (use your imagination)
- Project sponsor and project manager
- Project goal
- Business case for the project
- High-level results of the project

- Required resources for the project. (If you would like to find the prices of the new computers, software, and operating systems, you may, but they are not required for this exercise.)
- A basic timeline of how your team will implement the plan

Exercise Solutions

The following offer possible solutions for the chapter exercises.

Exercise 1-1: Analyze Objectives of Key Personnel

Person	Title	Objectives
Nancy Gordon	Chief Executive Office	All employees should be trained uniformly across all locations. All employees should have access to the website. Training should supplement existing training. Offer videos of common tasks. Site should be searchable. Users must be able to stop and resume lessons.
Cory Owens	Accountant	Speed issues need to be addressed. Computer needs to be brought current.
Sarah Sullivan	Claims Adjuster	Users should be able to pause and resume lessons. Need to take multiple operating systems into account. Include access from home and other remote sites.
Michael Bogner	Chief Information Officer	Upgrade computers to the organization's technical standard. Upgrade browsers. Replace 240 PCs as part of project. Systems require a new Windows operating system. OS training required. The cost of the solution should be considerably less than the estimated cost of live training.
Jill Vaughn	Web Design Manager	Include plug-ins for browsers.
Jackson Dahl	Web Integrations Team Leader	Need faster bandwidth. Confirm security issues. Need new web server for database.

Exercise 1-2: Write Project Charter

Project Charter

Project Name: Click and Learn: Web-Based Training Initiative
Project Sponsor: Nancy Gordon, CEO
Project Manager: Your name here
Project Goal: A new web-based training program will be created and implemented company-wide by January.

Business Case There's something new and exciting happening at Ogden Underwriters Insurance Company, and it's not a new life insurance policy. It's web-based training (WBT). WBT will enable us to replace and supplement traditional classroom training on all topics.

No longer will you have to enroll in the same four-hour class because you've forgotten how to do a single ten-minute task. No longer will you get hours and even days behind schedule because you needed to attend a class on the latest procedure for your department. No longer will you need to pester help desk staff, your neighbor, or your favorite computer nerd about how to write a macro.

Instead you'll just click and learn.

Our WBT service will enable employees from around the world to access the information it contains. That means each office, each mobile user, and even those who work at home will be able to log in to our site and access the information they need any time, day or night.

We'll provide forms, printable directions, and videos of various tasks for each department. Because this technology is web-based, it doesn't matter what operating system your computer is running. It's going to be great.

You'll be able to search for a specific topic or take an entire structured course. And thanks to our modular approach, you'll be able to pause your training should you get interrupted and then resume it minutes or even days later.

Required Resources Of course, with this technology there are fundamental changes that will affect all of us. For a start, each user will receive the latest version of their favorite web browser and the additional software required to view the videos and complete the WBT classes. We will replace 240 computers with new, speedy PCs running the latest Windows OS.

Our web server farm will grow to add a new database server. Additionally, our Internet access will become faster and more responsive.

Timeline

- **First 30 days** Replace 240 older PCs with new computers. Begin offering classes on Windows OS as part of rollout. Work with web developers to create schedule of course listings for each department.
- **Second 30 days** Continue development of web courses. Order faster throughput for Internet access. Install and work with Jill Vaughn and Jackson Dahl on integration for their database server and current web servers.

- **Third 30 days** Begin creation of videos, streaming software, and bandwidth utilization issues. Work with Jill and Jackson on links for Internet browser upgrade scripts.
- **Final 30 days** Go live with initial classes and test usage. Throttle servers and document results. As month progresses, continue to go live with additional offerings. Create forms to request additional courses, for troubleshooting, and for support.

Questions

1. What is project management?

A. The ability to complete a task within a given amount of time

B. The ability to complete a task with a given budget

C. The ability to manage a temporary endeavor to create a unique product or service—on time and within budget

D. The ability to administer a series of chronological tasks within a given amount of time and under budget

2. All of the following are project management process groups except for which one?

A. Initiating

B. Planning

C. Implementing

D. Monitoring and controlling

3. You are a project manager in your organization and would like to present a project to management. What organizational component is most likely to review your project to determine if it is worthy of investment?

A. Project sponsor

B. Functional management

C. Program managers

D. Portfolio review board

4. What individual has the authority over all of the project resources?

A. Project manager

B. Program manager

C. Project customer

D. Project sponsor

5. What is the purpose of the project charter?
   A. To launch the project team
   B. To identify the project manager
   C. To assign a budget to the project
   D. To authorize a project
6. You are the project manager for your organization, and management has presented a new project to you. This project requires that you adhere to several government regulations for the project deliverable. In this instance, what are the government regulations considered?
   A. Scope
   B. Requirements
   C. Assumptions
   D. Standards
7. Marci is a project manager in her organization and she is collecting requirements for a new project. She is working with Frances, an end user, to see how Frances is using the current software that Marci's project deliverable will replace. Marci asks Frances questions as she works, stops Frances to explore the purpose behind the software activities, and quizzes Frances on each step of the software activity. What is Marci doing in this instance?
   A. Gathering project requirements
   B. Exploring the learning curve
   C. Cross-training
   D. Active observation
8. You are the project manager for your organization, and you are starting an agile project. Which one of the following statements best describes agile projects?
   A. Agile projects plan the entire project and then work in short iterations.
   B. Agile projects utilize a prioritized product backlog and work in short iterations of planning and execution.
   C. Agile projects follow the logical phases of initiating, planning, executing, monitoring and controlling, and closing.
   D. Agile projects begin with a product backlog, and once all stakeholders approve, no additional changes can be added to the project.
9. Management has asked that you hire a moderator for a new focus group. What is a focus group?
   A. An interview process for elective team members
   B. An interview process by the team members to determine the success of a project manager

- **C.** A sampling of users affected by the proposed technology
- **D.** A sampling of management affected by the proposed technology

10. Why can a focus group be counterproductive?

- **A.** The participants may not understand the technology.
- **B.** The management involved may not like the technology being implemented.
- **C.** The participants may focus on the problems of the old technology rather than the goals of the project.
- **D.** Team members may have political agendas against the project manager.

11. A project manager would like to use an anonymous tool to gain a consensus on the needs of the project. Which tool is the project manager likely to use?

- **A.** A survey on an intranet site
- **B.** The Delphi Technique
- **C.** An e-mail message to all users within the organization
- **D.** A Monte Carlo simulation

12. You are working with management to evaluate and understand their goals for a new technical project you'll be leading. You want to use the SMART approach in your assessment of the project goals. What does SMART mean?

- **A.** Specific, measurable, achievable, relevant, time-bound
- **B.** Specific, metrics, action, relevant, time-bound
- **C.** Schedule, monitoring, action, risk, time-bound
- **D.** Scope, metrics, action, risk, time-bound

13. You are working with Sarah on developing a project charter for a new project. All of the following elements should be included in the project charter except for which one?

- **A.** Key project deliverables
- **B.** High-level milestones
- **C.** Stakeholders
- **D.** Company mission statement

14. You are working with your project team and management to determine how long the predictive project will take. One of your team members suggests padding all of the activities by 10–20 hours to ensure there's enough time in case there are problems with the work. You disagree with this idea. Why?

- **A.** Law of Diminishing Returns
- **B.** Moore's Law
- **C.** Parkinson's Law
- **D.** Pareto's Law

15. What is the primary difference between projects and operations?
    - A. Projects are temporary and operations are ongoing.
    - B. Projects are unique, while operations are administrative.
    - C. Projects are constrained by time and costs, while operations have budgets.
    - D. Projects are paid for by customers, while operations are paid for by management.

Answers

1. **C.** Project management is the ability to manage a temporary endeavor to create a unique product or service on time, within budget. Completing a project under budget is nice, but it reflects inaccurate planning of what the budget should have been at the project outset. In addition, the project goal must be met.
2. **C.** Implementing is not one of the five project management process groups. The five groups are initiating, planning, executing, monitoring and controlling, and closing. Some project managers remember the order of these groups by thinking about the syrup of ipecac, where each letter represents the order of the process groups (the "cac" means controlling and closing—not quite the same as monitoring and controlling, but similar).
3. **D.** Of all the choices presented, the portfolio review board is the best selection. This board consists of upper management and organization decision makers. They'll evaluate the worth of the project, its return on investment, and other factors to determine if the company should invest in the project or not. This is where "go/no-go decisions" originate.
4. **D.** The project sponsor is the person who has control over all of the project resources. This individual must have enough authority to deal with managers and authorize the resources needed on the project. The project manager has authority, to some extent, over the resources once assigned to the project. The project sponsor, not the program manager, has the authority over all resources needed for each project.
5. **D.** The purpose of the project charter is to authorize the project. The charter doesn't select or launch the project team, but authorizes the project manager for the project and authorizes the project to exist within the organization. The project manager is named and identified in the project charter, but the purpose of the charter is to authorize the project.
6. **B.** Government regulations are requirements and are never optional. You must always follow government regulations or your organization may face fines and penalties. Standards are optional and don't necessarily have to be followed by the project manager, while regulations must be followed.
7. **D.** Marci is participating in active stakeholder observation. She is working with Frances through the usage of the current software to understand better how the new software can be developed for the end users.

8. **B.** Agile projects utilize a prioritized product backlog and work in short iterations of planning and execution. Agile projects do not plan the entire project in its entirety before starting. While agile projects do move through initiating, planning, monitoring and controlling, and closing phases, this isn't the best definition of an agile project. Agile projects welcome change throughout the project.
9. **C.** A focus group is a collection of users your project will affect. A moderator is an impartial person who leads the conversation of the focus group to help gather project requirements.
10. **C.** The participants may focus on the problems of the old technology rather than the goals of the project. An improperly organized focus group can result in a gripe session about the old technology and its flaws rather than a discussion of the benefits and goals of the new project. A focus group requires a strong leader to help the participants focus on the future implementation rather than on their complaints with the current technology.
11. **B.** The Delphi Technique allows for anonymous input from participants and provides rounds of discussion for consensus building. While the Delphi Technique utilizes a survey approach, it requires anonymous input.
12. **A.** SMART is a method to assess project goals; it means that each goal should be specific, measurable, achievable, relevant, and time-bound. All of the other choices are incorrect meanings of the SMART approach to goal assessment.
13. **D.** The project charter does not need to include the company's mission statement. Project charters should include high-level milestones, high-level cost estimates, stakeholder identification, a general project approach, a problem statement, high-level constraints and assumptions, high-level risks, and the project objectives.
14. **C.** Parkinson's Law states that work will expand to fill the amount of time allotted for it. If you pad the project activities, they will likely expand to use all of the time allotted to each activity—something you don't want when it comes to time estimates.
15. **A.** Projects are temporary endeavors, whereas operations are ongoing initiatives. For example, a project could be to design a new software application, but the support of the application in the workforce would be an operational activity.

CHAPTER 2

Planning the Project

This chapter covers the following topics:

- How to plan a predictive project
- Planning in agile projects
- Creating a feasibility study
- Establishing project priority
- Considering IT concepts for project implementation
- Creating an approach
- Creating a strategy
- Managing with an agile approach

Picture this: you're an IT professional, and your manager informs you that the entire network, from the physical cabling to the network cards in each machine, has to be replaced with something bigger, better, and faster.

After you recover from choking on your coffee, you ask, "Something? What exactly did you have in mind?"

And what does your boss say? "I dunno. Something faster. Just figure it out and let me know what'll work. By the way, we can't spend too much. See ya."

While this may not be a typical way for a project to begin, you can see that before any charter, scope, or cost-estimating talks get underway, you have a ton of research and planning to do. Where does that planning begin, and how can you formalize your results? And what subject matter experts will you call upon to help? This chapter will answer these questions and help you streamline your efforts.

VIDEO For a more detailed explanation, watch the *Planning the Project* video now.

How to Plan

Don't laugh. Many IT project managers, executives, and professionals don't know how to plan. Oh sure, they think they do, but the reality is they don't. When these folks begin planning, their efforts consist of searching the Internet randomly, leafing through vendor brochures, and chatting with other professionals about similar problems they've encountered and how those problems were resolved. On the surface, this looks like a great effort. The Internet, vendor brochures, and interviews are all essential elements to IT research. The trouble, however, is there's little rhyme or reason, little approach, and, most important, few results to show for the effort.

The type of project management you use, agile or predictive, will also affect your planning efforts. Recall that predictive projects plan everything in detail early in the project. The goal is to predict what will happen. Agile projects also do some upfront planning, but the focus is more on the requirements of the project, not how and when everything will take place. Agile projects plan and execute in short bursts, while predictive projects plan and then execute with opportunities to return to planning as needed.

The goal of research is to come to a conclusion, a discovery, and hard-hitting facts upon which a decision, a plan, or an implementation can be based. Now here is the key: good research stems from an organized, concentrated effort. In order to do good research, you also have to know what you're researching for. This means you'll need to understand the business opportunity or the problem that the project will address—and you'll do that through stakeholder interviews. Some of these efforts in your organization may reside with the business analyst—but as the project manager, you still need to understand why a project is being initiated and what the project is to accomplish.

TIP In predictive and agile projects, planning happens throughout. While it's ideal in a predictive project to plan everything in advance of project execution, that's just not realistic for most projects. In a predictive project, there'll be changes, issues, and interruptions that'll cause you and the team to go back to planning and figure things out. Don't get too committed to really trying to predict everything that's going to happen in a project. Nobody knows everything that's going to happen.

In order for a predictive project to be successful, the project manager and the key stakeholders must know exactly what it is the project will create. Agile projects start with a general idea of the key requirements and then, over time, typically add more requirements to the project scope. Often, especially in information technology, the customers of the project don't know what they want or what your project should create. They may have a general idea of a scenario they'd like you to create for them. Through interviews, quantitative analysis, and in-depth research, you'll propose solutions to them. Project managers need to link project solutions to terms, scenarios, and conditions the stakeholders can understand.

When creating a solution for a customer, the project manager must have the same vision the customer has for the final product. While there will, no doubt, be iterations and revisions as the project progresses, it's better to understand up front what the project deliverables consist of. Root cause analysis allows the project manager and the customer to work together to find the solution for the problem, opportunity, or other condition the project is to resolve.

When you go about researching anything, from real-time transaction servers to wireless networking, you must possess a plan of attack, maintain laser-like focus, and document your efforts. How do you plan? Here is a sure-fire, six-step method that works:

1. *Define the purpose of the research in writing.* Writing a concise concept definition statement of the project helps form the research you are undertaking. The concept definition statement defines the intent of the project based on the high-level goals, problem to solve, or opportunity the project aims to capture. This document will help you develop the laser focus you'll need for success. Keep that statement in plain view as you research. Don't lose track of your purpose, or you'll meander through your research like a lazy walk in the woods. You might know the concept definition statement as a current and future state assessment, a needs assessment, or even a solution scope. It's a document defining the purpose of the research.
2. *Determine what resources you will use during this research.* Make a list of avenues of information you'll use. This is not to rule out any possible source of information, but to list your sources and then organize them in priority. Sources can include
   - Organizational process assets such as historical information, files from past projects, and documented experience with the related technology
   - Expert judgment such as consultants, subject matter experts, and other people in the organization
   - Qualified, quality Internet sites
   - Specific trade magazines
   - IT books directly related to the topic
   - Vendor brochures
3. *Delegate.* If you have team members in mind for this project, use them to help in the research. You'll need their expertise and experience to develop the best solution for the project purpose. Break down the planning into multiple components and then delegate portions of the research to team members. You might use a roles and responsibilities matrix to identify who'll research what area of the business problem or solution. Many hands may lighten the load, but accurate workers with knowledge develop the plan.
4. *Get to work.* Begin reading, evaluating, and taking notes on your discoveries. If you use the Internet, bookmark useful pages you've found. Few things are worse than knowing there's a great page out there somewhere, but you can't remember when or where you saw it. Record the names of books and magazines you've used and associated page numbers. This supporting detail will help you later when you present a solution to your stakeholder and formalize your project plan.
5. *Organize and document.* Compile all of the information you and your team have gathered. This is the start of a feasibility study. One key management skill is the ability to organize and recall the needed information at short notice. A knowledge management system is ideal for any project manager because it can help you quickly access information.

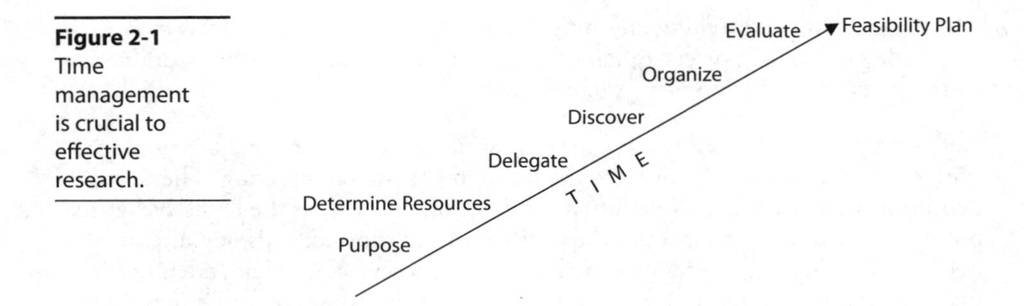

Figure 2-1 Time management is crucial to effective research.

6. *Evaluate and do more research.* Once your research has come together, determine if the collected data answers the research purpose. If it does, move on. If it does not, continue to research following these same six steps as your guide.

This method of research is simple and direct, but it will produce results. One key element is time; don't get bogged down in the research process. "Analysis paralysis" is the ongoing study of a problem to delay actually addressing the problem. Of course, quality takes time, but set a deadline to reach step 5. As you can see in Figure 2-1, the steps to successful research also follow a projected timeline.

Defining the Business Value

Projects aren't done for fun—there must be a business value behind the decision to invest capital in a project. It's no secret that IT projects are risky: they fail, come in late, run over budget, and sometimes are never completed at all. Even the best project manager can wrestle with risks, issues, and unforeseen delays in IT project management; this is why management and stakeholders need assurance that there is a genuine need and return on investment for their organization's efforts and funds.

Defining the business value is a project management research activity that overlaps with business analysis duties. If you're lucky enough to have a business analyst doing this bit of work for you, offer your sincere thanks. If you're the person responsible for linking the need to a project solution, you'll have some extra duties, but you'll also have a deeper understanding of what the project needs to accomplish. The business value addresses the business need and why the project should happen.

The business value will help the project manager define the project scope and the project management plan in a predictive environment. In an agile project, the business value is represented in the prioritized product backlog. To understand the business need fully, to write a feasibility study or business case, you need to experience the problem or interview people who have experienced the problem. It's also possible that the business need isn't a problem, but an opportunity, so you'll have to see the possible outcomes, good and bad, of a proposed solution that could create a loss or a return on investment. These elements help the project manager and the project team understand the stakeholder requirements and what the project must accomplish.

Determining the Business Objectives

A *business objective* defines the specific outcome the business wants to achieve. It paints a picture of what the organization should look like once your project is done. When you go about identifying the business goals and objectives, it's helpful to do a current state assessment: document what's happening right now in the environment that prompts your stakeholders to want a change. Once you've identified the current state, you'll next create a description of the desired future state; this is where the organization wants to go. The difference between the current state and the desired future state is the high-level view of the project scope.

The purpose of defining the business goals and objectives is not to create a solution. While your research, conversations with stakeholders, and organizational process assets from past projects may point you toward a solution, that's not the purpose just yet. When you define the business goals and objectives, you're simply defining the end results of what your project will create. Common business goals and objectives for IT are to

- Cut costs.
- Increase revenue.
- Secure the organization's technology assets.
- Improve customer satisfaction.
- Improve employee satisfaction.
- Seize a marketplace opportunity.
- Adhere to new or pending regulations.
- Become more efficient.

And there are probably hundreds of other technology-related goals. The purpose is to understand why a project needs to be created to get to how the desired future state can be reached. Business goals and objectives will also help you and the stakeholders realize the needed time and required costs for the project later in planning.

To create the business goals and objectives, you'll use six typical tools, as described in the following sections. Keep in mind that these tools aren't solo activities. You'll include your project team, the identified stakeholders, management, and maybe even subject matter experts. Determining business goals and objectives is crucial to understanding the problem so that you can create an accurate and cost-effective solution.

Benchmarking This tool compares one component to a similar component. In IT, you'll often use benchmarking when you have multiple choices between software packages or hardware solutions. For example, you might compare the benefits of Oracle software to SQL. The comparison could track technical metrics but also things like interoperability, support, learning curve, and maintenance costs.

Brainstorming This approach encourages participants to generate ideas about an opportunity or business problem. Brainstorming at this stage of research is useful to determine different types of outcomes for the project. You should encourage the participants to come up with as many ideas as possible, which then can be sorted and researched more in depth after the session.

Business Rules Analysis If the project outcome will likely affect the way your organization does business, you should study the business rules. Business rules define the internal processes to make decisions; provide definitions for operations; define organizational boundaries; and afford governance for projects, employees, and operations.

Focus Groups Focus groups are a type of stakeholder analysis. An impartial moderator leads stakeholders through a discussion about the opportunity or problem. A scribe or recorder keeps the minutes in the meeting, and then the business analyst, product owner, product manager, project manager, and other key stakeholders will analyze the results . An average focus group has 6 to 12 participants. The participants can be considered homogeneous if they all share the same characteristics, such as all salespeople. Or you can use a heterogeneous group, in which the participants are stakeholders with different backgrounds, such as users of a software product from different departments within your organization.

Functional Decomposition This method simply takes a large problem and breaks it down into smaller, manageable components. While it sounds easy, it's tricky. You want to break down the problem into subcomponents that are as small as possible so that each "subproblem" can be managed independently of the other problems. You will need to link the components together so that one solution to a subproblem doesn't adversely affect the solutions to other components in the decomposition.

Root Cause Analysis You'll see this approach often in project management. *Root cause analysis* involves studying the effect that's being experienced and then determining the causal factors of the effect. This is one of the purest forms of analysis, and the results are often graphed in a cause-and-effect diagram. You'll also use this approach in quality control.

Creating a Feasibility Study

A *feasibility study* is a documented expression of what your research has told you. It helps you determine the validity or scope of a proposed project or a section of a project. Feasibility studies can help solidify whether a problem is solvable or whether the organization can realize an opportunity. You might also be tasked with writing a feasibility study to determine the financial aspects of the project—including potential return on investment.

Feasibility studies are often written with upper management in mind, so they're direct, organized, and generally factual rather than opinionated. As you approach your project, keep in mind that the goal of any IT project is not technology for technology's sake, but to add value to the organization. The feasibility study determines if the proposed project can be feasibly accomplished.

As you draft the feasibility study, think like an executive and write with the business in mind, focusing on how the proposed technology will benefit the organization. If you approach any project as if it were a business venture and you were the proprietor of the business, you'll be much more successful in your work. As the "project proprietor," you assume ownership and responsibility of the project and its success or failure.

Organizing the Feasibility Study

To begin writing the feasibility study, refer to the concept definition statement and the business goals and objectives you used in the research phase. The business goals and objectives define why you initiated the planning process and should reflect the proposed project. As Figure 2-2 demonstrates, the concept definition statement is the foundation of the feasibility study structure.

For example, suppose an international company is investigating implementing a new, to-be-determined application that will need to manage multiple calendars, resources, and e-mail. The company's concept definition statement at the start of the research reads as follows: "To determine the selection of a web-based calendaring system that can provide resource management, e-mail, public and private meetings, and workgroup collaboration, the application must be proven, be able to integrate with our current network operating system, and address international time zones." This statement would introduce the feasibility study.

The feasibility study is divided into eight sections:

- Executive summary
- Defined business problem or opportunity
- Purpose of the study
- Options assessed
- Assumptions used in the study
- Audience impacted
- Financial obligations
- Recommended action

Figure 2-2
The concept definition statement is the foundation of the feasibility study.

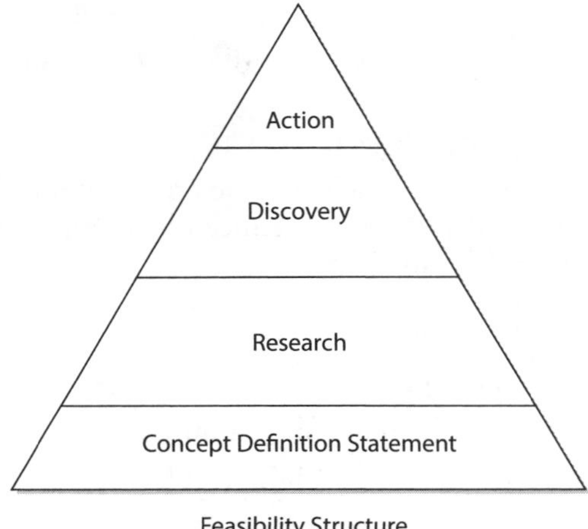

Each section is vital to the study and should be direct, be full of facts, and provide references to the historical information and supporting evidence you've used to create the study. The feasibility study is usually created when large projects with high priorities are proposed. You won't need to create a feasibility study for every project.

Executive Summary

The feasibility study should start with an *executive summary,* the purpose of which is twofold: to draw the reader into your findings and to define the key points of your analysis. As its name implies, it provides a summary of your findings so that the entire document doesn't have to be read. It should include a summation of each of the remaining sections in your feasibility study.

Defined Business Problem or Opportunity

This section describes the business problem and its effect on the organization or the opportunity the organization may seek. The business goals and objectives can be used in this section to link the proposed product or solution to the identified opportunity or problem to be solved. You'll also document the benefits of the technology you've investigated and are recommending. Ideally, you'll investigate several solutions, performing what is sometimes called *alternative identification,* so as not to marry the organization to any one solution before examining all options.

You should write this section to pinpoint the audiences impacted by the proposed technology. This section may also include

- Benchmark results of the alternative identification
- Support for the recommended product(s)
- How the recommended product(s) may dovetail with the current technology
- Vendor history
- Other organizations that have successfully implemented the product
- Any shortcomings or risks involved with the proposed product

Purpose of the Feasibility Study

Most feasibility studies are launched to determine if an identified opportunity is valid or to determine if an identified problem can be resolved. Feasibility studies can also be initiated for a number of other reasons:

- Compare hardware solutions.
- Compare software solutions.
- Determine buy-versus-build opportunities.
- Determine capability gaps with new technical solutions.
- Determine disaster recovery options.
- Investigate maintenance of the organization's technical status.

This section describes the intent of the feasibility study, considers why it was initiated, and connects the feasibility study to the business goals and objectives.

Options Assessed

Feasibility studies examine multiple options, often called alternative identification, for the business problem or opportunity to determine the best solution for the organization. The project manager needs to explain which options were assessed in the study, why the options were selected, and how the options differ from one another. For example, you might investigate for your sales reps four types of mobile phones and service packages from different companies. Each phone may have similar traits but may differ in application usage, costs, maintenance, and interoperability with existing technology.

Assumptions Used in the Study

An assumption is something you believe to be true but you have not proven to be true. In a feasibility study, you may have to make certain assumptions with the technology you're assessing for the sake of time and cost. For example, you might assume that your server configuration will work with each of the software options you're researching because it meets the hardware specs provided by the vendor. You'd choose this assumption rather than testing the software options on a live production server.

Audience Impacted

The feasibility study should address issues concerning the users, capability resource gaps, and who will be affected by the implementation:

- How much downtime will the audience experience because of the implementation?
- What is the learning curve of the new software?
- Will training classes be needed for all users?
- How will the recommended software transfer or work with your organization's existing technology?
- How long before this software will be upgraded again?
- How long before it will be retired, obsolete, or no longer supported by the organization?

Also in this section of the plan, you need to mention how the technology will be implemented. If one part of the organization switches to the new technology before other parts of the organization, for example, will the technology have an impact on work and communication between the two parts of the organization? How long will the technology implementation take?

Financial Obligations

This section of the feasibility study provides an overview of the cost of the technology rather than a full-blown budget. Consider these factors:

- Price of the technology product
- Necessary licenses

- Training for the implementation team
- Cost of labor to create or implement the solution
- Technical support from the vendor
- Outside talent and contractors to install the technology
- Monthly fees that may be associated with the technology (for example, service-related fees such as those for using a remote data backup service)
- Cost of not implementing the solution

The financial obligation section can also include return on investment (ROI) analysis. You should demonstrate how the technology will increase productivity, be easier to use, increase sales, or otherwise prove relevant. Of course, back up your facts with references from your research.

Recommended Action

Within this section of the feasibility study, you're ready to make your pitch for, or against, a technology to solve the problem. You should present a general overview of how the technology works, how it will be implemented, and what types of resources are required to make it work in your environment. You can also make a recommendation to investigate other options or newer technologies at this time—just be certain to explain why.

The solution and actions you recommend must be in alignment with the project purpose. A recommended action must address and satisfactorily answer the purpose of the project. Consider the reason why the project may be initiated, including

- To solve an existing problem
- To increase productivity
- To become more efficient
- To reduce costs
- To increase revenue
- To become more competitive in the marketplace

Now that you know the different parts of the plan, take a look at the executive summary of a sample feasibility study in the sidebar, "Executive Summary for Murray Enterprises." This company is moving to a new facility and will need to create a new network.

Executive Summary for Murray Enterprises

Written by Justin Case, IT Manager

Executive Summary The purpose of this feasibility study is to determine the type of cabling, wireless access, and related network devices required to improve the speed and reliability of our current LAN in our new facilities. Our current network is dated, sluggish, and unstable. A change of technology is required to increase the speed and reliability of our network. Our current LAN is a 10Base-T Ethernet network, and there is no LAN in our new office space. This is an opportunity to create a network that can support our current business needs, be scalable for growth, and be cost effective.

Defined Business Opportunity Our current network is nearly ten years old. The network's age, however, is not the problem; the problem is its capability to manage our company's growth into Internet publishing, online videos, and other Internet applications. As more users are creating and using Internet-based technologies, our network throughput is diminishing. As we prepare to move to a new facility, we have an opportunity to create our network with scalable, faster technology. The goal of this feasibility study is to determine what network options will best suit our current needs and allow us to continue to scale our networking demands.

Assessed Technology: Install CAT6a Cabling for Our Entire Network The current network is CAT5, but CAT6a is becoming the de facto standard for new cable installations. It provides better immunity from noise, which means less network traffic, fewer lost packets, and generally higher reliability than CAT5 networks.

- Install gigabit switches to segment and control network traffic.
- Upgrade wiring closet to gigabit equipment.
- Install 1000Base-T network cards in all compatible devices for faster throughput.
- Replace 850 PCs with new workstations that have gigabit-compliant hardware.
- Install CAT6a for speeds up to 10 Gbps.
- Install wireless access points throughout the facilities.

Audience Impacted The change would affect all users. The new network cabling would be created and installed, while the existing network would remain as is. The switch of the PCs to compatible NICs will happen by December 30. Users' logon processes and usual workflows will remain constant; only the speed will be faster and more reliable.

Financial Obligations

- Initial projected cost of the CAT6a installation: $19,800
- Cabling and connectors: $3,900
- Two switches: $11,800
- Wall-mounted patch panels: $4,800
- Network installation kits: $4,200
- Wireless access points: $2,550
- Electrician: $3,500
- 850 PCs: not included in this budget but will be coordinated with normal operations

Recommended Action Upon final approval, a project charter will be drafted and the team will be assembled. A plan of action will be created for the implementation. Upon arrival, the patch panels and switches will be installed and tested.

Cabling will begin at the top of the project. Next, our team will complete the testing of the switches and network cards, and then connectivity will begin. No PCs in production will be connected to the new infrastructure until the new technology has been proven reliable and has passed a quality audit. The workstations in production will cut over to the new infrastructure in waves, as the switches we choose will be backward-compatible for CAT5 networking. Wireless access points will be added throughout the facilities in discrete, secure locations.

Creating the Business Case

Another document the project manager may be required to create is the *business case.* Sometimes the business case is done in tandem with the feasibility study, and sometimes it's a stand-alone document. Either way, its purpose is the same: to help the organization determine whether it can justify the cost of the project in proportion to the return on investment. The business case links the value of the project's solution to the organization.

You'll need to do some research and analysis based on the business objectives and goals. Considering what the goals of the project are and the proposed solution to reach those goals, the project manager can predict the costs of the project, the duration until a break-even point is reached, and the anticipated ROI. It's not uncommon to include information on cash flow, related opportunities, and life cycle costing for the solution. *Life cycle costing* describes the cost to maintain the solution; usually management wants to know the maintenance costs for the solution for each year the solution is to be used.

EXAM TIP Depending on the organization, the business analyst, program manager, or some other stakeholder may write the business case. In some organizations, the business case isn't written at all. The business case is the documentation of the expected return on investment for the project.

The business case documents the quantitative value of the solution and the ROI, but it can also include an analysis of the qualitative values: morale, comfort level, and appreciation. In Chapter 1, I introduced the concept of the business case and its components. Now let's dig in a little deeper and explore four must-have elements you'll need for your business case.

Benefits of the Solution This section is the primary purpose of the business case. It defines the quantified benefits of the solution, justifies the costs, and defines the ROI for the project. The benefits of the solution can also include qualitative benefits, but there needs to be some evidence that the qualitative benefits will actually come into existence for the project. The solution benefits should be analyzed through SWOT analysis, a technique to determine a component's strengths, weaknesses, opportunities, and threats.

Cost of the Solution The business case should estimate the total costs of the solution. Cost estimates in the business case include the anticipated costs of the development of the solution, the life cycle cost of the solution, and the total cost of ownership once the solution is implemented. If you'll be outsourcing the solution to a vendor, you'll need to assess potential vendors and their abilities and potentially request and review proposals. You might also need to define the opportunity cost of the solution should your organization implement one solution over another viable opportunity. Opportunity cost is the total amount of the opportunity that could not be implemented because this solution is selected instead.

Risk Assessment The business case should include an initial risk assessment that defines the obvious risks in the project, the anticipated impacts and probabilities of the risks, and the risks of not doing the project. In IT project management, this initial risk assessment primarily focuses on the inherent technical risks such as downtime; data loss; delay in project deliverables; and risks specific to hardware, software, networking, or software development. It's not practical to include a full review and assessment of all project risks at this point of the project development. Once the project is green-lighted, risk identification and assessment happen at a much deeper level.

Results Measurement Everyone wants a good project, but what's good to the project customer may not be the same definition of good to the project manager. Metrics and key performance indicators (KPIs), such as cost, schedule, quality, and scope, need to be defined so that project requirements and goals can be established and agreed upon. When subjective terms are tied to the project requirements, too much is left to interpretation. Key performance indicators establish the project components that will be measured to determine the success of the solution.

TIP The business case and feasibility study happen before a project is launched. Every organization is different, but generally, the research to do a project is separate from the actual project launch. Once a project is deemed valuable, then the charter and other project planning commence.

Writing the Project Scope Statement

Now it's time to create one of the most important documents in your planning process: the *project scope statement.* This document defines all of the deliverables the project will create, the boundaries of the project, and the work that the project team will need to complete in order to create the project deliverables. This document is based on the project requirements, the feasibility study, the business goals and objectives, and the business case document. Note that project scope statements are used in predictive projects but not in agile projects.

The project scope statement usually passes through rounds of research and refinements before the project manager, the project sponsor, and the key project stakeholder(s) approve it. You might know this progression of scope refinement as *progressive elaboration,* which means that you start with a broad definition and then elaborate on the specifics of the deliverables through a series of refinements.

The project scope is a primary input to the remainder of the project planning and serves as a reference for all future project decisions. The project scope defines what's in the scope—but it also defines what's out of scope. For example, your project scope statement focuses on creating the new software, but the project won't include deploying it to the 10,000 users in your organization. Project boundaries are important to define so that stakeholders don't make false assumptions about what's in or out of the project scope.

One of the primary jobs of the project scope is to define how the project will be measured for its performance. You'll use KPIs to gauge how well the project is meeting the objectives of the scope; these indicators are usually linked to milestones at the end of project phases, schedule, cost, quality, and other objectives that you, management, or other stakeholders define. You'll use KPIs to determine what's important to the project stakeholders and to link project performance to satisfying those objectives by completing the project scope.

The project scope, in a predictive project, should be protected from change. Once the research, requirements gathering, and scope definition have been completed, the project scope statement should be averse to changes. Changes are allowed in a project, but all proposed changes should flow through a change control system so that they're documented, reviewed, and then (perhaps) approved for the project scope. Approved changes cause the project scope statement to be updated. (I'll talk more about change control for the entire project in Chapter 10.)

Your project scope statement will likely go through revisions on its way to final approval. It's important to involve the project stakeholders in the scope creation and to seek their approval of the project scope statement. This is part of scope validation—confirmation that what you've written in the project scope statement is what the project stakeholders expect from the project. Scope validation is needed to ensure that what's in the project scope statement satisfied the original business need of the stakeholders. Scope validation is also the customer acceptance of the scope once you've completed the project work.

The project scope statement has six components, as described next.

Product Scope Description The *product scope* is a description of the features and functions the customer will receive as a result of your project team completing the project. The product scope is the thing the project will be creating. For example, if a customer wants you to create a new piece of software, they'll describe the product scope, how it will be used, the functional components of the software, and the tangible components of the solution.

Product Acceptance Criteria The project scope defines either directly or by reference the technical requirements, expected deliverables, and/or the detailed design documents that constitute the product deliverables. The *product acceptance criteria* clearly define what the project must create in order for the project to be accepted by the customer and for the project to be considered completed. This portion of the project scope is important, as vague requirements and product acceptance criteria allow the project duration to linger. You and the project customers must be in agreement at the start of the project as to what constitutes the acceptance and closure of the project.

Project Deliverables The project will create *deliverables* that the customer will accept in the form of the completed product scope. The project may also create or purchase other deliverables such as tools, templates, reports, plans, and other ancillary deliverables that the organization will retain as a benefit for completing the project work. The project deliverables that the organization retains become part of organizational process assets so that other project teams can benefit from these deliverables in their relevant projects.

Project Exclusions The project scope statement must define the boundaries of the project to communicate what will not be included in the project deliverables. It's important to define what's excluded so that there's no confusion when the project manager wants to close the project and the project customers are expecting more deliverables.

Exclusions are items that are explicitly not in the project scope. You don't want the project customer to believe they are getting something that's not in the project scope. For example, your project may create the new software, but your project will not distribute the software as part of the project scope. Exclusions communicate exactly what's not included in the project scope and help to remove any assumptions about what the customer will receive at the project closure and operational transfer.

Project Constraints *Constraints* are things that limit the project manager's options. Predetermined budgets, deadlines, resources, preferred vendors, and required technology are all examples of constraints. Project management always has three constraints: time/schedule, cost/budget, and scope. These are sometimes called the *triple constraints of project management.* The project manager should identify and document all the known constraints. These are also called The Iron Triangle of Project Management. Each side of the triangle represents one of these constraints and all three sides must remain of equal length to the other sides.

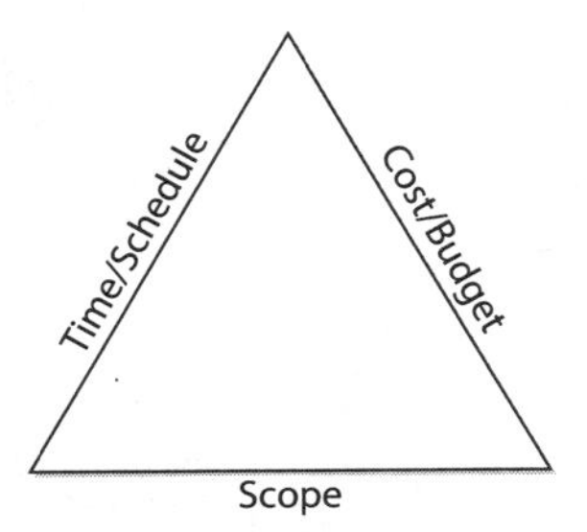

Project Assumptions As part of planning, there may be *assumptions* that must be made in order to plan effectively and timely. Assumptions about hardware and software compatibility, resource availability, longevity of the solution, and a stakeholder commitment to the project are all common assumptions. This portion of the project scope should also include information on the impact of the project's success should the assumptions prove to be false. All project assumptions should be evaluated later in planning to determine their risk for the project should the assumptions prove untrue.

Planning in Agile Projects

Agile projects use short bursts of planning and execution to move the project forward, but there is some upfront planning that takes place before or at the very launch of the project. The business case and feasibility can also be utilized in agile projects to help identify the business value and the features and functions of the product. Agile projects focus on creating business value and limiting the amount of time spent planning things that likely are going to change in the project. I'll talk in more detail about agile projects coming up.

One of the most common approaches to agile project management is called scrum. Scrum comes from the rugby term where the team huddles around the ball to get things in play and get the game moving again. Figure 2-3 shows the typical flow of a scrum project. The following list describes the common meetings, called *ceremonies*, and documents, called *artifacts*, that you'll see in a scrum project. These steps are repeated iteratively in this order throughout the project until the project is completed.

- **Product backlog prioritization** The product owner works with the customers, business analysts, executives, and other key stakeholders to gather, document, and prioritize all of the currently known project requirements. This prioritized list of project requirements is the product backlog. The product owner maintains the list, adds or removes features as needed, shifts the priority of requirements, and works with the development team to clarify any requirements, features, and functions.

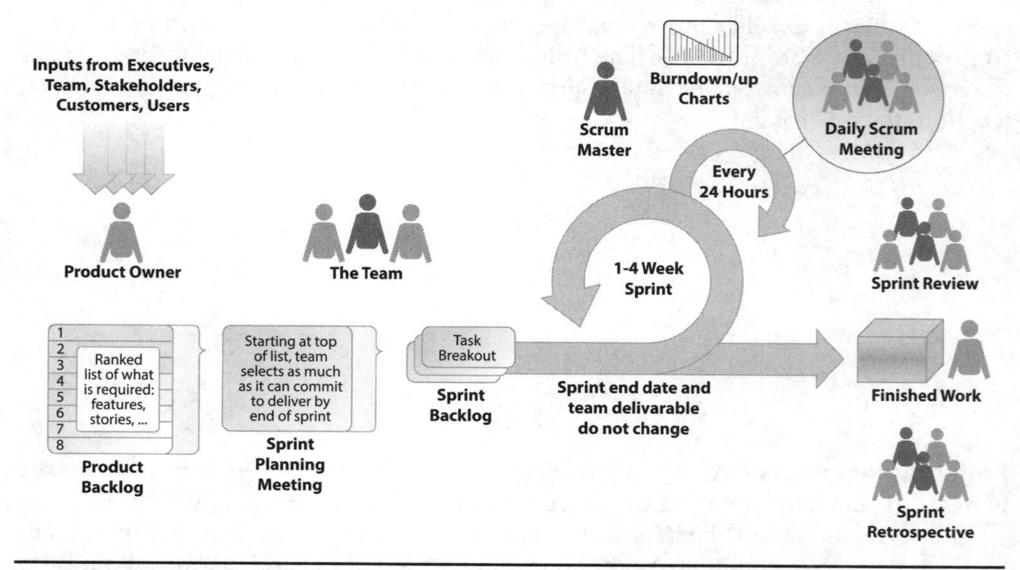

Figure 2-3 Scrum projects follow an iterative approach to project completion.

- **Sprint planning meeting** At the launch of the sprint, the development team, the product owner, and the project manager (called the *scrum master* in scrum projects) plan how much work the team can do in the next iteration, called a sprint. Sprints are time-boxed sessions of work that last from two to four weeks. Starting at the top of the product backlog, the development team selects a chunk of work that they believe they can complete in the sprint. The team always pulls items from the top of the prioritized product backlog so that the items with the most business value are being created first. The team also determines who will do what work during the sprint planning ceremony.
- **Sprint backlog** The items that are selected from the product backlog to be completed in the current sprint are called the sprint backlog. From the sprint backlog, the tasks and responsibilities are determined. Now the team can get to work.
- **Daily scrum** Every workday, at the same time and place, the development team meets for a 15-minute ceremony called the daily scrum. Although other stakeholders, such as the product owner, may attend, only the development team members speak. It's quick and the agenda asks the same three questions each day: What did you work on since we last met? What will you work on today? Are there any impediments that need to be resolved?
- **Sprint review** At the end of the sprint the development team demonstrates what it has completed during the sprint. The sprint review ceremony demonstrates results that are completely done (not nearly done) are demonstrated for the scrum master, the product owners, and the key stakeholders. This allows the stakeholders to see progress, get excited about the project deliverables, and offer feedback for any changes, approval, or clarifications.
- **Sprint retrospective** The sprint retrospective allows the team to discuss what did and did not work well in the past few weeks of the project. This is not a meeting to blame each other for faults but to discuss successes and failures and how to improve the project in the next sprint. Retrospectives are a kind of lessons learned meeting to immediately put the lessons learned into action during the next iteration of the project work.

These are the key steps of a scrum project, and they are repeated throughout the project until the project's conclusion. This example examined scrum project management, but the other types of agile project management have very similar concepts that we'll look at later in this book. For now, scrum is a good introduction to show the iterative planning that happens in all agile projects.

Establishing Project Priority

As a project manager, you'll likely find yourself managing multiple projects. You may also find yourself going head-to-head with other departments implementing similar projects or, worse, conflicting projects. Given that every organization has different approaches to project management, your odds of success will increase if you know your organization's approach.

Project priority may shift from quarter to quarter or year to year. Project portfolio management is a process an organization takes to pick and choose which projects are needed, are worthy, and should continue. Just as you might manage your financial portfolio, an organization has a responsibility to manage its portfolio of projects. The value of the project, the project sponsor, the current success rate of the project manager, and the purpose of a project are all factors an organization may use to determine which project takes the highest priority.

If your organization relies on a project management office (PMO), you may have some assistance in addressing project priority with senior management. Recall from Chapter 1 that the role of the PMO is twofold: it offers traditional project management services for an entire organization or portion of an organization, and it serves as a governing committee for all projects throughout an organization. If your organization were to participate in a PMO relationship, conflict resolution, budgeting, and the process of implementing projects and controlling projects would follow a system of checks and balances unique to your organization.

Your project sponsor should be as excited and motivated by the technology to be implemented as you are. The sponsor, hopefully, will be able to come to your defense, or more accurately, your project's defense, should the need arise. This is one of the fundamental reasons why the right sponsor is needed for the project. A sponsor who lacks the authority, commitment, and ability to protect and promote the project does little to move the project forward.

The goal of a project sponsor is to increase profits or reduce costs through the implementation of the technical project. The project manager acts on behalf of the project sponsor. While ideally project sponsors serve as mentors and guides through the project phases, a sponsor in many organizations is little more than a figurehead, and the project manager has no one to shield the project. Hopefully that's not the case in your organization.

Ideally, the project sponsor can help clear any roadblocks from the project's progress. A *roadblock* is anything that is preventing the project from moving forward, and the project manager can't resolve the issue alone. The project sponsor has the authority and power to clear the roadblocks and keep the project moving forward. It's important that early in the project the project sponsor is introduced to the stakeholders and it's communicated that the project manager is acting on behalf of the project sponsor.

One of your project management jobs is to relay updates from the project team to the project sponsor and from the project sponsor to the team, as defined in your communications management plan. A communications management plan defines all the required communications, scheduled meetings, and expected types of communications, depending on the project scenarios. I'll talk more about creating this plan in Chapter 6, but for now, know that you'll need the communications management plan to keep stakeholders informed. By keeping your sponsor informed of the project status, you personalize the project for him—as it should be. Figure 2-4 shows the path of communication among the project sponsor, project manager, and the project team.

The project sponsor typically has delegated the management of the project to you. You delegate some, or all, of the tasks of the project research and implementation to the team members. Your job, like that of the project sponsor, is not to micromanage, but to

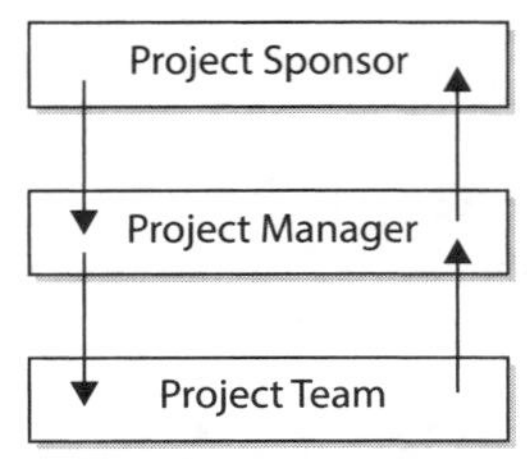

Figure 2-4 The project manager directs the flow of communication between the team and the sponsor.

organize and keep the team on track. As part of this delegation from the project sponsor to the project manager, and from the project manager to the project team, there needs to be a clear message that outlines the consequences of non-performance. Everyone needs to know the role they're assigned, what it means to be successful in that role, and what will happen for non-performance. Performing in the project means hitting goals, completing assignments correctly, and participating in the project.

Depending on your role in an organization, the project may be assigned to you or created by you. Should the project be assigned to you to manage, the role of the project sponsor is like that of the parent in a parent-child relationship. This is to say, the project sponsor is the parent of the project, which is the child, and is deemed responsible to complete and manage the tasks required to finish the project.

If you've created the project, the role of the project sponsor is similar to that of the investor in an investor-entrepreneur relationship. You, of course, are the entrepreneur. You've done the research, presented the facts, and then sold the sponsor or the organization on the project idea. The project sponsor has invested their credibility in your plan.

Internal Competition

IT projects can grow quickly and spin out of control and size. Imagine you are doing an operating system upgrade for the client workstations. Your plan calls for using Transmission Control Protocol/Internet Protocol (TCP/IP) for all of the hosts. You have decided to use Dynamic Host Configuration Protocol (DHCP) to assign all of the IP addresses for all of the workstation operating systems. (Without dynamically assigned IP addresses, users won't be able to access network resources.) Unbeknownst to you, another related team is working on segmenting the network and positioning switches and routers at key points.

EXAM TIP Agile projects use a project-centric approach, where the project team is on the project full-time and doesn't switch between tasks. Waterfall, or predictive, projects may use a matrix approach, where the team is on multiple projects and has responsibility for day-to-day operational work.

Can you see the trouble brewing? If you're not a network person, this scenario may look innocent enough; however, for either team to be successful, each team's plans must take the other team's plans into account! Figure 2-5 shows a map of the network. If the router and IP addressing information are not agreed upon between the two teams, the entire network can crumble.

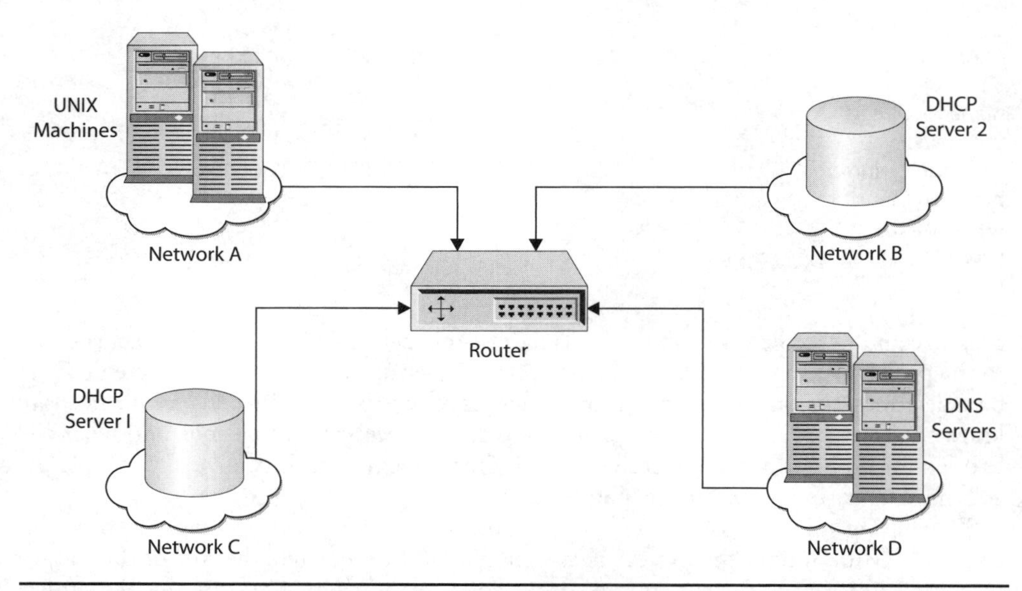

Figure 2-5 Teams must work together for projects to succeed.

The routers and switches team must agree on the network addresses, IP addresses for gateways, types of broadcasts that are permitted to pass through the routers, and more. The client OS team has to agree on which ranges of IP addresses to use, the positions of the DHCP servers and DNS servers, and the assignment of any static addresses for printers and servers within each segment.

This scenario is an example of *interproject dependencies.* Your project needs coordination with the routers and switches team, and they'll likely need information from you to cooperate. Interproject dependencies require communication, mutual respect for one another, and a spirit of working together. They can, however, cause problems such as scheduling issues, scope overlap concerns, and resource contention.

Resource contention happens when two projects need the same person or resource at the same time. This demand for a resource among projects creates an interproject resource contention that must be addressed as soon as the contention is discovered. Hopefully, the project managers can work together to resolve the issue with the needed resource, but in some instances, the contention has to be escalated to the project sponsors to help find a solution. This can often create frustration among those involved in the projects, as one project may have to wait before it can move forward with its execution.

Most organizations should have strategic tie-ins among departments, lines of communication between project managers, and ways to resolve differences and work together when conflicting projects arise. It is surprising, however, how many organizations do not.

At the basis of this problem are myriad issues that project managers find themselves dealing with: greed, personal achievement, personality conflicts, politics, and grudges.

These are all issues that can take the focus off of the success of the project and can ultimately throw the project off track and even bring any progress to a halt. When these situations happen, and they will happen, a project manager can take a few steps to bring the project back in line:

1. A meeting should take place with just the project managers from the conflicting projects. These individuals should outline their projects and discuss how both teams can work together and continue with their projects. The project managers should be diplomatic, willing to negotiate, and eager to find an agreeable solution.
2. If the project managers cannot find a solution between them, the next step is to conduct a meeting with the project managers and the sponsors of the conflicting projects. The project sponsors should lead the discussions and help the project managers find an agreeable, win-win solution.
3. If an agreeable solution can't be found among the project sponsors and the project managers, then the discussion can continue to work its way up the organizational chart until a decision or agreement between parties has been made. Ultimately, the good of the organization should win based on the priority of the projects.
4. The conflicting projects must be evaluated and weighed, with the goal being to determine which project will benefit the company the most. Once that has been determined, a solution must be created so the two projects can continue, one is dissolved, or one is put on hold. While that seems easy enough, it's not uncommon for a lower-priority project to need to be completed so that a higher-priority project can continue.

This entire struggle would be unnecessary if departments would simply communicate with each other. This internal struggle tears down morale, wastes time and finances, and hurts the organization. IT project managers must learn to work together, to be reasonable, and to communicate.

Every entity that has multiple project managers should create a system to share information on projects. An intranet solution would be easy to implement and manage. As new projects are being researched, a quick look into existing projects on the organization's intranet would allow teams to work together, accomplish more, and, again, be more productive. Far too often, unfortunately, projects are kept secret, fiefdoms are established, and domains are guarded among departments, managers, and lines of business.

Obtaining Budget Dollars

Research for any project must include information on the financial obligation required to implement the technology. Chapter 5 will focus on all aspects of budgets; this section introduces the financial planning stages of a project.

When you are considering implementing new technology, you must take a long, hard look at estimating the project expenses. Any organization can throw money at technology and hope for the best, though that free-for-all in IT is long gone for most organizations. Throwing money at a problem with little or no planning is as promising an investment as playing craps. The technology you recommend the organization to purchase needs to be the right power, the right size, and the right price.

In the technology world, it's easy to fall in love with the latest application, multiprocessor server, or network operating system. But is the technology good for the organization? Ask yourself these questions when making decisions on technology:

- How will this technology enable the organization to be more productive?
- Will the technology promise an acceptable ROI?
- How is the technology the right selection for the organization?
- How soon will this technology need to be replaced?
- What is the break-even point (also called the payback period) for this investment?
- When does it become profitable?

If you cannot answer these questions fully and accurately when selecting new technology, then you have not completed your research.

Value vs. Investment

Value and investment, when purchasing anything, often are incorrectly seen as the same concept. Value is a perception. Investments are a reality.

Here's a simple example: Imagine you are purchasing a bag of flour. The one-pound bag of flour is priced at $1.00. The three-pound bag of flour is priced at $1.92. Reason says that it's better to spend 92 cents more for the three-pound bag of flour, because it lowers the price to 64 cents per pound. The problem is, unless you actually use more than one-and-a-half pounds of flour, the purchase isn't a value, it's a loss. Use a little more than a pound-and-a-half of flour and you've used approximately $1.00 of flour at a better price.

How does this relate to IT projects? When you are making your technology selections for your organization, it's imperative that you know what technology will produce the desired results. The largest piece of technology is not always the best. Some consultants will advise you to buy the biggest chunk of technology you can afford, because it will be outdated in six months anyway. Wrong!

Always buy the right technology that will help your organization achieve the desired results of the project. And as for the consultants with the "spend big" mentality, question their credibility along with their advice. Take note that it's not their budget, their career, or their project. It's always wise to buy hardware and software that's scalable, but you must research how much larger the demand for a given resource will be in an organization. And you need to consider whether the value gained for speed, bandwidth, or any other resource is worth the investment.

A contributing factor in all projects is the expected level of quality. *Quality* is the ability and the completeness of the project deliverables to meet the requirements the stakeholders have defined for your project. Technically, quality is a conformance to the project requirements and a fitness for use. Within the confines of quality, you'll find grade. *Grade* is the ranking of service or materials. For example, you can purchase different grades of cables, monitors, computer equipment, and so on. You can also subscribe to different levels of software support: bronze, silver, and gold.

Within a project, you must evaluate the expected level of quality and then the correct grade of the materials and services needed to satisfy the requirements. Low quality is always a problem; low grade may not be. In addition, you must consider what happens if you overshoot the expected level of quality. While it's better to err on the side of caution, it can also be wasteful to deliver a level of quality that far exceeds what the customer was expecting. Some unscrupulous project managers consume their entire project budget by adding features, further testing, and other deliverables that were not called for in the project scope. They're trying to use all the monies in their budget by adding project extras—this is called *gold plating,* and it's a waste.

I'm sure some project managers will want to argue this point. They'll say that it's unreasonable to surrender funds in the budget. Or that management will cut future project budgets based on the performance of the current project. Or they'll look foolish with funds left over after they've committed the organization or customer to a set fee. This type of thinking is faulty for several reasons. First, the monies spent on gold plating extras is a defect in quality. Since quality is a conformance to requirements, anything that doesn't conform exactly to requirements is poor quality. Second, the monies wasted on gold plating are funds that may be needed elsewhere in the organization. Finally, the time and monies spent to deliver the gold-plated extras may affect the customer's ability to realize a faster return on investment, shorten the market window, or delay their usage of the project deliverables.

I do believe there's nothing wrong with presenting value-added changes to the project customer when there are funds and time left in the project. Gold plating consists of changes that the project manager adds without the customer's approval, simply to eat up the project budget, not to improve the project value.

Another budgeting concern you must consider during the research phase of your project is time—one of the largest and often overlooked expenses of project management. For example, if you have five members on your team and a project that will take three months to implement the technology, that's 15 months of combined labor time. If you want to see it on a more common, accessible scale, just look around the room during the next meeting that's a waste of everyone's time. Take a two-hour meeting with ten people, and there goes 20 hours of the cost of labor down the drain. Figure 2-6 shows how an increase in time must be tempered by specific goals and a deadline; otherwise, costs run awry.

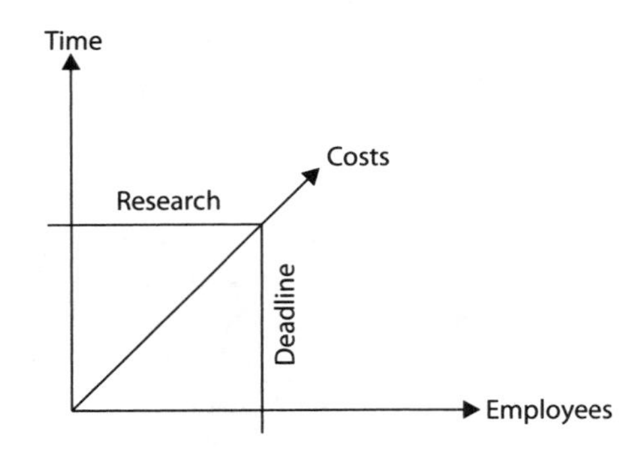

Figure 2-6
More time equals more expense.

So? If the project team is on the project full time, that's 15 months of time away from regular duties, 15 months of salary, and 15 months of time that can never be recovered. When implementing the technology, consider the time commitment required from each team member and yourself. While not all organizations assign team members to a project on a full-time basis, you still must account for the sum of the project team's time and its value to the project.

Should you decide to implement the plan through a third party, such as a value-added reseller (VAR) or the original vendor of the product, consider their costs combined with the time they'll need to complete the job. There are usually three different ways VARs will bill for technology implementations, and all can be costly if you don't stipulate all of the conditions:

- **Time and materials** Most technology integrators like to bill time and materials because there may be some additional problem discovered in the midst of the project that can result in the vendor working extra hours toward a solution. The trouble with time and materials billing is that some deceptive vendors take advantage of the situation and stretch the hours to increase the invoice. If you choose this billing method, you'll want a not-to-exceed (NTE) clause in your contract. You will also spend more time managing the contract to ensure you are getting the appropriate value for the time you are paying for.
- **Fixed-fee contracts** Some vendors know exactly what it will take to complete their technology installation and will offer a set fee. The trouble with set fees is that your organization may feel cheated when the installation takes very little time and is completed long before you thought it would be. Understand that most installers probably have developed a script or routine that automates much of the installation process. They can do the job in less time and with less frustration than you could on your own. Fixed fees are generally a low-risk solution for the buyer, as any cost overruns go back to the vendor.
- **Cost-plus contracts** A cost-plus contract represents a set fee for the procured work plus a fee for the work. For example, a vendor may charge $7,500 for a new server, rack, and cabling plus the cost of any materials used during the installation. Some unscrupulous vendors try to use a cost plus a percentage of costs contract where they expect you to pay for the cost of the materials plus 10 percent (or more!) for the materials. Cost-plus contracts are the pits for most buyers, as the vendor can drive the price up by actually wasting materials. There are some cost-plus contracts that include incentives and penalties if the vendor finishes early or late—though you can add these terms to a fixed-fee contract if you want.

With any contract method, consider the vendors' costs and calculate the ROI. Finally, make the companies completing the implementation guarantee their work in the contract you negotiate. Again, some vendors will want you to entertain a cost-plus-fee contract. These contracts demand a cost for the project materials, labor, or other elements of their implementation plus a fee for the project to be completed. The fee portion of the contract is usually fixed, while the materials and labor can fluctuate—giving the vendor an opportunity to drive the costs up and leaving you with more risk. As a general rule, you want to avoid cost-plus contracts. I'll talk more about procurements and contracting in Chapter 6.

Creating an Approach

When you begin to do your planning, you need a plan of attack. How much time will you commit to this phase of the project? What, or who, will be your resources? What is the goal of the research? Who else will be assisting you? These are all questions you should answer as your research begins.

The size of the project can help you determine how much time you'll need for planning. Of course, not all planning happens in one big chunk. You'll have some up-front planning, and you'll revisit the planning process throughout the project. For example, if your project team is creating an application, you'll meet with the stakeholders to determine their needs, create an approach to developing the application, and so on. As it moves into execution, your team may need additional planning time to solve problems within the development.

Here's a basic guide to determine how much planning time you can expect for the type of projects you'll manage:

Project Type	Attributes	Planning Time
Add/Move/Change	These are generally smaller projects that, as the name implies, add, move, or change some element within an organization.	Ten percent of the project time allotted to planning
Micro project	These projects take less than 2,000 hours of implementation and/or less than $250,000 to complete.	Twenty-five percent of the project time allotted to planning
Macro project	These projects take more than 2,000 hours of implementation and/or more than $250,000 to complete.	Thirty percent of the project time allotted to planning

It's easy to get bogged down in planning and hop from resource to resource rather than focus on a specific objective. While quality research takes time, an organized approach will allow you to find the information you're seeking in less time.

Create a Milestone List

One of the primary goals of planning is to determine how the project will be completed, what resources will be required, and what tasks will be involved in the project. Part of planning any project is to create a task list, which simply comprises the major steps required from the project's origin to its conclusion. You create a task list after the technology has been selected and before you create the implementation plan. Some projects use a task board that shows each task, who is responsible for completing the task, and the progression the task is making through the project.

There are multiple approaches to creating a task list, but some things must be accomplished before the task list can be created. In Chapter 4, we'll focus on creating the *work breakdown structure (WBS),* a cornerstone of project management. The WBS is a deliverables-oriented collection of the project. The WBS provides a true reflection of all the deliverables the project will create, and using this, you can create an accurate and complete task list. In this early phase of your project, you'll likely need a task list to ascertain how long the project may take, what types of resources are needed, and even how much the project will cost. One of the best and most direct approaches is a simple outline of what needs to be accomplished and in what order. For example, if you were creating a task list for the installation of new software on every workstation, your task list might look like this:

1. Test the software in our lab.
 - **A.** Test with current laptop hardware.
 - **B.** Test with Microsoft Office.
 - **C.** Test with Windows operating systems.
 - **D.** Test hardware with printers, scanners, and digital cameras.

2. Resolve and document any bugs or glitches from the testing phase.

3. Create installation methods for each OS.
 - **A.** System policies
 - **B.** Image deployment options
 - **C.** SMS server packages and database of hardware and software

4. Test rollout methods and document.

5. Roll out to pilot group of users.

6. Begin offering training of software and hardware to population.
 - **A.** Instructor-led training
 - **B.** Web-based training
 - **C.** Documentation for users

7. Finalize rollout plans and documentation.
 - **A.** Roll out software to users as they complete training class.
 - **B.** Work with help desk to answer support calls.

While the list is simple and direct, it allows the technical project to begin to take shape. It also allows the project manager to determine the type of talent, number of team members, and time commitment involved with the project.

The outline approach, as just demonstrated, is not always the best method, however. Some of the tasks in the preceding scenario can be completed simultaneously rather than sequentially. In these instances, a visual decomposition can really be beneficial. Using the same scenario, Figure 2-7 shows how the project would look in an organizational chart using parent-child relationships.

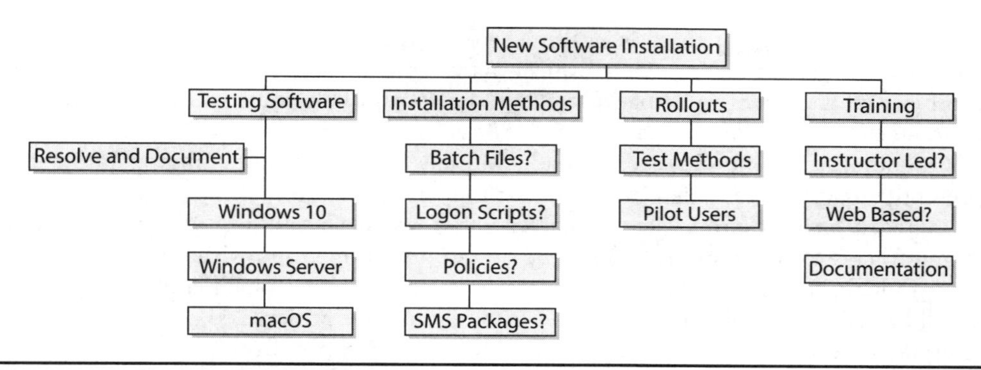

Figure 2-7 Charts can help you visualize the work to complete in a project.

Another approach is to use a software program such as Microsoft Project. Project enables you to enter tasks and then edit them in more detail as the project develops, as shown in Figure 2-8. If you are using Microsoft Project or other project management software, you may want to consider starting your entire project planning stages with

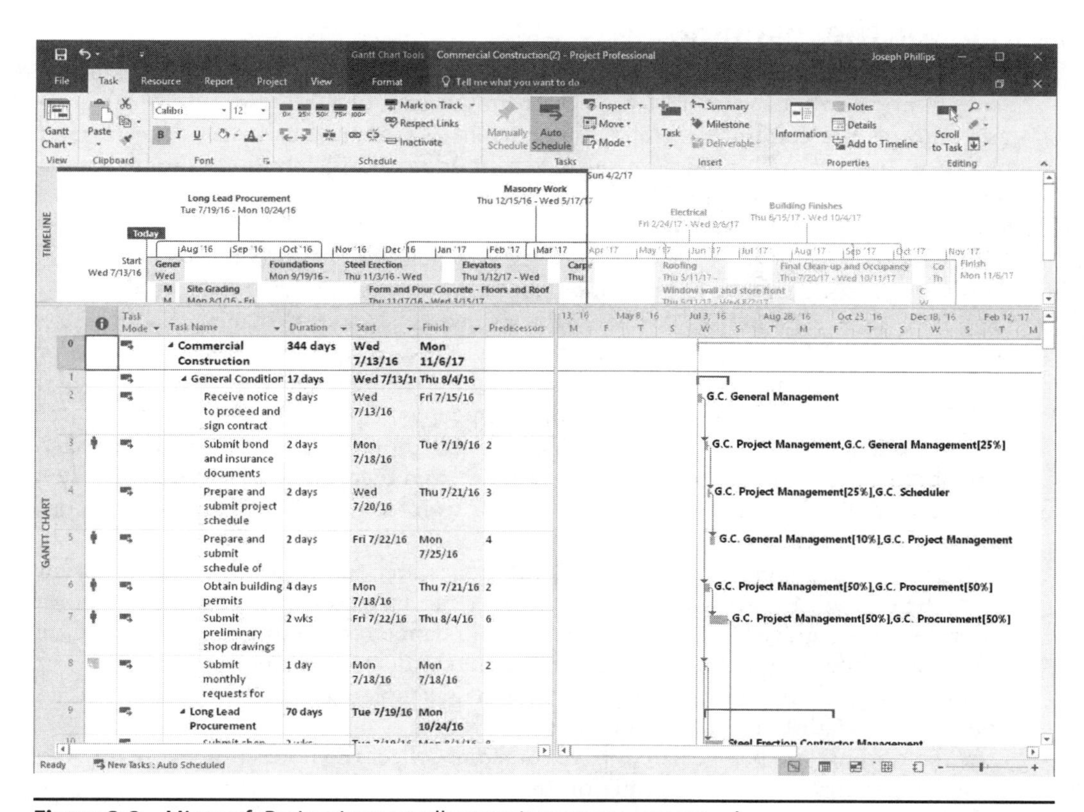

Figure 2-8 Microsoft Project is an excellent project management tool.

the software. Certainly, there is nothing wrong with creating an outline or flowchart and then transferring it into your project management software. Be cautious, however; a project management information system is a tool to help manage the project, not a guaranteed solution for project success.

Of course, planning starts very broad. You may have to do some initial planning and present your results to management so that they may determine if the project should be chartered. Once the project has a charter, you'll revisit planning to map out the milestones to complete the project. With the milestones in sight and with your WBS, you'll determine the activities required to get from milestone to milestone. Planning is iterative. You'll be visiting the planning processes throughout your project.

Agile projects use a product roadmap to show when releases from the agile project will be available. The roadmap shows milestones that identify the creation of key features. For example, a milestone in a software project might be to do a release when it completes three of four key features of the software. These features may take three to six increments to reach, which would mean a release would happen every 12 to 24 weeks depending on the software being created, the business cycles in the organization, and the desire from the customer to have the releases. The product owner is responsible for creating the product roadmap.

Manage the Planning

If you are completing the majority of the planning phase with your project team, pay close attention to the amount of time the team invests. Quality research doesn't come easily and takes a commitment of resource labor to produce quality results. However, too much time invested in research can lead to muddied results, meandering from topic to topic, and project burnout. It is always in your best interest to set a specific goal and deadline for the initial research.

To set a research goal, create a list of questions that you'll need answered to manage your research time. In the online resources that accompany this book, you'll find a document called Research Objectives. (If you need assistance accessing this file, refer to Appendix C.) While this worksheet is simple and direct, it can keep you focused on resolving fundamental issues and then branching into more detailed objectives.

If you are fortunate enough to have multiple people helping you research the project, don't be tempted to micromanage. Assign research topics to the team members, give them objectives that their research should produce, and then give them a deadline. There is no need to watch someone research. Let your team complete their assignments, and wait for the results.

Once your team members have completed the research, create a way for the information to be compiled quickly and easily so that it can be assessed and then you and the key project stakeholder can decide on what should happen in the project. If you have the resources, and depending on the project research, conduct a meeting and have the team members report their findings to the key project stakeholders. Face-to-face meetings are more productive than a report or an e-mail.

As the findings are being shared, have someone collect the notes and record any dialogue, controversy, or other information from the meeting. After the meeting, organize the collected data and disperse it to the team members. From the discussion on the collected research, the compiled report, and your own intuition, you should be ready to make an intelligent decision on how the project should move forward.

Contingency Plans

Every project needs at least one contingency plan. You may call these fallback plans, rollback plans, worst-case-scenario plans, or disaster-recovery plans. A *contingency plan* is a predetermined decision that will be enacted should the project go awry. If you ignore the creation of a contingency plan, you are tempting fate. A project that runs askew without a contingency plan will force your project to be late and most likely over budget.

Depending on the size of your project and its impact on your organization, your contingency plan may include a business continuity response. While *business continuity* usually addresses disasters and how the business can quickly react and continue to serve its clients, a large project failure can have an immense impact on an organization as well. You, management, and key stakeholders should examine the risk of failure and how it may affect the organization's continuity. This would include worst-case scenarios, disaster recovery, and any warning signs that the event is likely to occur.

As you complete the research and as the foundation for your project develops, think about how you would react if any phase of the project were to falter. As most IT projects will certainly have multiple steps to completion, there are plenty of opportunities for things to go wrong. And they will. Figure 2-9 shows how contingency plans are used and built into a successful project.

As part of the project planning process, record reported troubles, document any conflicts with other technology, and bookmark any articles or Internet sites that offer warnings on the technology you're implementing. By researching the negative possibilities of your technology, it can keep your love of the implementation from overriding reason and heighten your awareness that any technology can have flaws.

Use if-then statements such as the following to compose most contingency plans: "If the software conflicts with our video driver, then we'll write a driver that allows it to work." While this seems simple, a series of if-then statements can allow you to create a quick and concise contingency plan.

One of the primary reasons for creating contingency plans during the research phase of the project is in preparation of the next phase: dealing with management. Management loves to play devil's advocate. Some will swear they are the devil, but they aren't—usually.

Figure 2-9 Contingency plans are "in case of emergency" decisions.

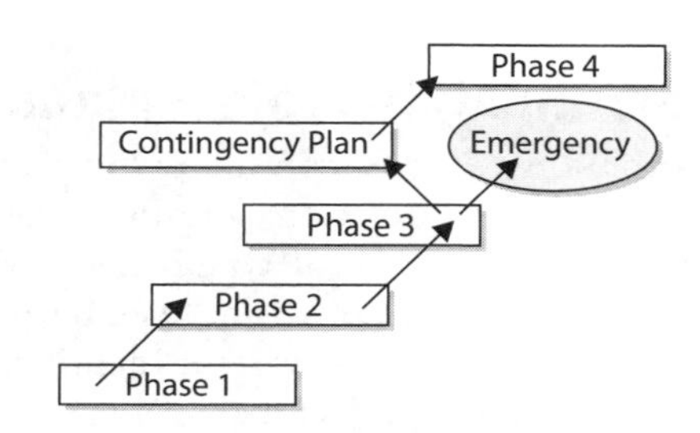

By having a documented, logical contingency plan for each facet of your project, you are working with management prior to the face-to-face meeting with them. This will build trust, confidence, and support of your project even before they say, "Make it happen."

Introducing Agile Methodologies

As I discussed in Chapter 1, traditional project management uses a predictive approach. Predictive project management is when you plan the entire project up-front. All of the requirements of the project are known, documented, and agreed upon. Lots of time is spent in planning at the beginning of the project, and additional planning time is sprinkled throughout the project life cycle. Predictive project management is also very averse to change requests. Changes are often frowned upon in predictive project management because changes can cause lots of effects in a predictive project.

Agile methodologies, however, use an iterative approach to planning and doing a project. You'll most likely find agile methodologies being implemented in software development. While there are many different types of "agile" methodologies to choose from, they all have some common characteristics—chief among these characteristics is that change is expected and welcome in an agile project.

The whole concept of agile projects stems from the "Manifesto for Agile Software Development," commonly called the "Agile Manifesto." This document was created in 2001 by 17 software developers. The developers were looking for a better approach to managing software development projects, because so much is unknown at the launch of a software development project when compared to a predictive project. It's also important to note that agile projects are also known as *knowledge work projects*—it's more brain than brawn power in the execution and creation of the project's scope.

The Agile Manifesto centers on four key values:

- Individuals and interactions over processes and tools
- Working software over comprehensive documentation
- Customer collaboration over contract negotiation
- Responding to change over following a plan

While the items on the right side of "over" in these statements are important in software development and knowledge work projects, the items on the left side of "over" are more valuable in an agile approach.

Introducing Scrum Agile Project Management

Scrum is widely accepted as the most common type of agile project management. Scrum has predetermined segments of iterations within the project. These iterations allow for planning, executing, team review, and a type of scope verification throughout the project. While there are lots of nuances of scrum (well beyond the scope of this book), there are five ceremonies within scrum that warrant identification. A ceremony within scrum comprises key activities with the project manager (called the scrum master), the development team, and the product owner. Let's take a quick review of these ceremonies now.

Sprint Planning

This first ceremony is based on a prioritized list of features and stories, called the *product backlog*, that need to be created in the project. The product owner, usually a representative of the project customers or end users, prioritizes these features and presents the items during the sprint planning session.

Sprint Iteration

The sprint iteration, usually just called a *sprint*, takes two to four weeks, while the development team creates the items selected from the product backlog. The selection of items from the product backlog is called the *sprint backlog*—it's the list of things the team has promised to complete in the current sprint for the customer.

Daily Scrum

The daily scrum is a daily meeting with the scrum master and the team. This is a 15-minute time-boxed meeting.

Sprint Review

Sometimes called the iteration review, this ceremony brings together the scrum master, the team, the product owner, and any relevant stakeholders, such as the project customer, to review what the team has, or has not, accomplished in the sprint.

Retrospective

This final ceremony is an opportunity for the team to discuss how the project is going. The team discusses any issues experienced in the past iteration, discusses what is working, and makes recommendations for team improvement and performance improvement in the next iteration of the project.

Reviewing Agile Management Principles

Agile aims to be flexible in its project management approach. Lots of formal overhead, forms, and documentation don't fit well within an agile environment. The overhead of traditional project management is something that project managers moving into a scrum environment often struggle with. The concept of project management in an agile approach isn't the predictive and directive approach many project managers are used to. Agile embraces the concept of *servant leadership* for the project manager. Servant leadership is based on four principles for agile projects:

- *Shield the team.* Servant leaders shield the team members from distractions that take their focus away from the requirements of the current iteration. To keep stakeholders at bay, hold minimal meetings and communicate on behalf of the team.
- *Remove impediments.* One of the topics from the daily standup meeting is to define what impediments are in the way of the project team's progress. The servant leader takes action to remove these impediments to keep the team working on the requirements of the current iteration.

- *Keep the team vision.* The vision of the project is communicated again and again by the servant leader to keep the team and stakeholders focused on the big picture of the project. The vision is a clear description of what "done" means for the project.
- *Carry food and water.* Okay, you're not really carrying food and water, but this servant leader characteristic means that you're providing the things the team needs to do the work. This can be access to resources, technical needs, and support from key stakeholders.

Another aspect of agile that is important is to keep things visual. A visual approach to project status and what's in the work queue are vital to success in an agile project. These visual items are sometimes called *information radiators*.

There are three visual items you should be familiar with when it comes to agile project management: burnup and burndown charts, dashboards, and Kanban boards.

Using a Burnup and/or Burndown Chart

Burnup and burndown charts are used in agile project management to show how many tasks have been completed by the project team during the current iteration. A *burndown chart* starts with the number of tasks assigned and "burns down" as tasks are completed in the project iteration. The goal of a burndown chart is to show the diminishing number of tasks still to do in the iteration. A *burnup chart* is similar but shows the accumulation of tasks completed in the iteration and "burns up" toward the goal or tasks to complete in an iteration. Figure 2-10 shows a burndown chart created in Microsoft Project.

EXAM TIP Burndown and burnup charts can also be used in a predictive project to show the consumption of the project budget.

Creating a Dashboard

Dashboards have become popular for project management over the past several years. A dashboard is like a quick report of all the important topics for a project. The dashboard organizes the information so anyone can quickly see the topics. Typical items in the dashboard can be activities in the queue, items in testing, items in the sprint backlog, number of defects found, and items currently being worked on.

Kanban Boards

Kanban boards are sign boards that visualize the flow of activities in a project. The development team writes features or tasks on sticky notes and then moves them across the board's categories as needed. For example, a task will start in the backlog, move into analysis, then development, on to testing, and finally move into the completed category on the sign board. It's like a to-do list of activities, but it shows what's in motion, what's done, and what's waiting to start. The category of "development" is called work-in-progress (WIP) in agile projects.

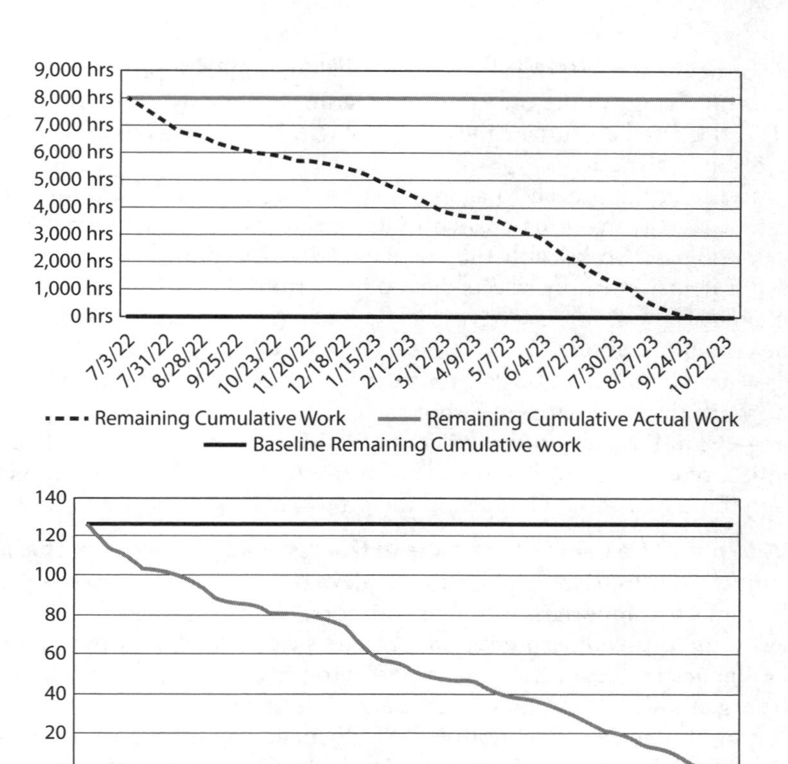

Figure 2-10 Burndown charts show the remaining work items for the project or for the current iteration.

CompTIA Project+ Exam Highlight: Planning Processes

This chapter doesn't actually map to any one chunky CompTIA Project+ exam objective—don't let that bother you one bit. You'll need all of the information in this chapter in your day-to-day world as an IT project manager, and it's the foundation for creating the project plans—something that's a big part of your CompTIA Project+ exam. You will need to be familiar with the terminology and concepts in this chapter to plan your projects successfully, which is something you can expect questions on during your CompTIA Project+ exam.

1.2 Compare and contrast Agile vs. Waterfall concepts Agile project management is a way to describe the iterative approach to managing a project; it's most often found in software development. Agile is based on the "Agile Manifesto," a document that defines the principles of agile, self-led teams; collaboration with project customers; and expectations for change in the project. While scrum is the most common flavor of Agile, Lean, XP, and Crystal are other agile methodologies that have a common iterative and change-is-expected attitude.

The scrum approach to agile project management is based on a product backlog of features and items to be created by the team. The product owner, a role in the scrum methodology, works with the customers and the project team to prioritize the backlogged items. The team selects the top items from the backlog and enters a *sprint* to create the items. The team may have a *sprint backlog,* which are items the team must create in the current iteration.

Waterfall projects, also called predictive projects, predict each phase of the project and the work that will happen within each phase. It's called "waterfall" because the project progression flows down through the identified phases of the work like a waterfall. These project types are plan-driven and are averse to change. Predictive projects like to predict everything that should happen in the project management plan.

2.1 Explain the value of artifacts in the discovery/concept preparation phase for a project Projects are launched to achieve business value. Business value describes why the project is important to the organization: it increases revenue or reduces costs. Business value is usually expressed in a business case. Recall that the business case examines the financial effect of the project, return on investment, the current and future state of the organization, and other financial considerations.

Project managers may not be the individual who do the research and write the business case. A business analyst may write the business case and gather requirements for the project before the project manager gets involved. It's not unusual for a project manager to receive the fully developed business case at the launch of the project and to work with the business analyst to refine the requirements or clarify the goals and expectations of the project.

2.3 Given a scenario, perform activities during the project planning phase Planning is important in both predictive and agile projects. While predictive projects plan in depth at the beginning of the project, agile projects plan iteratively as each iteration of the project begins. Planning in predictive projects is about predicting exactly what should happen. Planning in agile projects is more about the intent of the team and is always barely sufficient. For the CompTIA Project+ exam you'll need to be familiar with planning activities in both predictive and agile projects.

When it comes to planning, you'll often need to first consider the project constraints. Project constraints are things that limit your options. Common constraints are also usually tied to the key performance indicators: time, cost, and scope. Constraints can also include resources, quality requirements, and even the environment you're working in. Constraints create a boundary for the project and often can create problems for the project manager and the project team to work through.

The three big constraints of time, cost, and scope are often called the Iron Triangle of Project Management and the triple constraints of project management. The concept of the "iron triangle" states that time, cost, and scope must balance one another in the project.

You cannot have a tremendously large project scope and not have enough time or monies to create the scope. All three sides of the iron triangle must be in balance.

The other common constraints, such as quality and environment, are often set upon the project as part of the project requirements and may be seen as part of the project scope. The project scope defines everything that is included in the project and exclusions that are not included in the project. The scope in an agile project is represented by the product backlog. Items in the backlog can change throughout the project, in which case the product owner will reprioritize the product backlog so the most important items are always at the top of the product backlog list.

3.3 Given a scenario, analyze quality and performance charts to inform project decisions Agile projects utilize burnup and burndown charts to show project progress and performance. A burndown chart shows the consistent decline in the number of features left to create in the project against the number of iterations the project has completed. A burnup chart is a similar concept but shows the accumulation of features actually created in the project. Burnup and burndown charts are used primarily in agile projects, but you can also use them in a predictive project to show the consumption of the project budget and how much money is left in the project budget.

4.4 Summarize basic IT concepts relevant to IT project management IT project managers need to have relevant knowledge about the technology they are managing. Many IT project managers drift away from the technical side of the house as they get bogged down with managing the project rather than immersing themselves in the technology—and that's not necessarily a bad thing. However, it's important to maintain relevant knowledge, to continue to learn about the technology the project affects, and how the project solution may disrupt or affect other technical solutions and lines of business within the organization.

The project scope statement should address the computing services, architecture, networking, data, and software components of the technical project. You'll also need to identify and document how the project may affect any cloud services and software used by the organization. You'll need to see the big picture of the solution and work with stakeholders to fully understand the solution from their perspective to ensure that your project doesn't interrupt critical work or financial systems.

Chapter Review

Every project has constraints. Before you can begin to implement a project, assess budget dollars, or define a project goal, you must know the boundaries of the project. A project manager must know the business value the project aims to create. Based on the business value, the project manager and team can collaborate and plan. The initial goal of project planning is to answer questions in regard to the project scope. Planning allows decisions to be made, teams to be assembled, and the wheels of productivity to begin to turn. The constraints of the project are identified, documented, addressed, and managed throughout the project.

Create and evaluate a feasibility study to determine the project goal, the validity of the project, and the desired results of the project. Work together with your team to research, report, and develop the study. The feasibility study allows all parties to evaluate the project, its results, and its estimated ROI.

Establishing project priority, through researching the project, enables teams to work together for the good of the company. As conflicts, politics, and personal achievement arise, a set path of conflict resolution needs to be established between project managers, project sponsors, and upper management. The project sponsor should always win by putting the good of the organization ahead of the desire of others. In a perfect world, departments, management, and teams share information, work together, and strive to dovetail projects to create a powerful organization. It is achievable but not always probable.

Financial obligations to a project should be considered during the planning phase. A successful project manager evaluates the cost of the technology and the ROI. As a project manager, you must know the difference between value and investment and determine which technology will be the best investment. For each major phase of a project, you should create a contingency plan. A contingency plan enables you to make predetermined decisions in the event of project phases gone awry. A contingency plan also enables a project team to work with management, allows for different variables to a project, and adds a touch of realism to an expected flawless execution. As Henry Ford said, "Before everything else, getting ready is the secret of success."

In this chapter, we also discussed the popular project management approach that is largely used in software development projects: agile project management. Agile projects use time-boxed phases throughout the project to move the project work forward toward completion. Unlike predictive projects, agile projects expect and welcome change. The project manager and the development team work with the product owner to prioritize the product backlog, queue work for each iteration, present the completed work in a sprint review, and, before starting the next iteration of ceremonies, pause for a sprint retrospective to discuss the project.

Exercises

These exercises allow you to apply the knowledge you have learned in this chapter and are followed by possible solutions.

Exercise 2-1: Research Objectives Worksheet

In the online resources that accompany this book, you'll find a document named Research Objectives. (If you need assistance accessing this file, refer to Appendix C.) Print this document to begin working on this exercise. Based on the scenario that follows, complete the Research Objectives worksheet to reflect the required questions, sources of information, where you have found the information, and the information you've discovered. The first three questions on the Research Objectives worksheet have been provided for you. Try to come up with at least five more pertinent questions about the project. Use any resource you'd like—vendors, your experience, and Internet sites—to complete the questions and create new ones.

You are a project manager for Caulfield Educational Supply Company. You are about to research the validity of a project to implement an Internet server to host your company's own cloud applications. Currently the company's applications are hosted through a third-party hosting company. The average monthly bandwidth transferred is roughly 215 terabytes. Once the Internet server has been created, a new commerce-enabled site with database access will be built.

The Internet server you propose must be reliable and easy to implement, and it can be from any reputable software vendor. Your company has a T1 line to the local LAN, and Janice, your manager, is concerned that more bandwidth is going to be needed to ensure reliability for all the company's clients.

Exercise 2-2: Executive Summary for Caufield Educational Supply Company

Based on your findings for the Caulfield Educational Supply Company, create an executive summary of a feasibility study. Recall that an executive summary should include information on the recommended product, the audience impacted, financial obligations, and recommended action.

Exercise Solutions

The following offer possible solutions for the chapter exercises.

Exercise 2-1: Research Objectives Worksheet

The first three questions are included on the Research Objectives worksheet; the last two questions serve as example questions. Your research objectives will probably be different from those offered here.

Questions	Sources	Specific Locations	Information Discovered
1. What vendors provide easy-to-implement web server software?	Internet sites Personal experience	www.google.com www.amazon.com www.cisco.com	Google, Amazon, and Cisco offer cloud-based solutions that can mesh with our current network operating system.
2. What hardware is required to implement a web server?	Internet sites Personal experience	www.google.com www.amazon.com www.cisco.com	Each manufacturer will provide varying information needed to install and configure Internet sites on its solution.

(continued)

Questions	Sources	Specific Locations	Information Discovered
3. What type of network operating system (NOS) is our company using now?	Experience Network administrator	Ask the network administrator as part of expert judgment. Your sources of information can be, and often are, colleagues, management, and vendors.	Call upon your own experience, ask your company's network administrator, or decide for yourself what NOS is in place for the fictional company in this exercise. Knowing the type of NOS that is currently in place can impact the decision on what web server to implement, if it all, or to utilize a cloud-based solution, such as Amazon Web Services (AWS).
4. Where will the web server go in relation to the firewall?	Experience Books Internet sites	www.cert.org	Place the web server on a separate subnetwork. By doing this, traffic to and from the server will be isolated and will not traverse the private network. This will also ensure that the internal network's bandwidth is not hampered by traffic to the web server.
5. What is the yearly cost to host a web page through a third party?	Experience Magazines Internet hosting companies	Choose your favorite Internet hosting company.	For example, one company may charge $10 per month for startup Internet sites and 1500GB of data transfer. Another company may charge $7 per month for 100GB of bandwidth.

Exercise 2-2: Executive Summary for Caufield Educational Supply Company

Feasibility Study for Caulfield Educational Supply Company
Written by Justin Case

Executive Summary

The purpose of this feasibility study is to determine which cloud solution provides the best value for our organization. The executive summary is a portion of the feasibility study to give quick insight into the research and findings of the cloud solution.

Recommended Product

Windows Internet Information Services

Windows IIS is a robust web server. It will mesh with our current network operating system and will be familiar to administer, secure, and work with. Although there have been some security alerts for the product, these are well documented and repairable through service packs and security patches.

Audience Impacted The primary audience impacted by hosting our own Internet server will be our customers. Once the transition from an external Internet hosting company to our internal server has been made, it will be seamless to our Internet guests.

Financial Obligations A new server is required. The server will be based on a hardware array for fault tolerance. We will create a solution for fast, consistent data backup and restores. A UPS device is also required. This system will cost $18,500.

Although our company is currently using a Tl line, it is recommended that we replace the current connection with a fiber-optic line. This will keep our internal network from competing with the network traffic. The average estimated cost of the fiber line is $1,245 per month.

Recommended Actions

- Assemble a project team and begin developing the project charter and implementation plan.
- Order the fiber-optic line upgrade and equipment as soon as possible.
- Designate or hire an Internet server administrator and get this individual into IIS training.

We can minimize network downtime by first creating and securing our Internet site on the new IIS server. We will have to change the DNS information from our Internet hosting company to reflect the new IP address of our Internet server. This switch can take up to 72 hours to populate, so the transition should be made over a weekend or holiday period.

Questions

1. You are an IT project manager for the UYQ Organization and are working on a new project to create new software for manufacturers. You are just starting to work on the intent of the project, the high-level goals, and the opportunity for the organization to formulate a document that links the project objectives to the business needs. What type of document are you creating?

   A. The project scope statement

   B. The concept definition statement

   C. The key performance indicators

   D. The project charter

2. Jane is the project manager for her organization. She is preparing the business case and identifying the business goals and objectives for an identified problem that a project may solve. Jane is comparing two different applications for performance, cost, scalability, and the expected time for the end users to learn the software applications. What tool is Jane using as part of her creation and documentation of the business goals and project objectives?
   - **A.** Benchmarking
   - **B.** Brainstorming
   - **C.** Business rules analysis
   - **D.** Functional decomposition
3. What research technique breaks down large problems into smaller, manageable components?
   - **A.** Root cause analysis
   - **B.** Functional decomposition
   - **C.** Focus groups
   - **D.** Brainstorming
4. Mark is the project manager of the BGH Project. This project doesn't have much time for planning, as the project customer is urgent for a solution to be implemented. Mark has a limited amount of time to complete the research for the solution. When time is of the essence, what can a project manager do to increase research productivity?
   - **A.** Delegate the research topics among team members.
   - **B.** Use only one or two research outlets.
   - **C.** Limit the time spent doing the research.
   - **D.** Hire a vendor to do the implementation.
5. Gary is the business analyst for his organization, and he's created a proposed solution for the company to implement. Management has asked you to create a feasibility study based on Gary's findings. What is a feasibility study?
   - **A.** A plan for researching the project
   - **B.** A plan based on the project research
   - **C.** A plan that recommends the proposed technology
   - **D.** A plan that does not recommend the proposed technology
6. Of the following, which topic is not usually included in a feasibility study?
   - **A.** Executive summary
   - **B.** Market research
   - **C.** Defined business problem
   - **D.** Assumptions used in the study

7. You are the project manager for the JHW Organization. Management has come to you with a proposed solution to standardize all workstations and laptops in your organization. They would like you first to complete a feasibility study for the solution with a focus on the total cost of the project implementation. You agree but insist that you should also address the users of the project who will be affected by the change the project will create. Why must a project manager address the users impacted by the technology in a feasibility study?

   A. To determine their willingness to use the product

   B. To determine how many users will use the product

   C. To determine the downtime caused by the product

   D. To determine the validity of changing or adding technology

8. Why should a project manager demonstrate return on investment (ROI) in the financial obligation portion of a feasibility study?

   A. They should not; it will be determined by the project sponsor.

   B. They should include ROI to demonstrate the validity of the project.

   C. They should include ROI to demonstrate the initial cash outlay for the technology.

   D. They should include ROI to make certain his project is approved.

9. You are the IT project manager for your organization. Several projects being initiated in your organization have large financial requirements, entail extensive usage of human resources, and require access to network facilities. You suggest that project priority be established. What is project priority?

   A. It is the ability of a project manager to determine which project is the most valuable to an organization.

   B. It is the ability of a project sponsor to determine which projects should be implemented and which should be discarded.

   C. It is a process project managers and project sponsors must go through when conflicting projects arise within an organization.

   D. It is a process project sponsors use to determine which project is of greater importance when two projects conflict.

10. A new project has been launched by Shelly Dere, the project sponsor. Shelly insists that the project manager have total autonomy over the project decisions, but Shelly will retain the control of the project budget. What is the goal of a project sponsor?

    A. To manage the project manager

    B. To delegate duties to the project manager

    C. To increase profits through the project led by the assigned project manager

    D. To increase productivity through technical implementations

11. The "Agile Manifesto" values what more than processes and tools?

A. Completed activities

B. Customer collaboration

C. Individuals and interactions

D. Product backlog

12. What is project portfolio management?

A. It is the risk the project manager is taking when implementing a project.

B. It is the pool of available project managers.

C. It is the management and selection of which projects will be engaged, allowed to continue, or stopped based on conditions within the organization or project.

D. It is the relationship between the project manager and a third party that will implement the proposed technology.

13. Percy is the project manager of a large project that spans three countries. The project team is operating as a virtual team. There have recently been some arguments among the project team members over the technical implementation of the project. What causes internal conflicting IT projects?

A. Lack of communication

B. Lack of planning

C. Technology that develops too quickly

D. Conflicting technology

14. Beth is an agile project manager and she wants to create a dashboard for her team. A dashboard can also be known as a what?

A. Information radiator

B. Kanban board

C. Burnup chart

D. Queue

15. How long does a daily standup meeting last in an agile project?

A. One day

B. 15 minutes

C. As long as needed

D. One hour

Answers

1. **B.** You are creating the concept definition statement. This document defines the high-level goals and proposed solution for the project. It links the project objectives to the business needs.
2. **A.** Jane is using the benchmarking approach to compare the costs and benefits of one solution to another.
3. **B.** Functional decomposition breaks down a large problem into smaller, manageable components.
4. **A.** Many hands lighten the load. Whenever possible, a project manager should delegate the planning among team members. The project manager rarely completes planning alone. It may be tempting to hire a vendor to complete the implementation of the project, but the focus of the question is on the research of the project.
5. **B.** A feasibility study is a plan based on the project research. It contains a summary of the information you've discovered in an organized approach. It is a factual document that determines whether the project is feasible to complete.
6. **B.** Market research is not included in a feasibility study. Feasibility studies include eight sections: executive summary, defined business problem or opportunity, purpose of the study, description of the options assessed, assumptions used, audience impacted, financial obligations, and recommended actions.
7. **C.** A project manager must evaluate any downtime caused by the product implementation. Downtime for users, whether through a learning curve or lack of productivity, is an expense for the organization. Too high of a learning curve or long periods of inactivity due to lack of planning is unacceptable when there's a primary concern about the initial cost of the technical project implementation.
8. **B.** To implement the technology, an organization will have an initial cash outlay. The ROI will show how the technology can earn back the initial expense and more by increasing productivity. If the ROI is too little, the project may be scrapped.
9. **A.** A project manager may have multiple projects to manage. Project priority is the ability to determine which project takes precedence, as it is most important to the success of an organization.
10. **C.** The goal of a project sponsor is to increase profits through the proposed project. A project manager will carry out the implementation of the project. A project sponsor may manage a project manager, but it should not be the project sponsor's goal to do so. It's not unusual for the functional management to retain control over the project budget while allowing the IT project manager control over project decisions.
11. **C.** The "Agile Manifesto" values individuals and interactions over processes and tools.

12. **C.** Project portfolio management is a process that management uses to determine which projects should be engaged and which should not be. Project portfolio management is also used to determine if projects should be halted because of a shift in priorities, conditions within a project, or conditions within an organization.
13. **A.** Lack of communication causes conflicting IT projects. When a project manager is managing a virtual team, as Percy is in this instance, it's important to take extra steps to encourage communication among the project team members. If, within the organization, departments, teams, project managers, and project sponsors would effectively share plans, research, and needs, there would be fewer conflicts and more successful IT implementations.
14. **A.** A dashboard can also be known as an information radiator. Information radiators are visual boards or postings of key project information, such as work-in-progress, testing status, defects, and more.
15. **B.** Daily standup meetings should last 15 minutes. Standup meetings are short daily meetings to discuss what's been accomplished, what's in motion, and anything that may be preventing the project from moving forward.

CHAPTER 3

Working with Management

This chapter covers the following topics:

- Obtaining management approval
- Defining management's role
- Inventing a project kickoff
- Creating management alliances

Management. That very word conjures up so many different visions for project managers. For some, it brings up the image of the irate and belligerent boss who's always unhappy with someone about something. For others, it's the image of the boss who hides away, dreading to make a decision on anything. Still, for some, management is not a bad word at all. To these people, management is a mentoring, guiding presence that wants projects to succeed.

Whatever type of management you're dealing with, you have to work with them, and usually answer to them. If you're fortunate to have a good manager, count your blessings. There are plenty of people out there who would trade places with you any day.

The thing to remember about management is that it's not necessarily an "us-against-them" mentality. Management's job is to support the vision of the organization. Their role is to cut costs, increase productivity, increase revenues, and ensure that the requirements of upper management are met. The people you know as management are often in the uncomfortable position of being stuck between the executives and the day-to-day operational staff.

This chapter will examine how you, the project manager, can work with management toward project success. You'll learn how to present a plan to management, get management involved in the plan, and then work with management as the project progresses.

VIDEO For a more detailed explanation, watch the *Working with Management* video now.

Defining the Organizational Structure

The way your organization is structured determines the communication requirements, responsibilities, and reporting structure you have with management. The organizational structure, culture, and internal policies also determine the amount of authority the project manager will have in a project. Because all organizations and projects are different from one another (thankfully), they can be broken down into one of three different organizational structures: functional, matrix, and projectized.

Working in a Functional Organization

Functional organizations are fairly common. They are segmented by departments and their "functions." For example, you may have "Sales," "Accounting," "Legal," "IT," and so on, throughout your organization. In a true functional environment, all team members, including the project manager, report to their functional manager.

The project manager in a functional organization has very little power. Decisions flow through the functional manager—who is the one running the show. Project managers are sometimes simply called project coordinators or expeditors in a functional structure. The advantage of the functional organization, however, is that there's less anxiety than in other structures, communication demands are reduced, and team members stay within their departments to complete the project work.

Working in a Matrix Organization

A matrix organization structure allows a project team to incorporate resources from around the organization regardless of which department employees may work in. Project team members can be recruited from anywhere or any place within the organization. In contrast to the functional structure, this structure blends the project team based on team members' individual contributions and abilities.

Technically a matrix structure has three different flavors:

- **Weak matrix** The functional managers have autonomy and power over the project team members. The project manager has limited authority on project decisions, much as in a functional structure, except the project team members come from around the organization.
- **Balanced** The project manager and the functional manager have equal power and autonomy over the project team. While this sounds nice, it's usually pretty tough to implement. A good balanced matrix defines the decision types that management will make, such as costs and procurement, and leaves the core project management and technical decisions to the project manager.
- **Strong matrix** The project manager has autonomy over the project and the project team. This structure gives the project manager the most authority, but in reality it's still pretty tough to implement, as there are many functional managers to work with, budgets to approve, and personalities that affect project decisions.

On paper, the matrix organization looks pretty good, but there are some downsides. Communication demands increase for the project manager because they'll likely be required to keep all of the project team members' bosses up to date on how the project is moving and how the team members are doing on it. Team members have to report to at least two bosses: their functional manager and the project manager. In addition, team members can expect to be working on multiple projects at one time, which, of course, increases their responsibilities, communication requirements, and workload. When team members are working on multiple projects simultaneously, the project managers must coordinate schedules, share resources, and communicate even more. Further, team members may be expected to complete their regular job duties along with additional responsibilities on the project.

Working in a Projectized Organization

In this organizational structure the project manager works with complete autonomy over the project. The project team is on the project full time and reports only to the project manager. The project manager has the most authority in this structure. There are many advantages to the projectized structure:

- The project team is on the project full time.
- The project team reports to one boss for the duration of the project.
- The project manager has power over the project.
- Communication demands are reduced.

This can, however, lead to redundancy in some functions, such as tech support, accounting, purchasing, legal, and so on. Additionally, team members do not get the experience they gain when they work in an environment of their peers beyond the project team. This is particularly true for those working in technical fields. What can happen is that a needed resource is assigned to the project but sits idle for a few activities in the project.

Projectized organizations, while they can be efficient and create good teams, do have some negative aspects to consider. First, if a project team member leaves the organization, this can create a hole in the team, cause the project to stall, and require any new team member to ramp up quickly on the project. Projectized structures also face an increase in team anxiety as the project nears completion. Team members may begin to wonder (panic) about what assignment they'll be doing next as the current project enters its final stages of activity.

Presenting the Project to Management

It's been said that most people's number one fear is public speaking—and with good reason. Everyone has witnessed someone making a terrible job of delivering a presentation, so you know the dilemma. The fellow is sweaty, speaking low, and tripping over every word while he's searching for the next. Poor guy.

Chances are, though, if you were to talk to this person over a cup of coffee about the same ideas, he'd be rational, personable, and able to express his thoughts without a single "um" or "uh." What he's got is stage fright, and it's curable—with practice. The most important thing in a presentation? The message, not the messenger! If you know exactly what you want to say, you can say it much easier.

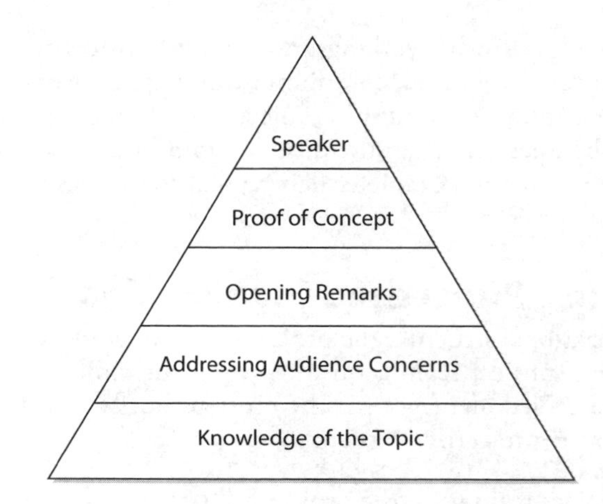

Figure 3-1 An effective presentation must sell the project through effective reasoning.

Presentations can be powerful, inspiring, and informative sessions. An effective speaker can captivate and motivate the audience to action. The core of an excellent presentation is the speaker's intimacy with the topic. The more familiar you are with the topic you are speaking on, the more convincing you will be. You must know your topic to speak effectively. Figure 3-1 shows the building blocks to an effective presentation.

Start at the End

When you begin a presentation, you want to capture your audience's attention. You want to hook them and reel them in to your project idea. One of the most effective ways to do this is to start at the end. Tell your audience first what the proposed project will deliver. Forget all the techno-babble—that only impresses geeks. Management wants to hear facts. If you are launching an agile project, you'll also need to sell stakeholders on how agile project management works differently than the traditional waterfall approach.

For example, Susan is proposing a real-time transaction server for databases across the United States. In her presentation, she could go on about how long each transaction will actually take, the number of processors in each server, and the network connecting each site—but who cares (other than IT folks)? To grab the attention of management, she needs to open with the end result: "Our company will be 33 percent more productive by implementing this technology. From coast to coast, our customers can buy more products in real time with a guarantee of when they'll ship. Of course, this means our company will be more savvy, more advanced, and more profitable than our competitors."

Wow! Now Susan really has management's attention. She now has to back up her statements with the proof she's already gathered from her initial research. Susan has sold the business value, not the technical solution. Once she has delivered the core benefits of the project, Susan should immediately show her supporting details regarding why she's so confident the implementation is a good thing. She can do this as a business case, which focuses on the return on investment (ROI) for the project. She could also use a slideshow or charts to show the expected growth.

NOTE Keep in mind that with slideshows it's easy to fall in love with the medium, not the message. Slideshows (ahem, PowerPoint) can be painful to sit through, boring, and overdone when they incorporate flashy animations that aren't relevant to the message. Keep it simple.

The WIIFM Principle

The WIIFM, or "What's In It For Me," principle is the ability to make a presentation touch the audience members so that they see how they will benefit from the proposal. When an IT project manager pitches an idea, they have to show the audience, typically management, how this technology will make customers' lives better, improve profitability for the company, and make the company superior. You can rarely go wrong if your proposal focuses on either cutting costs or increasing revenue—or sometimes both. Even not-for-profit organizations are concerned with the profitability of a project as better returns on investments mean better usage of the dollars saved and earned.

Professional salespeople describe this principle as selling the benefits of a product, not just the features. If you do present a feature, always back it up with a benefit. For example, "This new system has eight gigabyte of memory (feature), which will enable each user to increase productivity by 20 percent." You're selling the sizzle, not just the steak.

Obviously, whenever you get an idea to implement a new technology, you've got some personal reasons: productivity, an easier workload for you, an opportunity to work with technology, career advancement, and maybe some personal reasons you wouldn't say out loud and would deny even if you were asked. Management will have similar reasons for implementing, or not implementing, your plan. You have to look realistically at four major points in the WIIFM principle to determine if your plan is viable:

- **Profitability** If the ROI on your project is small or nonexistent, proceed with caution. Remember, all businesses exist to make money; make certain your project supports that goal.
- **Productivity** Examine how your project can increase the productivity of the company. Not all projects will increase productivity for everyone, but at the minimum it should not hinder productivity.
- **Personal satisfaction** At the core of WIIFM is the ability to personalize the project. Find attributes of profitability, credit for implementing the technology, new sales channels, and other benefits that will make management (and you) happy to implement the plan.
- **Promotion** Think of how the project can promote the company's products, but also think how it can promote careers—not only yours, but the decision maker who will see the advantage of the project and may become your project sponsor.

WIIFM is not all about greed. It's thinking for the other parties. It's showing them the need that they may not see. WIIFM is creating a win-win solution for the parties involved. There's nothing wrong with thinking about the interest of the parties you are pitching your presentation to—in fact, it's required if you want their approval.

In many instances, the project manager gets the project dumped in their lap, a slap on the back, and a hearty handshake. In other words, there's no pitch to management, or anyone else, to get the project off the ground. An effective project manager, however, should investigate why this project is needed, how it helps the organization, and what exactly the project must deliver.

Tailor the Presentation

If you are speaking to a group of executives who have but a few minutes to hear your proposal, you need to get to the point quickly using terms they can understand. Forget about processors and bandwidth altogether with this crowd—focus on the benefits. If you're pitching to a group of managers who have a background in technology, then show them the details of the technology and how it increases productivity, profitability, and sales. Figure 3-2 shows the overlapping scopes a project manager must consider when addressing an audience. The point is to know your audience ahead of time and tailor your presentation to that audience.

If you are presenting your solution to a group composed of upper management, middle management, and your immediate supervisors, always address your presentation to the decision makers—typically upper management. Whomever you are speaking to, tailor your presentation to what they want, and need, to hear to make a decision. Here are six tips to help you do so:

- *What is your track record?* Discussing the successful projects that you've completed gives confidence to management to invest in new projects you propose. It works the other way, too: If you've gained approval in the past but blew the implementation, that's valuable experience for you and them. If you've worked as a project manager before and failed miserably, management is going to be less excited to give you another opportunity until you can prove yourself again as a team player in other project implementations. Consider the lessons learned from the prior project and how they can help you on this new project. You may want to approach your immediate boss and seek her input.
- *Do they really want to listen?* You need to determine if this is the best time to even be talking about new ideas. If the organization is in turmoil financially, emotionally, or technically, your chances are diminished for implementing new projects—unless your project can resolve the turmoil.

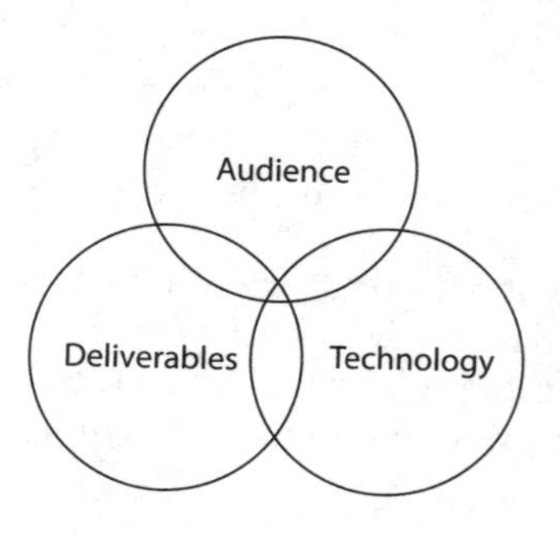

Figure 3-2 Project managers must address several factors in a presentation.

- *Are they listening?* Remember, you've got to speak the language your audience needs to hear. Describe your project and its deliverables in the business terminology your audience speaks, and you will be heard. A business case defines the problem or opportunity, the approach the project will take, and the anticipated return on investment.
- *And who are you again?* If you're low in the hierarchy, you may have big ideas but little track record. In this instance, your idea may be valuable but you aren't recognized—yet. Partner with someone in the organization who is a valuable leader and work with that person to pitch and manage the project. Use teamwork.
- *How does this project help?* If you are getting this question from your audience, then you have not started your presentation at the end, as previously discussed. Show them, in their language, exactly how this project helps. Show them how it increases profitability, what the ROI is, and why it should be implemented. In your organization, this task could be up to the business analyst, not the project manager. Know the rules before you start playing.
- *Are you following the rules?* You need to know the rules, and you need to follow the rules. Does your organization have formal or informal guidelines and procedures to pitch new projects? If so, follow these rules. Other organizations, and people, have enough on their plates that they won't dream of starting a new project until they can catch their breaths on their current assignments. Find the rules to make a project a reality and follow them.

For each presentation, it would behoove you to have handouts of your overheads, charts, graphs, and any other pertinent data. Your feasibility study is ideal for this collection of handouts if you've had the opportunity to complete one. Whatever you decide to disperse, always include an executive summary near the front of the study so that the reader can skim over the details and get the hard-hitting facts first. Of course, you should write the executive summary with your audience in mind. For formal meetings you should utilize a scribe. A scribe is a person that records the meeting minutes, business discussed, questions, and other important notes for your meeting. After the meeting, the recorded information, action items, and follow-up information are compiled and distributed to the meeting members.

Role of a Salesperson

Robert Louis Stevenson, the author of *Treasure Island,* said, "Everyone lives by selling something." No matter what you do for a living, you are selling something: time, wrenches, advice, a service, or a product. When it comes to presenting your idea to management, you have to slip into the role of salesperson.

When you think of sales, dismiss the high-pressure used-car salesperson stereotype. A good salesperson is someone who identifies a need and then helps to fulfill it. That's what you have to do here. You've identified a need through your research, and now you may have to show management where that need exists and how this technology will fill it.

When you're selling an idea, speak in direct, simple terms. If it takes you longer than five minutes to express the reason for the technology, you're taking way too long. In fact, an excellent summary of the technology and its benefits should take less than a minute. For example, if you've determined the solution to the problem you've been presented with is to implement Adobe InDesign for your web designers, provide the supporting evidence for your recommendation. You have done the research, interviewed the developers, and explored the cost of the decision, right?

To open the presentation, you'd say something to the effect of, "Adobe InDesign will increase productivity, streamline the efforts of our web designers, and allow for collaboration of in-house and external developers. The time saved alone will pay for the product in less than a month." (This is assuming you have the evidence to support this statement.)

Again, you're starting at the end, but the whole idea and pitch is done in less than a minute. You need to capture your audience's attention and draw their interests into the plan. Everyone wants to be part of an investment that works. Show them immediately how your plan will work and then invite them to participate.

Included in the presentation to management is an informal presentation. You may know this as the infamous "elevator speech." You know the drill: The CEO sees you in the elevator and asks, "So what are you working on these days?" And then you can sum up your project in 20 seconds or less on your elevator ride. This elevator speech is handy to summarize your project to anyone, anywhere. Agile projects may call this elevator speech the project vision statement.

The role of the salesperson in project management isn't really about sales—it's about marketing. You need to market your project to gain stakeholder support. Whenever a stakeholder first hears about a new project in the organization, especially a technology project, there's some immediate anxiety lurking beneath the surface. When an organization changes technology, it changes the way people work. People get accustomed to doing their work a certain way. People know that learning and mastering a new technology takes time. And people know that there are unknown risks in new technology: risks for how they'll work with the technology—not necessarily risks for your project, but from their perspective.

Marketing the project means you're helping stakeholders understand the vision of the project and why the project is needed. You address the negative aspects of the current state, and then you address how the new technology your project will implement will solve those negative aspects. Your goal is to help reduce fear, reduce perceived threats, and gain buy-in from stakeholders. You want stakeholder support, and you can get this by speaking with and involving stakeholders early in your project. Let's face facts: people want to be listened to, to feel that their concerns are being addressed, and feel that they are part of the project solution rather than just having a project solution forced upon them.

Defining Management's Role

Management, to some, is a necessary evil. Sure, there are plenty of bad, heartless bosses in this world, but not every manager is bad. The majority of managers want to be good bosses, they want to be well liked, and they want to do a good, thorough job.

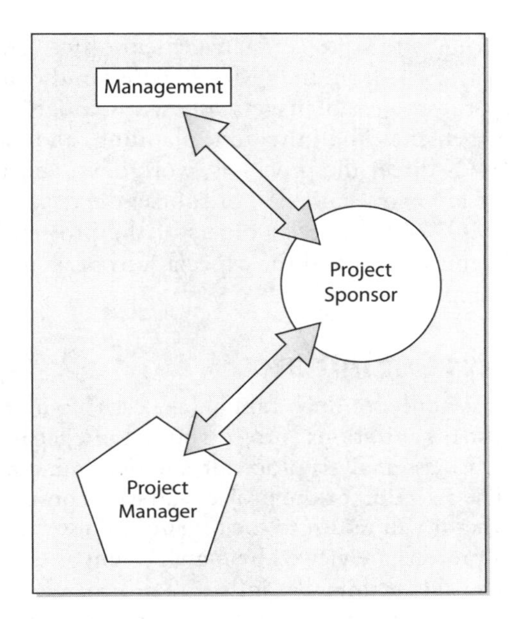

Figure 3-3 Project sponsors are mediators for project managers and management.

While management should show an interest in the project you're implementing, their role should be one of support, not one of implementation. Management should not be peering over the shoulder of a technician trying to install video cards and memory.

Project sponsors, however, do have an active role in the project management experience. Figure 3-3 shows the relationship between the project sponsor, the project manager, and management. Project sponsors need to be informed of the status of the project, of who is completing which portion of the project, and of how the project is doing on time and finances.

Project sponsors have invested their credibility in the implementation, and they are relying on you to report progress and to complete the work. Project sponsors, like management, should not be peering over technicians' shoulders, but should, in some cases, attend team meetings, be involved with the project planning phases, and have input on the project implementation. Don't be afraid to ask questions and share concerns with your project sponsor—the sponsor is on your side and wants you to succeed with the project.

In fact, if you are going to present the project, it is in your best interest to talk with the project sponsor ahead of time and get coaching on the presentation. Find out the hot buttons, allies, showstoppers, and so on. Then you can tailor your presentation to incorporate this information. If you are pitching a project that does not yet have a project sponsor, see if you can get some input from a likely sponsor or a friendly person in senior management. It always helps to stack the deck in your favor a little bit.

TIP In an agile environment, the product owner will have more interaction with the project sponsor and the stakeholders than the project manager. Agile projects have more autonomy for the entire team than what you may find in a predictive, waterfall project.

If you're working with a project management office (PMO), then your conversations and coaching may come from the PMO rather than the project sponsor. Recall that the PMO will help the project managers follow a standardized approach to most aspects of project management. The initiation, planning, and closing phases of a project are where most PMOs direct the processes, workflows, and involvement with the project managers. There are many variables, of course: the depth of involvement of the project sponsor, the PMO, and the stakeholders of the project. Each organization, like each project, is different; understand the expectations of your organization before venturing too deep into the project.

How Projects Get Initiated

Organizations have only so much capital (aka cash) to invest in new projects and opportunities. In most organizations, projects just don't get launched without first moving through some business analysis procedures to determine the worth of the project to the organization. The selection of the project can focus on its cost, its associated risks, or—most likely—what it will return to the organization for the investment.

While many projects are viewed by management as only a financial investment, there are usually many other factors that influence project selection: new laws and regulations, new technology, complaints from end users, competition, customer requests, and more. The formulas and mathematical approaches I'm about to share with you are rarely the only reasons why a project gets initiated, but these are almost always part of the conversation when it comes to launching a new project.

You can help yourself by being familiar with how your organization determines which projects are funded and initiated and which projects are declined. Knowing how your organization chooses a project will help you research solutions, prepare for management questions, and generally look like a genius. The project selection approaches described in the following sections are the most common and are often called benefit measurement methods for project selection.

Murder Boards

If your organization requires that project managers and technical gurus be interviewed before the project is funded, you're likely participating in a murder board. A *murder board* is a committee that asks every conceivable negative question about the proposed project. Their goal is to expose strengths and weaknesses of the project—and kill the project if it's deemed too risky for the organization to commit funds to. You might also know this approach as the kinder-named *project selection committee* or *portfolio management board.*

Scoring Models

Scoring models (sometimes called weighted scoring models) are models that use a common set of values for all of the projects up for selection. The common values can be profitability, complexity, customer demand, initial risk, startup costs, and expected return on investment. Each of these values has a weight assigned to it—values of high importance have a higher weight, while values of lesser importance have a lesser weight. The projects are measured against these values and assigned scores according to how well they match

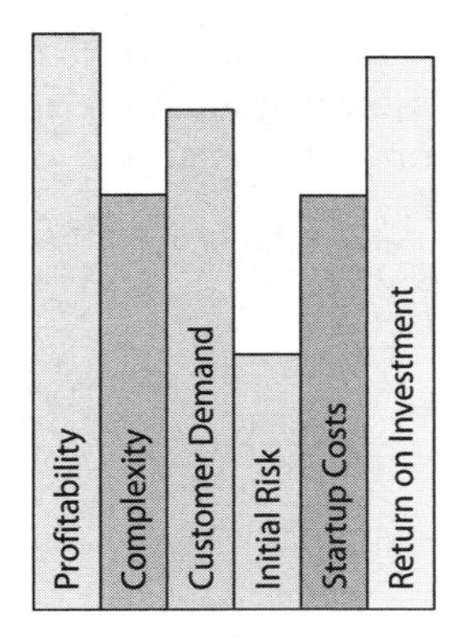

Figure 3-4 The weighted model bases project selection on predefined values.

the predefined values. The projects with higher scores take priority over projects with lesser scores. Figure 3-4 demonstrates the scoring model.

Benefit/Cost Ratio

Just like the name implies, a *benefit/cost ratio (BCR)* model examines the cost-to-benefit ratio. For example, typical measurements include the cost to complete the project and the cost of ongoing operations of the project product compared to the expected benefits of the project. Consider a project that will require $800,000 to create new software, market the software, and provide ongoing support for one year. The expected net profit on the product, however, is $1,500,000 in year one. The benefit of completing the project is greater than the cost to create the product. You can go into much more detail and consider costs such as the learning curve, competency levels required, market conditions, and even competition. BCRs are often used in the cost-benefit analysis of determining whether or not a project should be initiated.

The Payback Period

How long does it take the project to "pay back" its costs? For example, your project may cost the organization $500,000 to create over two years. The expected cash inflow (income) on the project deliverable, however, is $65,000 per quarter. From here, it's simple math: $500,000 divided by $65,000 is nearly eight quarters, or a little less than two years, to recoup the expenses.

This selection method, while one of the simplest, is also the weakest. Why? The cash inflows are not discounted against the time it takes to begin creating the cash. This is the time value of money. The $65,000 per quarter five years from now is worth less than $65,000 in your pocket today. Remember when sodas cost a quarter? It's the same idea. The soda hasn't gotten better. The quarter is just worth less today than it was way back then. You might also know the payback period as the *management horizon* or the *break-even point.*

Considering the Discounted Cash Flow

Discounted cash flow accounts for the time value of money. If you were to borrow $500,000 from an investor for five years, you'd be paying interest on the money. If the $500,000 was invested for five years and managed to earn a whopping 6 percent interest per year, compounded annually, it'd be worth $669,112.79 at the end of five years. This is the future value of the money in today's terms.

The magic formula for future value is $FV = PV \times (1 + i)^n$, where

- FV is future value
- PV is present value
- i is the interest rate
- n is the number of time periods (years, quarters, and so on)

Here's the formula in action with the $500,000 invested at 6 percent for five years:

$$FV = 500{,}000 \times (1 + 0.06)^5$$
$$FV = 500{,}000 \times (1.338226)$$
$$FV = \$669{,}112.79$$

$500,000 today is worth $669,112.79 in five years. So how does that help? If your organization invests $500,000 in your project and it is is worth less than $669,112 when it's complete, it's not worth doing from a purely financial standpoint. When four or five projects are up for selection, management can review the future values of their investment to see which project is worth investing today's dollars in.

While it's somewhat easy to determine what needs to be invested in a project, management may also want to see what a project will actually earn the organization. Sure, some of this is "blue sky" predictions, but you should have some evidence of what a project will earn the organization for investing its capital into your project. The earnings a project promises can then be evaluated to determine what the future value of the project is worth in today's dollars. In other words, if your project promises to be worth $750,000 in three years but it needs $420,000 to get started, management can examine what the $750,000 is worth in today's dollars. Management is looking for the present value of the promised future cash: $PV = FV \div (1 + i)^n$.

Let's take the project that promises to be worth $750,000 in three years and determine its present value. You can plug the values into the present value formula (assuming the interest rate is still 6 percent):

$$PV = 750{,}000 \div (1 + .06)^3$$
$$PV = 750{,}000 \div (1.191)$$
$$PV = \$629{,}714.46$$

So, $750,000 in three years is worth $629,714.46 today. This project needed $420,000 to launch, which is still considerably less than the present value of the project's completion. This is a pretty good investment for the organization. Now imagine if we had four different projects with various times to completion, costs, and expected project cash inflows at completion. We could calculate the present value and choose the project with the best present value, since it'll likely be the best investment for the organization.

Calculating the Net Present Value

The *net present value (NPV)* is a somewhat complicated formula but allows a more precise prediction of project value than the lump-sum approach found with the PV formula. NPV evaluates the monies returned on a project for each time period the project lasts. In other words, a project may last five years, but there may be a return on investment in each of the five years the project is in existence, not just at the end of the project.

For example, a retail company may be upgrading the facilities at each of their 1000 stores to make shopping and purchasing easier for their customers. As each store makes the conversion to the new point-of-sales (POS) software, the project deliverables will hopefully begin generating cash flow as a result of the project deliverables. The organization can begin earning money when the first store completes the conversion to the new software. The faster the project can be completed, the sooner the organization will see a complete return on their investment.

In this example, an interest rate of 6 percent per year is assumed. The following outlines how the NPV formula works:

1. Calculate the project's cash flow per time unit (typically quarters or years).
2. Calculate each time unit total into the present value.
3. Sum the present value of each time unit.
4. Subtract the investment for the project.
5. Examine the NPV. An NPV greater than zero is good and the project should be approved. An NPV less than zero is bad and the project should be rejected.

When comparing two projects, the project with the greater NPV is typically better, though projects with high returns early in the project are better than those with low returns early in the project. The following is an example of an NPV calculation:

Time Period	Cash Flow	Present Value
1	$15,000.00	$14,150.94
2	$25,000.00	$22,249.91
3	$17,000.00	$14,273.53
4	$25,000.00	$19,802.34
5	$18,000.00	$13,450.65
Totals	$100,000.00	$83,927.37
Investment		$78,000.00
NPV		$5,927.37

Management Theories

Your relationship with management, and how management sees their relationship with their employees, has been theorized and debated for years. As the project manager, these management theories can help you not only realize how management views you but also manage your own project teams more successfully.

EXAM TIP While the CompTIA Project+ exam objectives don't mention these management theories directly, it's a good idea to be familiar with these concepts as a project manager.

Maslow's Hierarchy of Needs

You've heard of Abraham Maslow, right? According to the psychologist's 1943 paper, "A Theory of Human Motivation," people go to work to satisfy their hierarchy of needs. Basically, if we satisfy our most basic needs, we can strive toward self-actualization, which enables us to contribute and use our skills and talents. Here are the five layers of Maslow's hierarchy:

- **Physiological** People need these necessities to live: air, water, food, clothing, and shelter. People need a place to work.
- **Safety** People need safety and security; this can include stability in life, work, and culture. People need a safe working environment—with job security.
- **Social** People are social creatures who need love, approval, and friends. People want to participate with their colleagues and peers—and to be liked at work.
- **Esteem** People strive for the respect, appreciation, and approval of others. People generally want to do a good job and complete their projects.
- **Self-actualization** At the pinnacle of their needs, people seek personal growth, knowledge, and fulfillment. People want to excel and work at something they enjoy and feel is valuable.

McClelland's Acquired Needs Theory

Psychologist David McClelland developed his acquired needs theory based on his belief that a person's needs are acquired and developed over time. These needs are shaped by circumstances, conditions, and life experiences for each individual. This is also known as the "Three Needs Theory," because he identified three basic needs for each individual. Depending on the person's experiences, the order and magnitude of each need shifts:

- **Need for achievement** People who have this dominant trait avoid both low-risk and high-risk situations. Achievers like to work alone or with other high achievers, and they need regular feedback to gauge their achievement and progress.
- **Need for affiliation** People who have a driving need for affiliation look for harmonious relationships, want to feel accepted by people, and conform to the norms of the project team.
- **Need for power** People who have a need for power are usually seeking either personal or institutional power. Personal power-seekers generally want to control and direct other people. Institutional power-seekers want to direct the efforts of others for the betterment of the organization.

McClelland developed the Thematic Apperception Test to determine what needs drive individuals. The test is a series of pictures, and the test-taker has to create a story about what's happening in the picture. Through the storytelling, the test-takers will reveal which need is driving their lives at that time.

Herzberg's Theory of Motivation

Frederick Herzberg, a psychologist and authority on the motivation of work, believed two agents affect people and their views toward their careers and work:

- **Hygiene agents** These elements are the expectations all workers have: job security, a paycheck, clean and safe working conditions, a sense of belonging, civil working relationships, and other basic attributes associated with employment.
- **Motivating agents** These elements motivate people to excel. They include responsibility, appreciation of work, recognition, the chance to excel, education, and other opportunities associated with work beyond financial rewards.

Herzberg's theory says the presence of hygiene factors will not motivate people to perform, because these factors are expected. However, when these factors are absent, workers are demotivated. The motivating agents inspire workers to strive for success. The tricky part, however, is knowing exactly what motivates the workers. Your motivating agents may not be anything like the motivating agents of the person working next to you.

McGregor's Theory X and Theory Y

In MIT psychologist Douglas McGregor's two theories, Theory X and Theory Y, management's view of their workers is divided into two categories: bad and good. X people are lazy, must be micromanaged, and generally cannot be trusted. Y people are wonderful people who are self-led, motivated, and can accomplish new assignments proactively. In reality, sometimes you, the project manager, must be both X and Y, depending on the scenario, the person you're managing, and what you want the outcome of the scenario to be.

Ouchi's Theory Z

Business management professor William Ouchi's Theory Z is based on participative management. His theory states that workers are motivated by the commitment, opportunity, and advancements provided by the organization employing the workers. Workers have a lifetime-employment mindset and learn the business by moving up through the ranks of the organization.

Vroom's Expectancy Theory

Yale Professor Victor Vroom's expectancy theory states that people will behave based on what they expect the results of their behavior to be. In other words, people will work in relation to the reward they expect for their work. If the reward is desirable to workers, they will work to receive it. People expect to be rewarded for their efforts.

Delegate Duties

In this discussion of how a project manager works with management, it also needs to be acknowledged that project managers are now part of management. If your organization has an "us-versus-them" mentality toward management, then it's up to you to bridge the gap. As a project manager, your team, especially a newly created team, may not fully trust you at the project's conception—which may be too bad, considering that your team will probably be made up of your friends and colleagues. You'll need to do your best to work with them—not against them—and earn their trust and respect.

One of your first challenges will be a delegation of duties. Delegation is necessary. You are the project manager, and you cannot do every task required of a project. Once the team has been created, you need to follow the path the management and the project sponsor have taken: put your trust in others so that they can do the job you've assigned to them. Predictive projects assign tasks to specific roles, while agile projects teams are self-led and self-organizing. This means the team will determine who'll do what tasks in each iteration rather than have assignments from the project manager.

Have you ever had the experience of someone asking you to do a task, only to stand over your shoulder and question every move you make? Or worse, have a boss watch you without saying a word? It's frustrating, to say the least. As a project manager, don't do this.

Once you have pitched the idea to management and your project has been approved, it's up to you to make it happen. It's easy to be tempted to do every piece of the project planning, or at least the exciting parts, but it's not wise to yield to that temptation. An effective project manager assembles the team, assigns tasks based on qualifications and credentials, and then trusts his team to perform.

Chapter 7 will detail the complete process of assembling and working with a team. For now, know that you are also considered management once you become the project manager. All those nasty thoughts and dislikes you have harbored for some managers can very easily be sent your way now from your team. As you start the project, consider these points to being an effective project manager:

- *Follow management.* Take management's lead and delegate as they've delegated. If you like the way certain managers have treated and challenged you in the past, follow their lead and do the same for your team. If you don't like the way some managers have delegated duties, find a role model and follow that manager's lead.
- *Delegation is necessary.* You can't do everything. By delegating duties, you are showing respect, trust, and wisdom to your team. As you move further and further into project management, you also move further and further from technology. Soon there will be a gulf between what you know and the present technology. A successful IT project manager must release the reins of the implementation to the project team members; they're closest to the project work.
- *You are in charge.* From the onset, as you delegate activities, be fair—but also remember you are in charge of the project. Establish the flow of communication from your team to you, not around you. Remember that your authority is related to the type of organizational structure you operate in (functional, matrix, or

projectized) and that the organizational structure affects the amount of authority the project manager has over the project team. Know what your authority is and then manage and lead to get the project done.

- *Remember the users.* As your project develops, don't forget to consistently address the needs of the users impacted by this technology change. Often it's easy to overlook the individuals affected by the project you are managing. At each phase of the project, remember to ask how this impacts the users of the technology.
- *Keep the big picture in mind.* As a project manager, you need to make decisions regarding trade-offs and resources that benefit the organization as a whole, not just you, your team, or your project. Develop the ability to see the macro environment and the details. This will take some getting used to, but it is an ability that will serve you well.
- *Learn how to speak different languages.* At times you will need to present the status of your project to senior managers. They speak a language of ROI, productivity, and competitive advantage. However, your team members communicate in techno-speak. Make sure you are speaking appropriately to the various audiences you address.
- *Delegate, delegate, delegate.* I mentioned this earlier and it bears repeating—as a project manager, you will have plenty of work to do: following each team member's successes and failures, tracking the status of the project, meeting with management, meeting with the team, and meeting with team members one-to-one. You need to delegate the tasks and leave them delegated. If a team member is having trouble with a task, then offer your assistance.

EXAM TIP Agile projects utilize a development team to describe the people who are doing the day-to-day work and creation of the project's deliverables. Development teams are self-led and self-organizing. This means they decide who'll do what in the project—somewhat different than the traditional project management approach. Many organizations don't break from waterfall directly to agile, but pass through a hybrid project management approach. Hybrid approaches use some of the predictive environment, such as planning, and some of the agile approach, such as the product backlog.

You are responsible for the project's success, the motivation of the team, and the communication with management. If this project fails, it is your fault. If the project succeeds, then everyone shares the glory. That's just the nature of the beast.

Focusing on the Results

Management's role is to help you, the project manager, focus on the end results. From the 1940s through the mid-1980s, management typically decided what task needed to be done, who would do it, when they would do it, and how it would be done. To complicate matters more, management would often supervise each step of the process to ensure that it was being done right. You know this, no doubt, as micromanagement.

Today's management philosophy is more laissez-faire—a hands-off, empowered approach to allowing teams to accomplish a goal. Management today is more concerned with results rather than the process of getting there. As a project manager, you, too, must adopt this strategy. You must recruit your team and then let the team do the work. Focus on the results, be available when you are needed, but allow the team to work.

Of course, this all sounds wonderful, but in reality it is hard to implement. It's tough to allow others to continue with a project you've created. It's difficult for others to have the same passion and drive that you do about a project. It's fearsome to put your future and your project's success in the hands of others. But, remember, you are not giving up ownership or control, you are allowing your team to do the job that you've asked of its members.

One of the best ways to create a team with drive and charisma is to, if at all possible, involve the team from the start of the project. By recruiting team members early and giving them responsibility early on, you have given them ownership in the project's success. A project manager, keeping the project results in mind, must have

- The ability to encourage participation from all members
- The ability to empower team members
- The ability to inspire team members and management

The project manager doesn't own the project or the project team. The project manager coordinates, directs, aligns, motivates, and leads the project team to accomplish the work. If you want to get people involved in your project, allow them to own the project, too. This means more than lip service about shared ownership and synergy; get the project team involved and let them make decisions about the project work, and then allow them to prove themselves by accomplishing activities and tasks. If you want people to do a great job on the project, delegate the work and then let the project team actually do the work. Get out of the way.

EXAM TIP Project managers in an agile project might be known as coaches, scrum masters, or mentors. Agile project managers still coordinate and help the team, but much of the decision-making rests with the team members rather than with the project manager.

Inventing a Project Kickoff

Every project requires a project kickoff. So what is it, how do you host one, and why is it needed? A *project kickoff* is a meeting—or an event—to introduce the project vision, the management backing the project, the project manager, and the team members. It should be friendly, yet authoritative; organized; and used as a mechanism to assign ownership of the project to the team.

A kickoff meeting is an opportunity for the key project stakeholders to hear and confirm the details of the project requirements, to get acquainted with the project goals, and, whenever possible, to review the approved project scope. A kickoff meeting isn't an opportunity to debate the project requirements, add new demands, or workshop new ideas for the project. It is a meeting to confirm the agreement of what the project will and won't do for the organization.

A project kickoff is needed to establish the launch of the project, who's in charge of the project, and who's in control of the project team. This event allows management to rally the troops, organize the team, and get everyone excited about the upcoming plans. It's also an opportunity to convey the expectations, roles, and responsibilities of all stakeholders. It's a bon voyage party—but, hopefully, not for the *Titanic.*

Set the Stage

Depending on your project or your organization, your team may comprise longtime colleagues or complete strangers. Use this opportunity to create a team—or at least the start of one. As the project manager, you are responsible for this collection of individuals, so you need to create a social environment of ownership and teamwork.

In the kickoff meeting, you can set the tone for the entire project. Most project managers want a sense of camaraderie but also a sense of formality to the project. Here are some recommendations of how you can create both for your project kickoff depending on the size, priority, and overall effect the project has on the organization:

- It's an event! Have some fun. Create a simple theme and have some prizes or handouts for the team that are relevant to the project. For example, using the theme "Together we grow," give each team member a small plant at the meeting, and tell the team that as the plants grow, so will the project. Remind them that the project needs daily nurturing, just like the plant.
- Get excited! It's okay to have fun at these meetings. If it's a major project, and management allows it, hire local professional cheerleaders to "cheer on the team." Have someone from the local zoo bring in some animals to jumpstart the event. Do something creative and unexpected.
- Invite a representative from the vendor whose technology you will be using, such as Cisco or Oracle, to give a pep talk to the team at the kickoff meeting.
- Have someone take candid photos of team members as they enter the room and then a group photo. Create an intranet page with each member's photo, bio, and contact information.
- If you can, schedule the meeting close to breakfast or lunch and order in food. Food has a wonderful way of bringing down walls and helping people to mix and mingle.

Project kickoff meetings can be boring and stuffy. Do something exciting, invigorating, and memorable. Create alliances between you and management and the project team. Invoke excitement, assign ownership of the project, and ask for a commitment to excellence. There is no reason why kickoff meetings can't be exciting. Anyone who says otherwise is a bore.

Timebox the Meeting

Timeboxing means that there is a time limit for the meeting. You do not want the project kickoff meeting to be a long-winded discussion about the merits of the project, how the project will take place, or who'll do what in the project. Timeboxing states how long the meeting will take place, something you should define in the agenda. You'll use timeboxing to keep everyone in the meeting on task and to the point. Deeper conversations can happen post-meeting or during project planning sessions.

How Management Fits In

At the kickoff meeting, invite all of the managers involved in the decision. Their presence signifies their commitment to the project. They don't have to stay for the whole meeting, but they should at least make an appearance to rally the troops and have a donut.

The project sponsor, however, should stay for as much of the meeting as possible. Ideally, the project sponsor should initiate the more formal part of the meeting by calling things to order and introducing the project scope. The project sponsor should speak for a few minutes on the value of this project and what it means to the organization. Then the project sponsor should introduce you as the manager of the project and clearly state your authority level on the project.

This approach signifies a role of authority among the team members without having to say who's in charge. Obviously, the project sponsor is the management most closely associated with the project, but the central line of contact between the team and the project sponsor is through the project manager. There needs to be a clearly defined path of who is in charge of the project. Projects are not a democracy; each team member should have input and some autonomy, but the success of the project rests on the shoulders of the project manager, so this individual must establish their authority.

Defining the Purpose

Once the project sponsor has introduced you to the team, you're on. This is your opportunity to define many things. Prepare yourself prior to the meeting regarding what message you want to convey to your team. Your opening remarks should do several things:

- Establish your role as the project manager.
- Clearly state the goal of the project.
- Define the objective of the project.
- Set the tone of how you'll manage the project.
- State the expectations of the different stakeholders in the meeting.
- Express the impact that the project will have on the organization.

In these opening remarks, you will confirm the purpose and importance of the project and assign that ownership to the team. Don't drone on and on about the project—the project's already been approved and there's no reason to continue selling.

If possible, present a slideshow of what the project will include. You can walk the team through a five- to ten-minute overview of the project's origin to the deliverables that signify the project has reached its end. There's no need or real possibility to have a detailed step-by-step plan yet. A simple timeline of each of the major milestones will be fine.

Once you've defined the purpose of the project, showed the team members the big game plan, and given them a sense of ownership, you can quit talking. You should be able to do all of this in 15 minutes or less. Yes, 15 minutes or less. Preferably less. The project team is already going to know much of what the project is designed to accomplish. This is an opportunity for the project team, management, the project sponsor, and you to all agree what the project should accomplish.

Finally, show how management fits into the plan. Show how a financial commitment has been made to the success of the project. Show how this team is responsible for the success of the project and how everyone is counting on them.

If possible, share the news of what failure would cost the organization and the impact any delays may have on the project. This isn't to scare the team into submission, but rather to create a sense of responsibility for the success. Of course, also share with them the benefits the organization will reap when the team does succeed.

TIP Kickoff meetings provide an opportunity to ensure that everyone understands the rules of the project management approach. Agile and predictive projects have a framework for risk, changes, communication, and other project characteristics that everyone must agree to follow.

Creating Management Alliances

You and management are also a team. Just like you want your project team to be dedicated, to trust you, and to work with you, the same applies in your relationship to management. You and management are working together for the good of the organization, striving in unison toward an obtainable goal. Figure 3-5 shows how your relationship with management can be reflective of how you manage your own project team. While it may not always feel that you and management are part of the same team, you are.

Figure 3-5 A working relationship with management is required for project success.

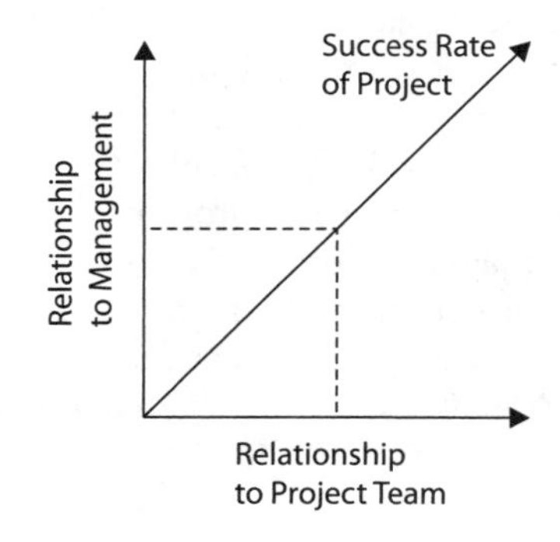

When management and you agree to implement a new project together, either by your choice or theirs, a team has been created. Hopefully, your management will be as supportive of you as you are of your own team. All of this is directly influenced by the organizational structure you're working in; functional, matrix, and projectized organizations affect the project manager–manager relationship, but overall the principles in this section are valid for any organizational structure.

Working Together

The first step in this team of you and management is to acknowledge the ability to work together. Whether you like the immediate management you are working with or not, you have to work with them. Keep in mind that your goal, the success of the project, can be impacted by the management you are working with. Likewise, the success of your project can impact the management you are working with. In other words, it's a symbiotic relationship—you both need each other to be successful.

The solution to working together is to create a channel of communication. You and management must be able to talk, to discuss the project, and to report on the status of the work, the finances, and expectations.

Your communications plan, which I'll detail in Chapter 6, will dictate how often you and management need to communicate. In some organizations, it's weekly; in others, it's monthly. How will you know? Ask management what their expectations are. In some instances, conditions within the project will prompt immediate communication. Basically the communications management plan defines who needs what information, when the information is needed, and the modality of the expected communication.

Intermediary communications in the shape of e-mails, an intranet site, or voice mail would be another avenue to keep management involved with the status of the project. By keeping a flow of communication open through you to management, you are ensuring management's involvement—but at a happy distance. Project managers must report both good and bad news. Don't candy-coat your findings; reporting both the good and the bad on an equal scale will build trust between you and management.

There are some problems that management and project managers together need to avoid. One of the largest complaints IT project managers have is that management will circumvent their position and go directly to the project team with instructions, input, and advice. In some instances, such as disciplining a team member, this action may be appropriate. The organizational structure will influence how project sponsors and other stakeholders communicate with the project team. Ideally, project sponsors should follow the same flow of communication through the project manager to the team.

While this flow of communication may require delicate handling, it's not impossible to achieve. At the conception of the project and prior to the project kickoff meeting, you, as the project manager, should express to the project sponsor that you would like to handle all avenues of communication and management of the project. If you're new to project management, this chain of command may not be granted, although it's not unusual. Most professionals respect the chain of command from management, to project manager, to the individuals on the team.

If, for some reason, members of management do bypass you and work directly with your team and this is disrupting the project, you must address the issue. Report to your project sponsor that this confuses the project team about your role as project manager and to whom they are to report, and it undermines your authority with the team. Don't be confrontational, but do be factual.

Following Organizational Governance

Organizational governance is a nice way of describing the rules and policies that you, as the project manager, must follow in your organization. Organizational governance describes the policies and rules that are unique to your organization—things like how you manage your team, forms you're required to use, processes you must complete, and how you get to buy resources for the project. You'll also encounter organizational governance when you complete a project phase and there's a review, sometimes called a phase gate, before the project can transition to the next project phase. Governance often comes from the PMO, if one exists.

Organizational governance for most organizations includes the internal process compliance—proof that you have followed the internal policies of the organization. This means your project will undergo audits for both process completion and financial reconciliation. Your project time, cost, and scope baselines will also be examined to see where changes entered the project, how the baselines were adjusted to reflect the approved changes, and how you completed versioning of each baseline. A *change log* is a document to record all changes to the project scope, schedule, and costs, and that'll help with any audits.

I'll talk more about change control in Chapter 10, as change is inevitable in IT projects. Changes, in a predictive project, must always follow a defined change control process, be documented, and be incorporated into the project plan for execution. Remember that in agile projects, change is welcome and expected. Most organizations use a *change control board (CCB)* to oversee significant changes to the project. Sometimes the change control board is known as a technical review board, an engineering review board, or a technical assessment board. These boards all do the same thing: review the change to determine its impact and value on the project deliverable.

Another aspect of organizational governance is how the project adheres to standards, regulations, and corporate compliance programs. A *standard* is a guideline that's appropriate for your industry—such as how network cables are punched down. No one's going to be fined for not following a standard exactly, but standards are the generally accepted practices for a type of work, service, or product. A *regulation*, however, is a law that your project must follow. Think of the Health Insurance Portability and Accountability Act (HIPAA) or the Sarbanes-Oxley Act. These are just two examples of laws that are now requirements for many project managers to contend with. As a rule, standards are optional while regulations are not.

Finally, your organization may participate in quality assurance programs such as Six Sigma or Total Quality Management. These are part of your organizational governance if your project is also required to participate in them. A common requirement of project managers is to map their projects to an ISO program such as ISO 9000:2000. ISO 9000:2000, from the International Organization for Standardization, is based on the Information Technology Infrastructure Library (ITIL).

NOTE The abbreviation is *ISO* and not *IOS* because of the language and translation differences around the world. *ISO* is Greek for uniform, and that's why it's used.

If your organization uses a program for quality and standardization, then the program is part of your organizational governance and your project will need to adhere to it. You might also have to deal with the CMMI—the Capability Maturity Model Integration. Don't expect many CompTIA Project+ exam questions on this process improvement program, but it's becoming a fairly popular system for IT project managers (especially if you're dealing with projects for the U.S. government).

Dealing with Challenging Bosses

Remember the boss who was a complete jerk? Or the one who would disappear for days and avoid any decision making? Do you still work for one of those?

While most counterproductive behavior is not tolerated in today's workplace, a fair amount of it still exists. Management has tended to shift into a more team-building, empowering, goal-oriented style of leadership than in past years. However, there are still plenty of managers who don't relate well to people.

Unfortunately, most of these managers stem from IT backgrounds, and they lack social skills. Or they're traditional managers and lack IT skills. As an IT project manager, it can be tough and confusing to deal with either type.

The manager who comes from an IT background may feel threatened that new technology is coming onto the scene to replace the work and implementation they did so many years ago. Due to their current position, they've lost touch with the rapid pace of technology and feel frustrated by it.

Other managers who stem from traditional roles often have no grasp of technology and of what it can or cannot do. These managers often hide from decision-making responsibilities, overanalyze every phase of the project, or immerse themselves in the project in an attempt to learn as much as or more than the IT project manager.

As a project manager, you will have to find a way to deal with different types of management. Here are six types of managers you'll likely encounter and advice on how to deal with each:

- **Managers who won't listen** Managers who won't listen are either not interested in what you are saying or have a general lack of respect for others. The best way to deal with these people is to document what you have to say. Often these managers put their confidence only in something that is in writing, as it's on record. Use e-mail, letters, and memos to confirm conversations you've had with the manager.
- **Managers who are aggressive** Managers who yell, stomp, and are outrageously rude have become less popular in today's workplace; however, these bullies still exist. The best way to deal with these managers is to befriend them, as much as you can, and let them know that when they act the way they do, it offends you. Don't cower before them, and if the behavior persists, seek help from the human resources department.

- **Managers who avoid decisions** These managers are afraid of making the wrong decision, so they make no decision. They request more research, cancel meetings, and delay their way out of any forward progress. The best way to deal with these managers is to set deadlines with them on when the next phase of the project will commence. These deadlines don't have to be exact dates; they can even be the accomplishment of key milestones within the project. Put the deadlines in writing and try to get a commitment from them. As an alternative, present them with the decision you suggest, and let them know if you don't hear from them by a certain date, you will implement your recommendation. Make sure this is documented and that you give the manager a final heads-up before going ahead with your recommendation.
- **Managers who micromanage** These managers are typically perfectionists, feel that no one else can do the job as well as they can, or don't trust anyone else to do the task at hand. The best way to deal with these managers is to let them know politely that they are micromanaging. They just need to be told they aren't allowing you to do your work. Many of these managers don't realize that they are guilty of micromanaging and need to be told to back off. Of course, you'll then complete the task proficiently and with excellence to show the manager you can do the activity without their hovering.
- **Managers who hog the credit** These managers step into the spotlight as the project is created and again when the project is finished. In the meantime, their presence is not seen or heard. The best way to deal with this type of manager is to document the progress of the project publicly on an intranet site. Allow everyone to see who has done what work, when the work was completed, and how it was done. The glory-hungry manager may be there at the kickoff and at the event's end, but their contributions aren't recorded on the intranet site. If this solution does not work, tell the manager that they need to credit all of the team members and their hard work in the project's success.
- **Managers who rotate the discipline** These managers think someone always has to be in trouble at any given time, and they will discipline someone once a week just to remind everyone else that they are in charge. Your department may refer to it as "being in the doghouse," "called on the carpet," or "your turn." The best way to deal with this is to confirm the cycle of discipline and then confront the manager about it. You should always follow your organization's human resource practices concerning confrontations in the workplace. You don't want to create more trouble.

Working with Good Bosses

Just as plenty of bad bosses exist in the world, a large number of truly good bosses are out there. These individuals are caring, hard-working, goal-oriented individuals. They have the good of the organization in mind, know how to lead, and treat people fairly. If you are fortunate to have a good boss, let that person know that you appreciate the way they offer advice, listen to what you have to say, and treat you with respect.

Working for a good boss, however, can often be mistaken as working for a passive boss. If you can imagine another project manager working for a boss with a temper, that person's motivation to work hard is not to get yelled at, publicly embarrassed, or put in the doghouse. On the other hand, some who have a kind boss may be tempted to become more lax because they know their manager would never yell at or embarrass them. If you have a good boss, don't take advantage of them. Continue to work hard, to work persistently, and to lead your team.

Learn from your boss. As an IT project manager, you can learn from either type of boss that you may have. A bad manager is showing you how not to manage, while an excellent manager is showing you how it's really done. Find the attributes of your manager that work and then repeat those skills with your project team. Not only will you become an effective manager, but you'll also become an effective leader.

CompTIA Project+ Exam Highlight: Management Processes

There are several exam objectives covered in this chapter—and you'll need a solid grasp of these objectives to pass your CompTIA Project+ exam. Dealing with management is an integral part of a project manager's job, and knowing these objectives can help you become a better project manager, operate more efficiently, and boost your career. But let's start with first things first—passing your CompTIA Project+ exam.

1.1 Explain the basic characteristics of a project and various methodologies and frameworks used in IT projects All project management approaches include a kickoff meeting to communicate the project vision and selected project framework. A kickoff meeting is a formal event for the key project stakeholders, such as the project manager, the project team, the project sponsor, and managers to walk through the project objectives, project management approach, goals, approved project scope, and project requirements. It ensures that all of the key stakeholders have a clear understanding of the project, how the project will operate, and any rules that everyone in the project must agree to abide by. Kickoff meetings aren't opportunities to debate the project requirements, add new demands, or workshop new ideas for the project. A kickoff launches the project work based on the agreed-upon requirements.

Agile and predictive projects utilize project kickoff meetings. A project kickoff meeting communicates the effect the project will have on the organization and sets expectations for all of the stakeholders contributing to the project success.

1.9 Given a scenario, apply effective meeting management techniques Meetings are a necessity in all project management approaches, so you want to have a good grasp on meeting management techniques. This means you'll want an agenda that maps out the intent of the meeting, a thoughtful presentation that quickly addresses the meeting purpose, and you need facilitation skills that keep the participants on track and to the timeboxed duration of the meeting. A scribe can record the minutes, notes, questions, and other details so the information, action items, and follow-up questions can be distributed to the attendees after the meeting.

1.10 Given a scenario, perform basic activities related to team and resource management A major exam objective is understanding all the organizational structures. There are three primary structures that affect the project manager. When you're answering exam questions, pay attention to the organizational structure if possible—the structure of the organization will often affect the actions the project manager is allowed to take. Here's a quick recap of the organizational structures:

- **Functional** Gives the project manager the least power and gives the functional manager the most power over project decisions. The project resides in one department or line of business, such as sales, manufacturing, or IT. Resources aren't shared throughout the organization, and the project team members have regular operational duties in addition to their project duties. The project manager might be called a project coordinator or expeditor and must do whatever the functional manager requests on the project.
- **Matrix** A matrix structure uses resources from around the organization, and the project team members may be working on multiple projects at once. Matrix structures require additional communication, scheduling, and trade-offs for resources. There are three types of matrix organizations:
  - **Weak matrix** Little authority for the project manager; functional management has the decision-making power over project decisions.
  - **Balanced matrix** The project manager and the functional managers share project power, though there can be a power struggle over who's in charge of project decisions.
  - **Strong matrix** The project manager has more project authority than the functional management.
- **Projectized** This organization assigns a project team for the duration of the project, the project manager has total authority over the project, and communication demands are lower than in matrix structures because the team is working on one project full time.

4.1 Summarize basic environmental, social, and governance (ESG) factors related to project management activities. Organizational governance describes the internal processes, policies, standards, and regulations that the project manager must follow. This exam objective can be a little tricky, as organizational governance can vary based on the project, organizational structure, and the existence of a PMO. You won't have the same rules and processes at Company ABC as you might at Company XYZ. You'll find, however, that certain industries use the same standards and follow the same regulations, such as in manufacturing, health care, or construction. IT project managers must be aware of the standards and regulations that affect project planning and execution for the industry in which their project resides.

Chapter Review

To begin working on a project, you need two fundamental things: dedication from you and approval from management. Often to gain the approval of management, you will have to conduct a presentation in which you sell management on the idea of implementing your business case. The business case has to be condensed into the language that all management speaks: return on investment. Once you've got the business case condensed, create and tailor your presentation to the audience.

Just as your opening lines on a first date are crucial, so are the opening remarks at a presentation for new technology. By starting at the end of a project and exposing what the project will deliver, you'll capture your audience's attention and have them clamoring for more—hopefully. Once the project has been approved, you must continue to work with management and keep them informed of the project's progress. Management's role in the project is that of support, not implementation.

The organizational structure will determine your level of autonomy on a project. A functional structure restricts the amount of power the project manager has. This structure assigns the power to the functional manager. The matrix structure has three levels of power for the project manager: balanced, weak, and strong. When the project manager and the functional managers work in a matrix structure, they may struggle for project power and control. The projectized environment affords the project manager the most power and authority.

At the onset of the project, you, the project manager, must bring management and your project team together. At the kickoff meeting, you'll create a sense of camaraderie and excitement for both management and your assembled project team. Have fun! Invite management and project clients to press the flesh and snack on a donut. Create excitement to make the kickoff an event and build immediate morale. While the kickoff meeting can be an event, its purpose is to set expectations, formally authorize the project work, and explain the project goals and vision.

Project managers and the project team must follow the organizational governance for your organization and industry. Understanding how your organization does business, how the organization allows projects to move forward, and what's expected of you as a project manager is paramount to project success. This also means the project manager must be aware of the standards, regulations, and best practices that will affect the planning and execution of the project.

Finally, as the project begins to move forward, you'll need to work with your management. Just as there are different types of people, so are there different types of managers. Learn how to work with your manager, not for your manager. Mutual respect must be present, or the project will be grounded. No matter what type of boss you have, good or otherwise, learn from them. Mimic their good attributes and avoid their bad ones. Bosses come and go; leaders endure.

Exercises

These exercises allow you to apply the knowledge you have learned in this chapter and are followed by possible solutions.

Exercise 3-1: Preparing a Presentation Opening Statement

In this exercise, you will create an opening statement for a presentation to management. Recall that an opening statement should start by describing the end goal of the project—focusing on the project deliverables instead of the process to get there.

You are a network administrator for TriStar Manufacturing. Your network currently consists of six Windows Servers and 300 Windows workstations. Most of the workstations have a 1 GHz or faster processor and 1GB of RAM. Users are constantly complaining that the network, their computers, and new applications are very slow and often crash.

Management has asked for your opinion on the technology and what can be done to improve the situation. You've decided it's past time for the company to upgrade. You have done some initial research and would like to recommend the following hardware and software upgrade for your company:

Hardware	Features
Six Windows servers (rack servers)	Each server equipped with two 3.7-GHz, eight-core, 45M-cache processors, 32GB UDIMM memory, and five 1TB RAID 5–configured solid state hard drives
300 Windows workstations	Each workstation equipped with a 2.0-GHz processor and 8GB of RAM
Three network switches	Used to segment the network
Two network storage devices	10TB storage each

Based on this information, create an opening statement for your presentation. Your opening statement should be snappy, captivating, and focused on how this upgrade will increase productivity. (As technology pushes forward, these hardware requirements may seem dated to you. No worries—do some research and update these to your ideal specifications.)

Exercise 3-2: Creating a Kickoff Meeting Event

Congratulations! Your opening statement and presentation to upgrade the company's hardware is a hit, and management wants you to proceed. They've agreed that you will be the project manager for the rollout, and you've handpicked 15 people to be on your project team.

You've completed the feasibility study and are now ready for your project kickoff meeting. The following table will help you create an event for your kickoff. Management has allotted a whopping $500 for the event. Have fun, and make it exciting and inspiring!

Questions	Action
Who should attend this meeting?	
When will the meeting take place?	
What is the theme of the project?	
What type of project-related handouts will you have?	
What fun things do you have planned?	
How will you inspire the team?	
Who will speak first?	
What are your opening remarks to the team?	
What are some of the topics you'll cover in your team presentation?	
How long will your presentation take?	
What will you spend the $500 on?	

Exercise 3-3: Dealing with Negative Stakeholders

Excellent kickoff party…er, meeting. The project team and your project sponsor had a good time—everyone thought it was great. Well, everyone except Amar Abbot, the company crank, and, unfortunately, your immediate manager. Amar is less than thrilled about your new project and thinks technology generally gets in the way of any real work getting done. Amar doesn't believe that technology has made the company more efficient or more productive.

Your assignment in this exercise is to record how you would react to different scenarios regarding your manager, Amar Abbot.

Scenario	Your Response
Amar Abbot is getting increasingly cranky about the time you'll need to be spending on your "pet" project. He's insisting that you focus less on technology and help more users clean their workstation's mouse, monitor, and keyboard.	
Now you need Amar's help. He has to make a decision about when you can have the testing lab for the new workstations and operating systems. You need to create workstation images, test software compatibility, and work with your new servers. The equipment has arrived, but Amar just doesn't want to make a decision on when you can use the testing lab for your project.	
Today is Frank's turn to be in Amar's doghouse. It was your turn last week, and next week it'll be Mary's. Every week someone on your team is in trouble with Amar. This whole cycle of Amar being angry with someone every week is demoralizing to your team, prohibiting progress, and generally frustrating team members.	
Amar Abbot is becoming more rude and belligerent with you and your project team. He's yelling at anyone who comes into his office, threatening to "write you up," and acting very unhappy with any task you try to do. Yesterday he embarrassed Sam and made Jane cry.	

Exercise 3-4: Calculating Time Value of Money

In the online resources that accompany this book (see Appendix C) you'll find an Excel spreadsheet called Time Value of Money, and you can use it to complete this exercise. The spreadsheet includes the future value and present value formulas, and Excel will calculate the results for you. Follow these steps to use the worksheet's formulas:

1. You can access comments about the formulas by hovering your mouse over the red dots in cells A4, A5, A6, A17, A18, and A19.
2. Navigate in the spreadsheet to cell B4.
3. Enter **$890,000** as the present value in cell B4.
4. Navigate to cell B6 and set the number of time periods to **3**.
5. What is the value displayed in cell B11?
   Your answer: ________________
6. What is the future value of $890,000?
   Your answer: ________________
7. Navigate to cell B17.
8. Enter the future value as **$625,000**.

9. Change the interest rate to **0.05** in cell B18.
10. Enter the number of time periods as **2**.
11. What is the present value of $625,000?
    Your answer: ________________
12. You can use the formulas in the Excel spreadsheet to answer this scenario's question. Assume that you present a project to management that will last for five years and will require an investment of $775,000. If the rate of return is 6 percent, what must your project be worth in five years to be worth the investment to management?
    Your answer: ________________
13. You can use the formulas in the Excel spreadsheet to answer this scenario's question. Assume that Shelly has a project that promises to be worth $550,000 in three years. If the rate of return is 6 percent, what is the maximum amount management should invest in the project if they want to break even in three years?
    Your answer: ________________
14. You can use the Excel spreadsheet to test your comprehension of the formulas. Try to solve the formulas using just a pencil and calculator and then test your results in the Excel spreadsheet. Complete the following table:

Present Value	Interest	Time Periods	Future Value
$300,000	0.06	4	
$550,000	0.06	2	
$1,245,000	0.05	4	
Future Value	**Interest**	**Time Periods**	**Present Value**
$789,000	0.06	3	
$500,000	0.05	4	
$1,922,001	0.06	5	

Exercise Solutions

The following offer possible solutions for the chapter exercises.

Exercise 3-1: Preparing a Presentation Opening Statement

Your opening statement may be fashioned like this one: "By upgrading our servers and our workstations to current technology, we can increase productivity by 47 percent and increase morale by 100 percent. This implementation would allow all of us to work smarter, faster, and with less frustration."

You'd then continue your presentation to include the details and evidence of your opening statement. You would need to include a statement as to how productivity will be measured. You should also link the performance improvement to a cost savings. While the organization may experience less frustration with the solution, you should back up the claims through a quantitative statement, such as hours saved, number of help desk calls placed, or productivity opportunities. Throughout the presentation, you should link the technical specifications back to the claims you made in your opening remarks.

Exercise 3-2: Creating a Kickoff Meeting Event

Here is a sample of how a fun, informative project kickoff meeting could go:

Questions	Action
Who should attend this meeting?	Management, the project sponsor, the project manager, key stakeholders, and the entire project team will attend.
When will the meeting take place?	Early morning; breakfast will be provided.
What is the theme of the project?	"Working on the Future"
What type of project-related handouts will you have?	Everyone will receive a hammer with the theme "Working on the Future" printed on the handle. Construction hats, breakfast food, and drinks will also be provided.
What fun things do you have planned?	Construction hats will be handed out at the door. Orange cones and sawhorses will lead the way into the meeting room. "Construction" workers from the computer and software vendor will be on hand to meet and greet the attendees.
How will you inspire the team?	In the opening statement, the project manager will relay the theme of "Working on the Future" by laying out the goals of the project.
Who will speak first?	The project sponsor
What are your opening remarks to the team?	"Construction sometimes has a bad reputation, but everyone is happy to see the end results of the work."
What are some of the topics you'll cover in your team presentation?	Goals of the project Project deliverables Flow of communication
How long will your presentation take?	One hour from start to finish
What will you spend the $500 on?	Hammers, plastic construction hats, food

Exercise 3-3: Dealing with Negative Stakeholders

The following table presents possible responses to the different scenarios with unhappy Amar Abbot:

Scenario	Your response
Amar Abbot is getting increasingly cranky about the time you'll need to be spending on your "pet" project. He's insisting that you focus less on technology and help more users clean their workstation's mouse, monitor, and keyboard.	"Amar, I understand how you feel in regard to the computer upgrade process. There really isn't a pressing need to clean the workstations, as we'll be replacing those within a few weeks anyway."
Now you need Amar's help. He has to make a decision about when you can have the testing lab for the new workstations and operating systems. You need to create workstation images, test software compatibility, and work with your new servers. The equipment has arrived, but Amar just doesn't want to make a decision on when you can use the testing lab for your project.	Document and date that the room is needed for the project rollout testing and planning. Send the document not only to Amar Abbot, but also to the project sponsor. By documenting the problem, you've created a paper trail that Amar may be more responsive to. You might also negotiate with other users of the room, prepare an agreed-upon schedule, and just ask Amar to approve the solution you and the other employees have created. This involves presenting management with a solution rather than asking them to take action.
Today is Frank's turn to be in Amar's doghouse. It was your turn last week, and next week it'll be Mary's. Every week someone on your team is in trouble with Amar. This whole cycle of Amar being angry with someone every week is demoralizing to your team, prohibiting progress, and generally frustrating team members.	Take Amar to lunch and explain that the cycle of discipline is unnecessary. You are all adults, and his behavior is interrupting the progress of the project and demoralizing your team. Be sure to pay for his lunch.
Amar Abbot is becoming more rude and belligerent with you and your project team. He's yelling at anyone who comes into his office, threatening to "write you up," and acting very unhappy with any task you try to do. Yesterday he embarrassed Sam and made Jane cry.	You, Sam, and Jane should all take Amar to lunch and discuss his behavior. Acknowledge that he is your manager, but his actions will not be tolerated. After lunch, document the conversation in an e-mail and send it to Sam, Jane, and Amar.

Exercise 3-4: Calculating Time Value of Money

The following list shows the steps to use the worksheet's formulas along with the correct answers.

1. You can access comments about the formulas by hovering your mouse over the red corner dots in cells A4, A5, A6, A17, A18, and A19.
2. Navigate in the spreadsheet to cell B4.
3. Enter **$890,000** as the present value in cell B4.
4. Navigate to cell B6 and set the number of time periods to **3**.
5. What is the value displayed in cell B11?
   Your answer: 1.191016
6. What is the future value of $890,000?
   Your answer: $1,060,004.24
7. Navigate to cell B17.
8. Enter the future value as **$625,000**.
9. Change the interest rate to **0.05** in cell B18.
10. Enter the number of time periods as **2**.
11. What is the present value of $625,000?
    Your answer: $566,893.42
12. You can use the formulas in the Excel spreadsheet to answer this scenario's question. Assume that you present a project to management that will last for five years and will require an investment of $775,000. If the rate of return is 6 percent, what must your project be worth in five years to be worth the investment to management?
    Your answer: $1,037,124.82
13. You can use the formulas in the Excel spreadsheet to answer this scenario's question. Assume that Shelly has a project that promises to be worth $550,000 in three years. If the rate of return is 6 percent, what is the minimum amount management should invest in the project if they want to break even in three years?
    Your answer: $461,790.61

14. You can use the Excel spreadsheet to test your comprehension of the formulas. Try to solve the formulas using just a pencil and calculator and then test your results in the Excel spreadsheet. Complete the following table:

Present Value	Interest	Time Periods	Future Value
$300,000	0.06	4	$378,743.09
$550,000	0.06	2	$617,980.00
$1,245,000	0.05	4	$1,513,305.28
Future Value	**Interest**	**Time Periods**	**Present Value**
$789,000	0.06	3	$662,459.61
$500,000	0.05	4	$411,351.24
$1,922,001	0.06	5	$1,436,230.96

Questions

1. You are the project manager for your organization, and you are preparing a presentation for management. This project will help your organization become more efficient, and it will help your career as a project manager in your organization. Your audience will include the CEO, customers, and end users. Considering this information, what is the most important thing in your presentation?

   A. The audience

   B. The message

   C. The technology being presented

   D. The length of the presentation

2. Marty is the project manager, though her manager calls her the project coordinator, for her organization. Her manager, Tom, meets with Marty and the project team weekly to assign project work and to review the project work completed so far. The project team members have regular operational work to do in addition to the completion of the project work. What type of an organizational structure is Marty likely a project manager for?

   A. Functional

   B. Weak matrix

   C. Strong matrix

   D. Projectized

3. Harold is the project manager for his organization. He believes that project teams can be self-led and want to accomplish the project as long as they understand the work they are to complete. Harold believes in challenging the project team and pressing their abilities to learn and do more. What management theory does Harold most likely believe in?

   A. Herzberg's theory of motivation

   B. Maslow's hierarchy of needs

   C. McGregor's Theory X and Theory Y

   D. Vroom's expectancy theory

4. You are the project manager for the NHQ Organization. You and several project managers have been invited to present your potential projects to a management committee to determine their worthiness, return on investment, and potential risks. What is the name of this committee that may select your project to be initiated?

   A. Murder board

   B. Organizational governance board

   C. Technical assessment board

   D. Project management office

5. You are the project manager for an organization that is a balanced matrix. Your project team members come from many different areas of the organization. Which one of the following is likely a negative component of working in a balanced structure?

   A. Lack of project team resources

   B. Increase in team productivity

   C. Anxiety for the project team as the project nears completion

   D. Higher communication requirements

6. When creating audience handouts for a presentation, what information must be included in the handouts?

   A. Information on the profits

   B. Information on the project manager

   C. An executive summary

   D. An implementation plan and timeline

7. Frank is proposing a project to management. His project will require $300,000 to be initiated and will last for two years. Considering that the current rate of return is 6 percent, what must Frank's project be worth in two years at a minimum for management to consider this project?

   A. $300,000

   B. $300,001

   C. $337,080

   D. $266,988

8. What role does management play in project management?
   A. Hands-on implementation
   B. Authoritarian
   C. Support
   D. Financial watchdog
9. If you subscribe to Herzberg's theory of motivation, which one of the following must exist as a hygiene agent before you may offer motivating agents to your project team?
   A. Paycheck
   B. Reward
   C. Bonuses
   D. Physiological needs
10. Your project is expected to be worth $575,000 four years from today, with a 6 percent rate of return. What is the present value of the project?
    A. $455,453
    B. $609,500
    C. $725,924
    D. $34,500
11. You are the project manager for your organization. Your current project is a database upgrade project, and your project team has completed the first phase of the project. Before your project can move forward, management must review the first phase and determine whether the project is allowed to transition to phase two. What is this review process called?
    A. Organizational governance
    B. Phase gate approval
    C. Quality control
    D. Step funding
12. You are the project manager of the GHR Project for your organization. This project has just recently been initiated, and the requirements of the project have been agreed upon. Your sponsor would now like you to host a meeting that will review the agreement of the project objectives and establish the direction of the project. What type of a meeting does management want you to host?
    A. Workshop
    B. Scope decomposition session
    C. Kickoff meeting
    D. Bidders conference

13. Complete the following sentence: Standards are ________________, while regulations are ________________.
    A. guidelines, penalties
    B. rules, requirements
    C. optional, penalties
    D. guidelines, requirements
14. Management is considering a new project for your organization. This project will likely last for three years with the realization of benefits each quarter the project is in existence. Knowing that the benefits will be available to the company at each quarter, management would like you to examine the value of the project to determine its true worth for the required investment. What type of benefits measurement approach does management want you to complete?
    A. Net present value analysis
    B. Future value analysis
    C. Cost-benefits ratio analysis
    D. Management horizon analysis
15. Management is considering several projects for implementation in the organization. To determine which projects are the best for the organization, management has created seven categories to measure each project: profitability, risk exposure, schedule, marketplace conditions, competency required, experience, and efficiency gained. Each project will receive points in each category, and the project with the most points will be selected. What type of benefits measurement approach is being completed in this organization?
    A. Scoring model
    B. Weighted analysis
    C. Time value of money analysis
    D. Cost-benefits analysis

Answers

1. **B.** The most important thing in a presentation is not the audience; it is the message. The presentation must be clear, concise, and to the point. The technology being presented and the length of the presentation are both important elements, but they are not the most important things in the presentation.
2. **A.** Marty is working in a functional structure. Your clues in the question include that Marty is called a project coordinator—something that's common in a functional structure. Tom's involvement with the project team is another clue that this is a functional structure. Finally, the project team members have regular operational duties to complete, so you can reason that this is not a projectized organization.

3. **C.** McGregor's Theory X and Theory Y say that X workers are lazy and unwilling to work. Y workers are self-led, have initiative, and are willing to work. Harold demonstrates this belief by putting confidence in his team to complete their project work.
4. **A.** A murder board is a management-driven board that reviews potential projects to determine which project will likely be the most successful for the organization.
5. **D.** Matrix structures have higher communication requirements than the projectized or functional structure. This is because the project team members come from across the organization and are likely to work on multiple projects at once. The project managers must communicate with the functional managers, the other project managers, and the project team members to coordinate schedules, activities, and resource use.
6. **C.** Always include an executive summary. As you give your presentation, the executive summary allows the audience to read over the quick facts of the project and get an idea of where the project will end. It also documents the goals and overview of the project for individuals who may not be able to attend your presentation.
7. **C.** You can use the future value formula to determine the minimum amount that Frank's project should be worth: $FV = PV \times (1 + i)^n$. In this instance, the formula is \$300,000 × (1.1236), as the interest rate is 6 percent and the project will last for two years. In two years, if Frank's project is worth less than \$337,080, it's not a good financial decision for the organization to make.
8. **C.** Management should be supportive but not authoritarian or hands on.
9. **A.** Hertzberg's theory of motivation defines hygiene agents that must first exist as the expectation of the employee. A paycheck must exist for the project team as a hygiene agent before they are likely to be motivated for other rewards and incentives.
10. **A.** The present value (PV) can be determined by using the formula $FV \div (1 + i)^n$, where i is the interest rate and n is the number of time periods. In this instance, the answer is \$575,000 ÷ (1.2624), for a value of \$455,453.
11. **B.** The review at the end of a phase has many names, but of all the choices presented, only the phase gate approval is the correct choice. While it's true that the phase gate approval is part of organizational governance, organizational governance includes many other components and is not the most exact of all the choices.
12. **C.** This is a kickoff meeting to discuss the project objectives and establish the project's direction. A kickoff meeting is not an opportunity to request scope changes or debate the project's merit.
13. **D.** Standards are guidelines, while regulations are requirements. You can change or deviate from some standards in a project, but you must always follow the related regulations or there will be fines and penalties.

14. **A.** Of all the choices, net present value analysis is the best choice, because net present value allows you to consider the cash inflow and time value of money for each time period of the project.
15. **A.** A scoring model is a benefit measurement method that assigns points to defined categories of projects. The project that receives the most points is the project that is selected, funded, and initiated. Incidentally, project managers can use a scoring model for vendor selection as part of the procurement process.

CHAPTER 4

Managing the Project Scope

This chapter covers the following topics:

- Creating a project scope
- Defining a work breakdown structure
- Obtaining management approval
- Establishing communication channels

Remember when you were a kid and you bought your first model car? You opened the box, sorted all the pieces, put the decals aside for safekeeping, and gathered all your tools. Of course, you read the directions completely and carefully assembled each piece of the model with just the right amount of glue, patiently waited for it to dry before proceeding, and then finally applied the decals with a pair of tweezers.

Doesn't sound quite right? Were you more like the kid who ripped the box open, tossed the directions aside, and ended up gluing your fingers together? What's the lesson here? With experience, you became more like the kid in the first example: meticulous, careful, patient, planning, and savvy with a tube of glue. The same is true with project management—except for the glue thing. You, the project manager, need a detailed plan of the work, what phases are required in the work, and then what tasks are required within each phase. Just as it was when building the model car, taking the proper steps won't be easy, but if you plan for success, you will reach your goal.

VIDEO For a more detailed explanation, watch the *Managing the Project Scope* video now.

Creating the Project Scope

Consider all the IT projects that you've ever worked on. Add to that pile of projects all the IT projects that fall outside your realm of experience, expertise, and concern. That's a huge amount of hardware, software, databases, and network projects that have been defined, financed, planned, and completed. IT project management is a huge arena of endeavors, possibilities, and conditions. Each project is different from the next, but all projects must have boundaries that establish what the project will accomplish and what the project will exclude.

The requirements of a project define the boundaries. Requirements are the elements, conditions, services, and expectations that the project customers, stakeholders, and management expect the project to create. You can generally visualize the project requirements when you consider the current state of an organization and then examine the future, post-project state of the organization.

As a project manager, you may be responsible for gathering the requirements, though this responsibility often rests with a business analyst. It's healthy for you to understand the process, techniques, and outputs of collecting requirements so that you can form a complete picture of how the project is defined and what the project stakeholders are anticipating from your project.

Gathering Requirements Through Communications

One of the most effective approaches to collecting project requirements is to get out there and talk with your stakeholders. This is true in both predictive and agile projects. Communication is a major part of project management, and it really starts when you and the stakeholders discuss the desired future state of the organization—the future state that your project will help create. Your organization may have a different approach to collecting requirements, and that's fine; always follow the governance of your organization.

If you're lucky enough to have a business analyst who will complete the requirements collection process for you, you'll want to discuss the requirements in depth with the business analyst and the stakeholders to confirm that you understand the requirements. This discussion focuses first at a high level of what the customer wants, and then you can drill down into the individual components that contribute to the overall solution the business analyst has documented. You and the business analyst should create a partnership and identify your roles and responsibilities on the project, work together, and stay out of each other's way.

If you're not so lucky to have a business analyst, then it's all up to you, the project manager, to collect and document the project requirements. One advantage of this approach, however, is that you'll have an in-depth understanding of what the customer expects, and that will help you create the project scope, maintain quality, and address any threats or stakeholder concerns. Through interviews, focus groups, workshops, and requirements analysis meetings, you can discover, define, and document what the stakeholders expect the project scope to achieve.

If you're dealing with a large mass of stakeholders, such as the end users of the software your project may create, then meeting with each stakeholder isn't feasible. In these instances, the project manager can use surveys to interview the stakeholders. With web technologies, it's a snap to compile and share responses. You might also elect to use representatives for large groups of stakeholders. Representatives of the stakeholder group and the project manager meet to discuss the requirements that will affect the large mass of stakeholders, given the representative's input.

Finally, as introduced in Chapter 1, you might need to use observation, sometimes called job shadowing, to see how the stakeholders complete their work. *Passive observation* happens when the project manager quietly observes the work being completed to understand how the stakeholder performs the duties and processes that the project deliverable will affect. *Active observation* allows the observer to interact with the stakeholder—often stopping the work, asking questions, and even trying the work to fully understand how the stakeholder completes the work.

Whatever approach you take to document the requirements, you must ensure that the stakeholders are in agreement that you've captured the project requirements. You don't want to create the project scope without having captured the project requirements or base the scope on requirements that are wrong.

The requirements documentation usually includes all of the following information:

- Business need to solve or the opportunity the project will seize
- Project objectives and goals that can be traced to the requirements
- Functional requirements of the project deliverable, such as features and functions of the thing or condition the project will create
- Solution design is needed for technical projects. This can be for software development, network infrastructure projects, and databases.
- Nonfunctional requirements of the project deliverable, such as the level of service, performance objectives, security, interoperability, and support
- Quality requirements
- Factors for project acceptance
- Effect of the deliverable on the organization, departments, or lines of business
- Effect of the deliverable on entities outside of the organization
- Need for education, training, and ongoing stakeholder competency support
- Identified assumptions and constraints

One method that can help track the requirements from the requirements document to the finished product is a *requirements traceability matrix (RTM)*. An RTM identifies the project requirements, documents when each requirement is to be created in the project life cycle, and records when the deliverable was actually created. The RTM can also help control and evaluate changes to the project scope. At the end of a project phase and at the end of the project, the project manager can examine the RTM as part of scope verification to confirm that the requirements that were expected were created.

You can also use the matrix to record attributes about each deliverable, such as a globally unique identifier, a description of the requirement, the requirement's owner, the versioning information, and the status of the requirement. All of this information can help with stakeholder management and communication throughout the project. You and the project team can use this information as you plan each phase of the project and to confirm that you've completed all of the requirements that the customer asked for. And then you're everyone's hero.

Creating the Product Backlog

In an agile project, the project scope is represented as the product backlog. The product backlog is a prioritized list of all the features, functions, and requirements that the final product should have. The product owner represents the project customers and is responsible for sorting, prioritizing, and maintaining the product backlog. The product backlog items are gathered by a business analyst, customer representative, the product owner, the project manager, the project team, and other stakeholders depending on the size of the project and the formality of the organization.

Throughout the project, any stakeholder can add items to the product backlog, but only the product owner may prioritize the items in the backlog. The team is always developing the most important items in the product backlog. Items that are not as important are shifted to the bottom of the list, and it's possible, due to time and financial constraints, that not all items in the product backlog will be created in the project. This isn't a bad thing, as the most important items in the product backlog are created first, and the items of lesser importance, and lesser value, are created last.

TIP Eat your dessert first. This is the fundamental idea of an agile project, where the dessert in a project is the most valuable things the project will create. The prioritized product backlog means the most valuable items in the project are what the team will create ("eat") first.

Working with the Balanced Scorecard Approach

Many organizations use a strategic tool called the *balanced scorecard* to align their projects to the organization's vision and strategy. If your organization uses this approach, you'll work with management to ensure that the project requirements and its scope are in alignment with the organization's balanced scorecard.

The balanced scorecard has four categories that are scored based on key performance indicators (KPIs):

- **Financial** Increase profits by lowering costs and increasing revenue.
- **Customer** Reduce customer wait times and improve customer retention for the organization.
- **Internal process improvement** Increase organization efficiency and lower the process cycle time to complete activities.
- **Organizational capacity** Improve the organization's knowledge and skills and improve tools and technology.

A project in a balanced scorecard organization would address how the project will be in alignment and supportive of these four specific factors as part of its project scope statement.

Writing the Project Scope Statement

In Chapter 2, I detailed the process of writing the project scope statement. This is one of the most important documents you'll create in the project. Once you've captured the project requirements, you're ready to write this important project document. Remember that the project scope is all of the work, and only the required work, that the project must complete in order for the project to be done. In this document, "work" doesn't refer to the physical activity but to the deliverables that the project team will create. Think of work, in this sense, as you would the works of Mozart.

The details of the project scope statement are based largely upon the work you, the project team, and the business analyst have already done through requirements gathering. The project scope is often based on forms and templates that your organization or project management office (PMO) uses to capture these requirements. You might also use expert judgment through your project team, consultants, and other experts internally in your organization. The goal is to take the requirements document and elaborate on the requirements to define the exactness of the project deliverables.

EXAM TIP Agile projects don't have a project scope statement, but they do have a set of requirements in the product backlog. Requirements help define what the project must create and are often written as a list of must-have, should-have, could-have, and would-be-nice-to-have. This is known as MoSCoW, where the capital letters correlate to the level of necessity for each requirement.

Sometimes the project requirements are wide and broad—for example, to create a website that allows customers to log in to a secure area of the site. While this requirement does define the functional attributes of the deliverable, the intricate details, the behind-the-scenes programming, and the interface are left to interpretation. The creation of the project scope statement allows the project team to elaborate on this information and define exactly what the customer would like. Through the project scope statement, you can define the web services software, the database type, and the security expectations of the web server. It's typically easier to get customer approval of a concept than a physical, working deliverable. Figure 4-1 depicts this concept: changes early in a predictive project aren't as difficult and costly as changes late in the project.

The project scope statement defines these items:

- **Product scope description** Remember, the project will ultimately create the product scope, or the deliverables. It's essential that the project scope statement define the product scope either directly or refer the reader to a specific document where the existing and approved product scope resides. You're basically confirming that your project will create the product, service, or condition defined in the product scope documentation and the requirements documentation.
- **Product acceptance criteria** You need to know what it will take for this project to be considered done. You and the customers, the project sponsor, and any other key stakeholders need to define in the project scope statement the conditions that must be met for the project deliverables to be accepted and the project to be closed.
- **Project deliverables** These are the products, services, or conditions the project will create for the project customer and for the organization. For example, the customer may be expecting new software from the project, but the creation of the software will require your project team to buy certain resources, acquire hardware, and the like. You'll be creating all sorts of project plans, templates, and reports that can be used by other project managers in your organization. Not all project deliverables go to the project customer.

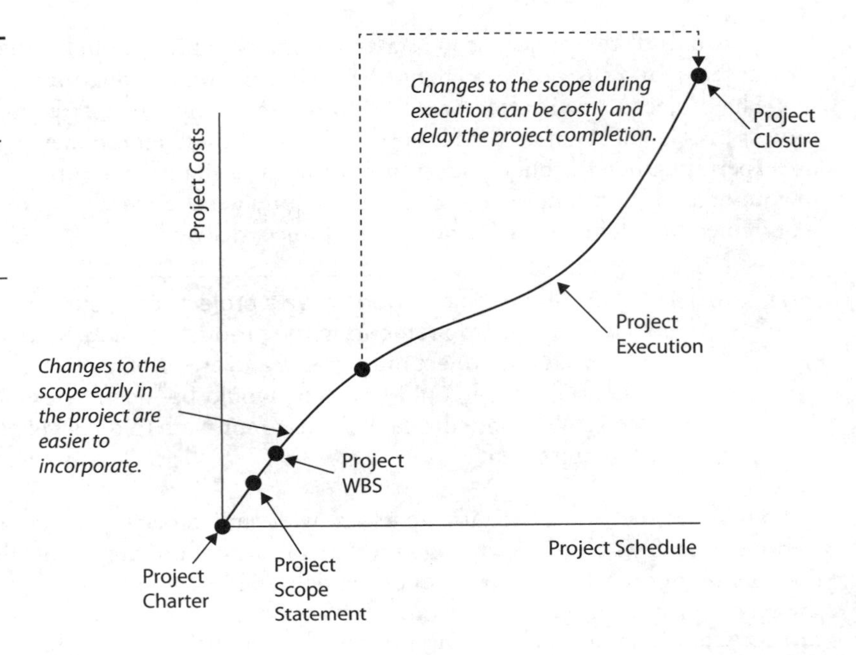

Figure 4-1 Changes early in a predictive project are easier and less costly than changes later in the project.

- **Project exclusions** The project scope statement defines what's included in scope and what's out of scope. By establishing the boundaries of the project, you'll eliminate many assumptions and disappointments when your project is near closure.
- **Project constraints** A constraint is anything that limits the project manager's options. You can usually see a constraint when the word "must," "required," or "mandatory" is included in the description. Common constraints are budgets, schedules, and contractual terms.
- **Project assumptions** An assumption is anything that you believe to be true but that hasn't been proven to be true. Assumptions can have huge impacts on the project's success if they prove false. For example, you might assume a vendor will be available for the duration of the project or make assumptions about software compatibility and data security. These assumptions are difficult to prove in IT, but you often have to accept them to get moving on the project work.

Creating the project scope statement is not an easy process, but organizational process assets, in the form of historical information, also called artifacts, can be helpful. If you've completed similar projects in the past, just adapt past scopes to the current project to save time, to ensure accuracy, and to follow a proven approach. The project scope statement creation may also require that you create a constraints log and assumptions log rather than listing all of the constraints and assumptions directly in the project scope statement. You might also need to update or create a stakeholder directory that includes all of the stakeholder contact information, project details, and other information about each stakeholder and the project expectations.

Agile projects utilize the product backlog to represent all of the requirements prioritized from most valuable, at the top of the list, to least valuable, down at the bottom of the backlog. Early in the project, the project manager and stakeholders do establish the project vision; it isn't as formal as the project vision for a predictive or waterfall project, but it helps everyone understand what the end result of the project should be. Along with the vision, the team works towards the *DoD*—the Definition of Done. The DoD is utilized throughout the project to communicate when the project is complete, when each requirement is complete, and when each work activity is complete. The DoD is a great tool to communicate what constitutes done in the project. It's a clear, simple concept that keeps everyone committed to the project work and the desired future state of the project.

Defining the Work Breakdown Structure

Once the project scope statement is approved, you're ready to move forward with project planning. An approved project scope statement is needed in order to begin creating the project's *work breakdown structure (WBS)*. A WBS is a deliverables-oriented collection of project components used in predictive projects. It is a categorization and decomposition of the project scope statement. And, yes, it's called "decomposition," because you're breaking down the massive project scope statement into smaller, more manageable components. Each component is subdivided again and again until you reach the smallest element of the WBS: the work package. A *work package* is a WBS element that you can schedule in your project, estimate the costs from, and monitor and control.

Consider a project to establish a new network. The WBS could offer high-level deliverables such as LAN, WAN, extranets, and intranets. Each of these high-level deliverables would be broken down into more defined deliverables that make up the high-level components. At the lowest level of the structure, you have the work packages. Work packages will be further decomposed into *activities* in the project schedule. If you think about the project, each assigned activity should relate to a work package. The sum of all the work packages, when they're created by the project activities, will equate to the project scope. The project scope, when it's completed, will fulfill the product scope. And all of that equates to a completed project.

A WBS is important in all projects. It is necessary because it serves as input to five key project management activities:

- Cost estimating
- Cost budgeting
- Resource planning
- Risk management planning
- Activity definition

Later, in Chapter 10, we'll get into the juicy business of managing changes to the project scope. When there are changes to the project scope—and they do happen—you'll also have to update the WBS. When you update the WBS, it often triggers changes in all areas of the project, but especially in costs, schedule, resources, risks, and activity definition. They're all related.

Working with a WBS

There's no right or wrong way to create a WBS. You can draw an elaborate decomposition on a whiteboard, sketch it out on a cocktail napkin, or be more technical and use software such as Microsoft Project or even Excel, Visio, or PowerPoint. It is best, however, to use some common terminology when addressing your WBS.

A *project,* of course, is a temporary endeavor that has a definite end date, produces a defined set of deliverables, and is an investment by an organization. For example, the software application that allows web users to search a database requires a scope, a defined deliverable, a commitment of resources, and a targeted end date.

EXAM TIP Agile projects don't utilize a WBS, but rather the product backlog. The items in the product backlog are small enough that anyone on the team can quickly understand the item and its intent in the project. Incorporating changes is easier in a product backlog than in a WBS because the new items are simply added and prioritized in the big list of features.

Within the project there are phases. A *phase* is a portion of the project that typically must be completed before the next phase can begin. Phases make up the project life cycle, and the completion of a phase usually creates a *milestone* that shows progress in the project. For example, a database project could have four phases: creation of the database, creation of the application, creation of the web interface, and troubleshooting and implementation. Typically phases do not overlap each other in execution, but it is possible that phases could overlap to save time or if the nature of the project work allows phases to happen in unison.

EXAM TIP Phases are segments of work that describe the labor in that segment. Usually, when you complete a phase you don't return to that phase, but move onto the next chunk, or phase, of the project. CompTIA, frankly, gets a little loose with their definition of phases. In the commonly accepted practices of project management, a phase is unique to the project work, while knowledge areas, such as initiating and planning, describe the project management work that takes place therein.

The *work* within each phase is linked to the *work package*s of the WBS. The project's activity list is derived from the work packages that the decomposition of the project scope creates. For example, a phase to create a database encompasses several work packages required for completion. A database administrator needs to create the database with the application designer to ensure consistency, a system needs to be built to enter the different attributes of the database items for searching, and there could be connection hooks between this database and existing databases that production uses.

The point is that it's often easiest and most logical to create a WBS based on the nature of the project work. In a project where it's easy to see and relate to the phases of the project, you should decompose the project scope accordingly. In other projects, where the focus of the project execution isn't as clear, it may make more sense to categorize the project components and then decompose. The right way to create a WBS is what's right for the specific, unique project.

Coordinating WBS Components

Some project managers would recommend that you continue to break down each component until you cannot possibly break down the deliverable any further. However, conventional wisdom contradicts a continual decomposition of any deliverable, as it eventually leads to units of work that are too small to measure, assign, and manage. While some control over the work to be completed is required, a project manager needs to put faith in their team to complete the tasks necessary to finish the job.

As a rule, find an acceptable amount of time that will serve as the smallest increment of work. For example, with a small project, you may only break work down into what equates to days of labor. With a larger project, you may choose to break work down into weeks. The key is not to continue to break down each deliverable into tiny, unmanageable work packages but to break them down into assignable, realistic chunks of deliverables. A heuristic you can rely on is the *8/80 Rule,* which suggests that the smallest work package should take no less than 8 hours and no more than 80 hours to complete. While this rule can apply to most projects, it's not always applicable. For example, you may have a work package that represents a subproject or part of your project that you'll be outsourcing to a vendor. The group completing the activities will likely decompose these work packages into their own WBS components.

As you plow through the decomposition of the project scope to create the work packages, you'll likely find that you need to rearrange elements, add deliverables, and refine the WBS several times. This is expected—don't get frustrated with your efforts. Over time, the refinements you and the project team add to the WBS will help you plan the execution of the project, create more accurate time and cost estimates, and effectively close the project.

One element you can use through the WBS is a code of accounts to help keep things organized. A *code of accounts* is a numbering system that shows the different levels of WBS components and identifies which components belong to which parts of the WBS. For example, suppose that I have a project named Data Center and it's been assigned a project code of 507 in my organization. Within this project are four major categories of deliverables: database, network, application, and hardware. Each of these categories would append the assigned project code and would look like this:

- 507.1 Database
- 507.2 Network
- 507.3 Application
- 507.4 Hardware

When my project team and I decompose these elements, we'd continue to branch off the elements within the code of accounts. For example, I could decompose 507.2 Network a bit more and created this:

- 507.2 Network
  - 507.2.1 Routers
  - 507.2.2 Switches
  - 507.2.3 LAN
  - 507.2.4 WAN

You can see how each element can be decomposed again and again into related but separate deliverables. Each deliverable, in this instance, can be subdivided again, and I can continue to append to the code of account numbers. Finally, I might end up with a work package of J-hooks for the LAN deliverable, which could have a code of accounts identifier of 507.2.3.2.1. That identifier would be used only once in the entire project, and it would reference only the J-hooks that are needed for the LAN. There's no confusion as to which deliverable I'm discussing with my project team, and it's easier to track time and costs for just that one particular project element.

Defining a WBS Approach

There are two broad methods used to create a WBS: top-down and bottom-up. The *top-down* approach uses deductive reasoning because it starts with the general and moves to the very specific. *Bottom-up* moves from the very specific toward the general. Figure 4-2 depicts the difference between the top-down and bottom-up methods.

Both methods have their advantages. The bottom-up method is ideal for brainstorming a solution to a problem. Imagine that a project team is trying to find a solution to connect a network in Chicago to a network in Phoenix without having to spend much money. The bottom-up method would call for very specific solutions without delving

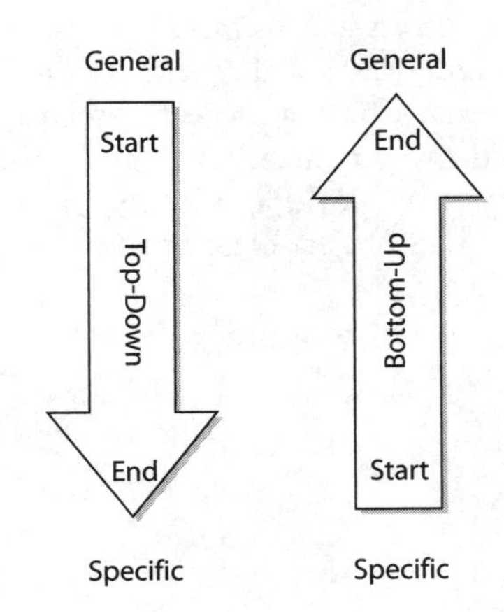

Figure 4-2 There are two general methods for creating a WBS.

into all of the details of each solution. The method could investigate the use of new software, a new service provider, or practically any implementation that is still open for discussion on the actual work to be implemented.

You might also use the bottom-up method when a stakeholder with lots of influence and power over your project is demanding a very specific feature in the project. With the bottom-up approach, you could start at the specific deliverable the stakeholder is demanding and work backward to the rest of the project scope.

The top-down approach requires more logic and structure, but it is generally the preferred method for creating a WBS. A WBS using the top-down approach would identify a solution first and then dissect the solution into the steps required to implement it. You probably use the top-down approach in your daily life. For example, when making a decision to purchase a car, you'd decide what kind of car to buy: SUV, sports car, sedan, minivan. And then what can you afford? What color? What about the bells and whistles? This process of thinking begins with a broad approach and then narrows to specifics.

The Mechanics of Creating a WBS

The process of creating the WBS is not a solo activity. Typically it requires the involvement of the project manager and the project team. On some occasions, it may require the project sponsor or other stakeholders—though typically by this juncture the team has been given the go-ahead and management supervision is not required. It is, however, a good idea to involve your key stakeholders to create shared ownership of the project and to keep them involved. Depending on the size of your project team, gathering around a computer screen to build a WBS is not ideal. What is ideal is to assemble the team and lead them through the process of creating the WBS together. Here's how.

For starters, make certain you have a whiteboard, plenty of markers, sticky notes, and control of the meeting. You'll start with the top-level requirements first. These are usually the phases of the project, but they can also be the major deliverables, categories of the project that may span the calendar phases, and so on.

For this example, we'll assume that the deliverables are contained within each logical phase. If your budget has not been created or did not include project phases, determine what the project phases are by asking these questions:

- Are there logical partitions within this project such as deadlines, milestones, or activities based on the calendar?
- Are there business cycles within your organization that need to be considered?
- Are there financial obligations or constraints within this project that could signify phases?
- What processes are currently in place for system development within your organization?

Once you have the phases identified, write each one on a sticky note and attach it to the whiteboard in the order of the phases. Now within phase 1, you'll decompose the components into smaller deliverables. You'll continue to decompose the project deliverables until they are at a manageable work package.

Decomposing the project deliverables requires some fine-tuning; you do not want to get too granular with the tasks, but you do want to break down the components so that you may allocate time and resources to the activities that must be completed to create each component. This is where you'll reference the 8/80 Rule: no task should take less than 8 hours or more than 80 hours to complete. If you remain general and acknowledge the work to be completed rather than describe the actual activities required to create each component, you'll be fine.

After you finish phase 1, or the first major deliverable, move on to the next component and repeat the process, and so on, until all of the deliverables have been broken down into work packages and the work packages have been broken down into the necessary tasks. What you'll have on your whiteboard may appear to be a very messy collection of sticky notes, but in reality, it represents your project from start to finish.

From here, document the WBS and let the structure cool for a day or two. Chances are you and the project team will think of other elements that need to be added to the WBS or you'll think of a better approach to organizing the project work. You can create the WBS in a software package such as Microsoft Visio—but Word and Excel can work just as well. Some project managers go immediately to Microsoft Project, but Project is really good for organizing the activity list and not so much the WBS deliverables that you're concerned with here. Whatever approach you take to create the WBS, it's important to document and structure the WBS to reflect all of the project requirements.

Why You Need a WBS

You may be tempted to skip the process of creating a WBS—especially on smaller projects. Don't yield to that temptation. By creating a WBS, even on small projects, a project manager can help accomplish several things in a project:

- *A WBS defines the work required to complete the project.* How many times have you started a project only to uncover deliverables that you had totally forgotten about? Or worse, you realized that a component was needed that didn't exist and had to be created before your project could continue? A WBS ensures that a project manager knows all of the required deliverables that must exist for the project to be considered complete.
- *A WBS creates a sense of urgency.* By creating a WBS, a project manager and the team are working toward the project deliverables. Because the WBS is broken down at its lowest level into work packages—and the activity is derived from the work packages—the tasks can then be assigned start and end dates. The WBS is needed to ensure proper scheduling and sequencing for the identified activities to create the project deliverables. The project can maintain its momentum—and its schedule—if all members complete their tasks on time. A WBS allows a project manager to track the success or failure of team members based on the completion of activities, which in turn creates the deliverables the WBS identifies.
- *A WBS can help prevent scope changes.* When management and departments try to add new features for an existing project, a WBS can ward them off. Because a WBS defines all the project deliverables based on the project scope and identified requirements, it becomes easier for a project manager to rule out additions and

new features to a project that has already started. It is possible, however, to add new features to the project scope, but there are ripple effects into the project schedule, cost, quality, risks, and other areas of the project that must be considered and managed.

- *A WBS provides control.* As a project manager, you may be in charge of several different IT projects. A WBS can allow you to graphically view the status of any project and how much progress is being made. You can easily home in on a particular phase, work unit, or task and make adjustments, counsel team members, or adjust the schedule as needed. Control is good.
- *The WBS is a portion of the scope baseline.* The scope baseline is the combination of the project scope statement, the WBS, and the WBS dictionary (discussed next). Your goal is to manage the project team to create all of the things defined in the scope baseline. Basically, deliverables that are not in the WBS are not in the project. The scope baseline provides a point of agreement between the project manager, the customer, the sponsor, the team members, the vendors, and other stakeholders on what is and is not in the project.

Creating a WBS Dictionary

The WBS *dictionary*, as its name implies, is a dictionary that defines each work package in the WBS. You'll use the WBS dictionary to identify clearly what the components of the WBS are and how they relate to the project scope. The WBS dictionary is a great place to define each element of the WBS in simple-to-understand terms that all project team members and stakeholders can reference. Think of all the acronyms you use in your organization or as an IT project manager and the terms your project team members use as part of their work. By defining these terms, you assure that if they're part of the WBS, there's no confusion as to what's what.

The WBS dictionary typically defines, at a minimum, all of the following about each WBS element:

- Code of accounts number
- Description of the WBS element
- Person, vendor, or other organization responsible for the WBS element
- Scheduled creation of the WBS element
- Resources required to create the WBS element (resources are people, materials, and facilities)
- Cost to create the WBS element
- Quality requirements
- Criteria for acceptance of the specific deliverable
- Technical references, drawings, and other supporting detail
- Contract terms and information
- Milestone schedule

You'll use the WBS dictionary throughout the project for reference. You'll likely need to update the WBS dictionary as more information becomes available during the management of the project or when changes are introduced and approved for the project scope. Updates to the WBS and the WBS dictionary are called *refinements.*

Obtaining Stakeholder Approval

Once the WBS has been initially created, it must pass through management for a final sign-off. In some instances, such as when the project manager and the project team are consultants or vendors integrating the technology into an existing enterprise, the project manager probably won't have to pass every work unit within the WBS through management for approval. Stakeholders will need to approve the WBS in most projects. They'll need guidance from the project manager and often experts to guide them through the WBS, but they're usually just looking for confirmation that your WBS is linked to the requirements they're expecting from the project.

Because agile projects don't use a WBS, there is less formality with the sign-off of documentation and project planning. But this doesn't mean that agile projects don't have rules and obligations for the stakeholders and team members. The product backlog is managed by the product owner. Anyone in the project can add items to the product backlog, but the product owner approves or declines the change, prioritizes the added items, and maintains the integrity of the backlog throughout the project.

Presenting to the Project Sponsor

The project sponsor, the advocate of the project, will be the first stop on the road to approval of the WBS. The project manager must be prepared to explain any component of the WBS or WBS dictionary to the project sponsor. If the project manager has fully researched and skillfully planned the WBS with a focus on the project requirements and has identified deliverables and business cycles, there should be few revisions to the schedule.

EXAM TIP The project sponsor can be the project champion of the project, but not always. The champion is an informal role of the person who is campaigning and cheering the project's success. The champion wants the project to exist and be successful in the organization.

However, if the project manager has failed to work with the team, to consider business needs, or to take into account other implementations within the organization, the project sponsor can (and should) work with the project manager to correct these issues. You can imagine the frustration that could ensue if the WBS had to be re-created because of poor planning and an inadequate understanding of other activities within the IT realm of an organization—not to mention that failure in creating the WBS does nothing to gain the confidence of your project sponsor and the project team.

Presenting to Key Stakeholders

Once the project sponsor has approved the WBS, the project manager should present the WBS to the key project stakeholders. Depending on your organization, this may be the customer of the project, another department within your organization, or management. As always, tailor the presentation for the audience you are speaking to. The WBS presentation does not need to go into great detail for each deliverable represented within each phase or component. To begin the presentation of the WBS, start with the deliverables of the project. By again reminding the stakeholders what the project will produce, you'll be reinforcing their decision to move the project forward. The whole point of presenting the WBS to the stakeholders is to confirm that your project includes all of the requirements they've requested for the project.

Once you've established the deliverables, reveal the phases required to reach them. It would be most effective to show a milestone schedule or a high-level overview of the project schedule. Figure 4-3 is an example of a milestone chart for a project that's already in motion. As each phase is revealed, you can superimpose an arrow over the timeline to show where each phase will put you in relation to the project completion. You could also detail the requirements traceability matrix along with the WBS to show the relationship between the project requirements and the components in the WBS. This will allow your audience to visualize how your plan has been well conceived and how each phase will produce a deliverable, moving your team, and the organization, toward the end result of the project.

Within each phase, you may wish to show a few of the highlights to convey the core activities that are required. It is not, however, necessary to illustrate every task required to complete each phase unless the stakeholders explicitly ask for it. You should be prepared to discuss each phase in detail, and it would serve you well to have an alternative presentation that does include each task of each phase of the project if your project is planned to that level of detail.

Generally, stakeholders do not want to know about each activity associated with installing a network, replacing workstations, or upgrading a dozen servers. You should, however, always share with management any phase of the project that may require any

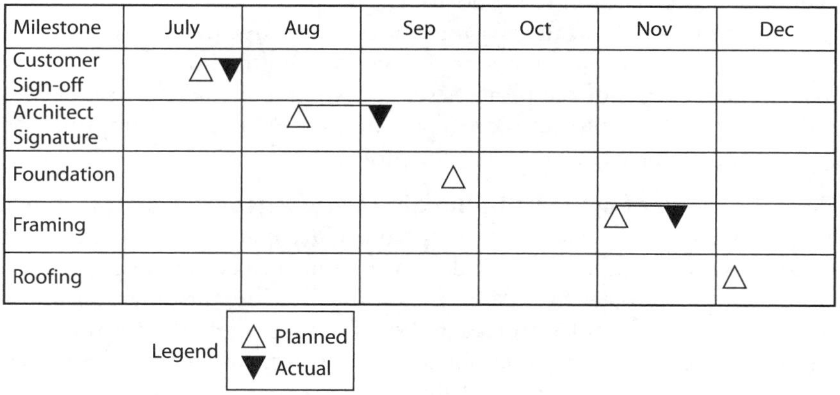

Milestone	July	Aug	Sep	Oct	Nov	Dec
Customer Sign-off	△▼					
Architect Signature		△	▼			
Foundation			△			
Framing					△ ▼	
Roofing						△

Legend: △ Planned ▼ Actual

Figure 4-3 Milestone charts track planned targets and actual fulfillments of milestones.

downtime of IT components, even if it is over a weekend. The WBS dictionary should address these downtimes, if they exist, so that the stakeholders are aware of the impact. Always take into consideration, through audits and logging, the type of activity that occurs over weekends or at night by remote users, international users, and users who work late and long hours. Don't assume anything.

If management does not approve the WBS, the project manager should immediately address any areas of concern. In some instances, management may approve on the condition of a few revisions. In other circumstances, management may delay your approval in order to study the WBS in detail. Plan your management approval meetings in accordance with what the norm is in your environment. If management always delays decisions to begin projects, don't let their delay infringe on the implementation. Plan the WBS approval meeting with ample time for management to review and revise the WBS.

If your WBS needs to be approved immediately, stress to the stakeholders that the schedule must be implemented within *x* number of days. If there are no concerns with the first phase of the schedule, then perhaps at least that part of the plan can begin while management reviews the later details.

CompTIA Project+ Exam Highlight: Managing Project Scope

The primary focus of this chapter is the project scope and planning the project work. It's been said that projects fail at the beginning, not the end, so it's no surprise that you'll need to clearly identify what the project is intended to create as early as possible. The project scope, for predictive and agile projects, maps to the project vision and the business value of the project. Each item the project creates as part of the scope must have business value. Business value correlates to why the project work is being done. If you can't identify the business value for an item in the project scope, the item likely doesn't belong in the project.

You wouldn't create a project scope document without a charter—there's no authority for the project to exist without the charter. The project scope document, or what's more commonly known as the project scope statement, defines the boundaries of the project. Recall that the boundaries of the project establish what is included in the project scope and what's left out of the project scope. This helps to alleviate assumptions, establish ground rules, and set the expectations of the project.

2.1 Explain the value of artifacts in the discovery/concept preparation phase for a project You can utilize artifacts to help you manage and plan the current project. Artifacts are sometimes called historical information, project archives, organizational process assets, or past project files. Whatever you call it in your organization, the concept is the same: leverage similar projects to better manage the current project. With the historical information, you can better create the business case, create the project scope statement, and plan for the cost expectations of the current project.

All projects should define the current state of the organization and what the future state of the organization will look like when the project is done. A current state analysis can be time-consuming, and a future state analysis is really just a prediction of what you think may happen. Historical information, however, is proven information, so it can help you better predict what will happen in the project and throughout the project. Of course, no one really knows what will happen, but the historical information can increase the likelihood of project success.

2.2 Given a scenario, perform activities during the project initiation phase During project initiation, much of the focus is on launching the project and creating the project charter. However, this exam objective includes three subobjectives discussed in this chapter:

- **Identify and assess stakeholders** Stakeholder identification is done early in the project and the assessment of stakeholders is done to determine who has the influence, power, and authority for the project decisions.
- **Review existing artifacts** The historical information from past projects to better manage the current project.
- **Determine solution design** Technical projects require a good technical design to meet the expectations for performance, reliability, scalability, and other success factors.

Applying these three exam subobjectives can help the project manager and the team complete the project successfully. The identification of the key stakeholders will directly affect the solution design of the project scope. Existing project artifacts can save time, cost, and frustration as this historical information is proven information for the project team to utilize.

2.3 Given a scenario, perform activities during the project planning phase In this chapter, I talked about the final project acceptance criteria and how you and the stakeholders need to establish early in the project how you'll know the project's done. The details of the project scope, the detailed objectives, and the project acceptance criteria all need to be documented and agreed upon between the project manager, the project sponsor, and the key stakeholders. Recall that the project scope statement actually has six components:

- **Product scope description** The features and functions of the product, service, or result your project will create
- **Product acceptance criteria** The conditions that must be true for the project deliverables to be accepted by the stakeholders and for the project to be considered done
- **Project deliverables** The things your project will create; remember, project deliverables can be more than just the product scope fulfillment
- **Project exclusions** The things and conditions your project won't create

- **Project constraints** The limitations the project manager must deal with in the project; time, cost, and resources are common constraints
- **Project assumptions** The things that are believed to be true but have not yet been proven to be true

The project scope must be approved by the project sponsor and the key project stakeholders. When changes enter the project, the project scope likely will need to be updated to reflect these changes. Changes to the project scope will cause changes in the project's WBS and WBS dictionary, too.

The WBS is a visual decomposition of the project scope statement. The WBS dictionary is an indexed definition of every component of the WBS document. You'll need the WBS and WBS dictionary throughout the project, not just during project planning. The process of creating the WBS dictionary can happen by identifying project phases, milestones, or major deliverables within the project and then subdividing these elements into smaller, more manageable components. The smallest element in the WBS is the work package; work packages can be cost-estimated, scheduled, monitored, and controlled.

The WBS is not an activity list, but consists of deliverables that in turn will equate to the project scope. The work packages in the WBS will help the project team create the activity list. Each element of the WBS, from the largest to the smallest work package, can be identified by using a code of accounts numbering system. The code of accounts uses a sequential approach of appending a globally unique identifier to each element of the WBS. For example, a project that has a project code of 431 and includes a deliverable of three print servers could identify each print server as 431.1, 431.2, and 431.3. Each print server could be subdivided again if there were additional deliverables in the server that you needed to account for.

The subdivision of project elements should generally follow the 8/80 Rule, which states that deliverables in the WBS should not be decomposed smaller than the equivalent of 8 hours of labor or larger than 80 hours of labor. This is a general rule and doesn't, of course, apply to all situations.

Chapter Review

Every predictive project demands a project scope statement, a work breakdown structure, and a WBS dictionary. These three project documents compose the scope baseline, and the project manager and project team will rely on the scope baseline to consider every major project decision going forward. At the heart of the scope baseline is the WBS—it's a detailed look at the things your project will create for the stakeholders and for the organization. Agile projects are less rigid in planning and utilize the product backlog.

Decomposition is the process of breaking down the project deliverables into a logical order. As a rule, work packages do not need to be broken down into granular step-by-step tasks, but rather tight, individual units that can be scheduled, can have costs estimated, and can be monitored and controlled. The 8/80 Rule can help the project team gauge how large or small the work packages should be in the WBS.

A *code of accounts* is a numbering system that can provide order to the WBS. The code of accounts is a unique identifier for each element in the WBS to pinpoint each element in each of the major deliverables of the WBS. The code of accounts should also be included in the WBS dictionary for each WBS element so that it's easier to match up the conditions, characteristics, and other attributes of an element from the WBS to the physical creation of the deliverable.

There are multiple ways to create a WBS, and any combination can be used as long as the end result depicts the exact project deliverables and expectations of the project stakeholders. A WBS, once completed, needs approval from the project sponsor first to confirm the scope decomposition and how the scope meets the project scope requirements. Once you and the project sponsor are in sync on the scope, WBS, and WBS dictionary, the key stakeholders need to sign off on the WBS to ensure that all of the project deliverables are accounted for as they're defined in the project scope statement.

Exercises

These exercises allow you to apply the knowledge you have learned in this chapter and are followed by possible solutions.

Exercise 4-1: Writing a Project Scope Statement

In this exercise you'll prepare to write a project scope statement. In reality, the project scope statement can be a lengthy, complex document that provides a lot of information about the project requirements and objectives. In this exercise, your goal is to create a summation for a hypothetical project scenario. Your high-level project scope statement should include

- Product scope description
- Product acceptance criteria
- Project deliverables
- Project exclusions
- Project constraints
- Project assumptions

You are the project manager for the POQ Organization (POQ) and you're working with Sam, the business analyst for your organization. POQ traditionally sells running shoes through retail stores but now wants to take a new approach and allow customers to access POQ's inventory and order shoes online. Sam reports that he has captured the requirements from the project stakeholders and they've approved his final documentation of the requirements. The stakeholders are excited about this new project, which will create software for their online web presence.

Your project must create a website that meshes with your organization's color scheme, branding, and feel of the POQ's existing online presence. The website will allow users to access an online catalog of running shoes, purchase shoes from the online store, and keep a history of shoes they've ordered in the past. The website should be secure, should calculate shipping costs, and should recommend related items to the customer based on what they've added to their shopping carts.

Management has approved the requirements but stipulated that the solution not cost more than $75,000 and must be done within three months. Sam also tells you that some of the retailers are concerned that people will just order the shoes online rather than visit their locations. You need to create a monthly calendar for each shoe retailer in your company that encourages people to visit the stores in addition to your online presence.

To help you write the project scope, answer these questions based on POQ's shoe project:

Question	Answer
What's the primary goal of this project?	
What are some characteristics of the product scope?	
What are the primary deliverables of the project?	
What type of constraints will this project need to plan for?	
What assumptions are in the scenario?	
What elements could become part of the project scope that wouldn't necessarily be part of the product scope?	
What conditions must be true for this project to be considered done?	
What type of skills would you need on your project team to help create this project scope?	
What additional information would you need from Sam in order to write the project scope?	
Are there any initial risks that you see in the project?	
Which approach would you recommend for the project, agile or predictive? Why?	

Exercise 4-2: Create a Work Breakdown Structure

In this exercise you will complete a work breakdown structure. To assist you in the WBS and the research phase, questions will prompt you to complete the work decomposition. In addition, you can use the exercise solution in the next section as a guide to complete the WBS.

You are the IT project manager of a network upgrade project. The network will consist of 187 workstations, 5 servers, and 17 network printers. The network will be segmented through switches. The 187 workstations and printers will be on one segment and the servers on the other.

The five servers on the network will be replaced with five new servers; four servers will be domain controllers, and the final server will be an e-mail server.

In addition to the creation of the network infrastructure, each workstation will be replaced with a new PC. Each PC will be configured identically with Windows and will use DHCP to receive its IP address.

Complete this table to begin the creation of the WBS:

Question	Answer
Are there any major deliverables within this project?	
What are the deliverables you see? (Hint: There are at least three.)	
In what order should the phases take place, and does it matter at this point of the WBS? Why or why not?	
Which components are within the first deliverable?	
Which components are within the second deliverable?	
Which components are within the third deliverable?	
How do you break down the components within the first major deliverable into work packages? Break down a few.	
Do your work packages conform to the 8/80 Rule?	
Can you further decompose the existing structure and still be within the 8/80 Rule?	
What other deliverables can you identify in this project?	
Assuming the project has a code of 675, number the components of the WBS using a code of accounts.	

Exercise Solutions

The following offer possible solutions for the chapter exercises.

Exercise 4-1: Writing a Project Scope Statement

On most projects, many people contribute to the project scope statement. It's rarely an activity that the project manager completes alone. Templates and standards in your organization can help streamline this process. The point of this exercise is to prepare you for the types of questions you'll need to ask when writing the project scope with others and to think about all the information you'll need to write an effective project scope. Here's a possible solution to this exercise:

Question	Answer
What's the primary goal of this project?	The primary goal of the project is to increase revenue for the organization by adding a sales channel through an online presence. The website will help the organization realize sales that may have been lost to competitors and gain sales of customers who aren't within a reasonable distance to the local retail stores. The project will also create goodwill between the organization, its customers, and its retail stores by promoting interaction between web users and calendars and features that web users will have access to only in the retail stores.
What are some characteristics of the product scope?	The product scope describes the features and functions of what the project stakeholders are expecting from the project. The product scope in this scenario has several characteristics: • Web interface that meshes with the organization's marketing presence • Database-driven customer history • Catalog of shoes and prices • Inventory connection • Secure shopping cart • Recommendation of additional items based on purchases • History of shoes the customer has purchased • Calendar feature for each retail store

Question	Answer
What are the primary deliverables of the project?	Usually the deliverables of the project are slightly larger than the product scope. This is because the project may need to purchase additional things, such as software and hardware, to create the product scope. In this instance, there's not enough information to know if the project manager will need to purchase new servers, software, Internet connectivity, and other resources to make the product scope a reality. The project manager would also need to consider the competence and skill sets of the current staff to create the website that management is asking for. Finally, the project deliverables can also include the project management plans, schedules, charts, and other outputs of the planning processes the project manager and team will complete.
What type of constraints will this project need to plan for?	A constraint is anything that limits the project manager's options. In this instance, the requirements of the project will serve as scope constraints. Management has also set a cost constraint of $75,000 and a schedule constraint of three months. More research and information are needed to determine whether these constraints are feasible. It's possible that these constraints are fine given conditions within this organization. However, it's also possible that these constraints are unreasonable and that they may become project risks.
What assumptions are in the scenario?	The scenario assumes several things: • Skills and resources for the project already exist in the organization. • The technology capabilities are present or can be created. • Stakeholders believe there will be additional software rather than using web technologies for the website. • The online catalog of shoes exists or will be created in the project. • The online stores and the website data will communicate with one another if users' in-store purchases are part of their purchase history. • The cost constraint could be an assumption that the project can be completed for that amount of funds. • The schedule constraint may also be an assumption of a realistic timeframe for project completion. • The online retailers will maintain or participate in the monthly calendar section of the website on an ongoing basis.

(*continued*)

Question	Answer
What elements could become part of the project scope that wouldn't necessarily be part of the product scope?	Because the project scope may need to acquire additional resources that the company could retain, it's often possible that the project scope has slightly larger requirements than the product scope. Consider these deliverables that aren't part of the product scope: • Software • Hardware • Internet connectivity • Training materials • New employees • Project management plan • Project documents, such as supporting detail, execution plans, and lessons learned documentation
What conditions must be true for this project to be considered done?	The project scope will need to be defined and approved. Once the scope has been approved, the project's WBS can be created. Each of the work packages in the WBS will need to be created according to the quality requirements of the project and organization. By completing the project scope, the product scope will also be fulfilled. A future state assessment could be created by the business analyst as part of the business case, and there'd be more details in this document as to the expectations of the project customer. Basically, customers should be able to access the website, order shoes, have recommendations on related products, securely check out, and receive their purchases via a delivery service. The website should calculate the cost of the shoes, shipping, and any taxes, as well as create an estimate for the customer when their order will arrive. In addition, a history and profile of the user should be created in a database that would record past purchases, store information about the customer, and welcome the customer back on their next visit.
What type of skills would you need on your project team to help create this project scope?	For this project to be complete, the project team would need, at a minimum, developers, writers, database engineers, graphic artists, web designers, photographers, network engineers, and the project manager.

Question	Answer
What additional information would you need from Sam in order to write the project scope?	You'll need to see the requirements document in detail. Sam, as the business analyst, should have consulted with you, the project manager, much sooner in the requirements gathering. It's possible that Sam has already used experts from your organization or outside your organization to determine the skills, costs, and schedule for the project, so you'll need to examine the requirements and supporting detail for the project. The requirements documentation will help you determine if the exact solution for the project has already been selected, such as the database, software, and connectivity issues; or you and the project team may need to research and define the solution that will satisfy the product scope the customers are expecting.
Are there any initial risks that you see in the project?	A risk is anything that can have a positive or negative effect on the project. Most project managers focus on negative risks, such as the predetermined costs and schedule in this scenario. Other negative risks are the skills needed, the security of the data, the acceptance of the site by the customers, interaction with the retail store owners, and the promotion of the site for the customers. Positive risks could be the opportunity for the project team members to learn new skills, the increased revenue, and the benefits of connecting the inventory with retail stores.
Which approach would you recommend for the project, agile or predictive? Why?	A predictive approach could be utilized as all of the project requirements have been identified and there is a clear vision of what the desired future state should be. Predictive projects utilize upfront planning and the WBS. An agile project could be utilized to quickly deliver the most important items first in the project. Agile projects bypass much of the front-loaded planning of a predictive project and instead plan in short iterations throughout the project. Agile projects are ideal for projects that will have lots of change and uncertainty in the final result of the project.

Exercise 4-2: Creating a Work Breakdown Structure

Here is a completed worksheet that demonstrates the possible answers for the WBS process. Your chart may be slightly different from the answers presented here.

Question	Answer
Are there any major deliverables within this project?	Yes, there are four major deliverables.
What are the deliverables you see?	The four major deliverables are • Network infrastructure • Servers • Printers • Workstations
In what order should the phases take place, and does it matter at this point of the WBS? Why or why not?	The phases would logically follow the installation of the network infrastructure, the servers, printers, and then the workstations. There are multiple parts to each phase that could be simultaneously completed with the other phases. During WBS creation, however, activity sequencing is not crucial.
Which components are within the first deliverable?	The first deliverable has many requirements to produce the desired results: the creation of a network topology to map out the path for each network cable and drop; the installation of a suitable and speedy network cable that follows installation code, such as plenum-grade cable, installation J-hooks, and wall jacks; the termination and testing of each cable to ensure reliability; and the installation of patch panels, switches, and connections between the network elements.
Which components are within the second deliverable?	The second deliverable requires several primary components: • Planning and implementation of the five servers • Planning and implementation of the domain controllers • Planning and installation of the mail server and their secured role in the enterprise • Planning and implementation of the user account names and auto-configuration of the associated mailboxes • Planning of domain security, policies, group creation, and access permission to resources such as home folders, printers, and data

Question	Answer
Which components are within the third deliverable?	The third deliverable requires these components: • Workstations • Network access • Applications • User profiles • Development, testing, and implementation of a system to automate the rollout of the workstation operating system images
How do you break down the components within the first major deliverable you've identified into work packages?	The installation of suitable network cables, plenum installation procedures, and the creation of wall jacks could be broken down into three separate work packages. The installation of patch panels, switches, and connections between the network elements could be broken down into three separate work packages as well.
Do your work packages conform to the 8/80 Rule?	The WBS work packages should not take more than 80 hours or less than eight hours to complete; if they do, the work packages should likely be further decomposed.
Can you further decompose the existing structure and still be within the 8/80 Rule?	Depending on the approach you've taken with defining the WBS elements, you probably can continue to subdivide the WBS components into smaller work packages. While you can subdivide these elements, it's not always the best choice to do so. You want the work packages to be at a manageable level for the project's time and cost estimating, for project execution, and for monitoring and controlling. If you break down the packages too small, even if they follow the 8/80 Rule, you may be creating more work for yourself and the project team. The objective is to find a good balance of control of the project work; to trust in the project team to complete the work; and to take the time to plan, monitor, and control the work packages.
What other deliverables can you identify in this project?	You may have identified server and workstation licenses, patch cables, spare parts for failed hard drives, data redundancy, and other components.

(*continued*)

Question	Answer
Assuming the project has a code of 675, number the components of the WBS using a code of accounts.	A code of accounts identifies the components of the WBS and the WBS using a numbering sequence. In this example, I've continued the "Network infrastructure" deliverable to a second layer of decomposition. Your code of accounts could look something like this: 675 Network Upgrade Project 675.1 Network infrastructure 675.1.1 Domain controllers 675.1.2 Mail servers 675.1.3 User accounts 675.1.4 User mailboxes 675.1.5 Domain security 675.2 Servers and configuration 675.3 Printers and configuration 675.4 Workstations

Questions

1. You are the project manager for a project that will develop in-house software used to monitor a computer parts inventory. Your project sponsor asks that you begin working on the WBS. What is a WBS?
   A. A breakdown of the project work activities
   B. A decomposition of the project scope
   C. Weekly deadlines for the project
   D. A topology of the project team's responsibilities
2. You are the project manager for the NQQ Project. You have been working with the project team to create the WBS and have now decomposed the project down to work packages. What is a work package?
   A. A unit of work that must be completed before the next unit can begin
   B. The smallest unit of work that can be performed by the team as a whole
   C. The smallest decomposed object in the WBS
   D. One of the three parts of any project: the introduction, the implementation, and the project wrap-up

3. You are working with your project team to decompose the project scope down to the work packages. Some of the project team members are concerned that you'll want to subdivide the deliverables to a very granular level rather than trust the project team to do their work. You assure them that this won't happen, because you're using the 8/80 Rule. What is the 8/80 Rule?

   A. How long a phase should last

   B. A heuristic that says a project should not last more than 8 months or less than 80 days

   C. A heuristic that says a task should not last more than 80 hours or less than 8 hours

   D. A description of a collection of tasks within one phase

4. Grace is the project manager of a small project for her organization. You are serving as a project management consultant to this project. Grace tells you that because her project is so small, she doesn't feel the need to create a WBS. You tell Grace that it's in the project's best interest for her to follow through and create the WBS. Why must Grace and the project team create a WBS?

   A. The WBS allows the project manager to work backward from the targeted date to assign tasks.

   B. The WBS allows the project manager to assign resources to tasks.

   C. The creation of the WBS ensures that all of the project deliverables are fully identified and decomposed so that the necessary resources may be obtained and assigned to the work.

   D. The WBS allows the project manager to assign multiple team members to multiple tasks to speed up the implementation.

5. You are leading a project to create a new application. Todd, a key stakeholder, wants to review your project's work breakdown structure. You report that there is not a WBS in this project. Todd is upset and demands that you create a WBS. Why would you not create a WBS?

   A. To ensure maximum billable hours.

   B. Because the project is too small to warrant a WBS.

   C. All of the requirements haven't been gathered yet.

   D. You are utilizing an agile approach in the project.

6. You are the project manager for your organization and are reviewing the requirements for your new project. The business analyst has completed the requirements and is consulting with you on what each requirement is and why it's important to the project stakeholders. You would like to create a table that maps each requirement, its characteristics, when the requirements will be created, and other information. What type of a table would you like to create?
   - **A.** RACI chart
   - **B.** Roles and responsibility chart
   - **C.** Requirements traceability matrix
   - **D.** WBS dictionary
7. You are the project manager for your organization, and you and the project team are creating your WBS for a software development project. You are mapping the WBS to phases within your project. Of the following, which one is the end result of a phase that can help in the WBS creation?
   - **A.** Milestones
   - **B.** Project management life cycle
   - **C.** Deliverables
   - **D.** Project funding
8. All of the following components are part of the scope baseline except for which one?
   - **A.** Project charter
   - **B.** Project scope statement
   - **C.** Project WBS
   - **D.** WBS dictionary
9. You are the project manager for your organization and are working with your project team to create the project scope statement. Part of the scope statement is to define the constraints and assumptions that you must work with in the project. Which one of the following is an example of a project assumption?
   - **A.** A predefined schedule
   - **B.** A predefined budget
   - **C.** Interoperability of the software and existing hardware
   - **D.** A requirement to include scalability in the hardware you add to operations
10. What is the name of the numbering sequence a project manager can use to identify the components of a WBS?
    - **A.** Code of accounts
    - **B.** Chart of accounts
    - **C.** WBS dictionary index
    - **D.** Project glossary

11. Nur is the project manager for the NAA Project in her organization. She is about to create the WBS based on the approved project scope for her project. She would like to include several key stakeholders in the creation of the WBS. What is the primary benefit of including the stakeholders in the WBS creation process?
    - A. Including stakeholders ensures that all of the requirements are identified and captured.
    - B. Including stakeholders ensures that the stakeholders know who the project manager is and who the project team members are.
    - C. Including stakeholders ensures that the stakeholders see that all of the requirements have been identified and are accounted for in the project.
    - D. Including stakeholders helps to promote shared ownership of the project.
12. Which one of the following is not needed when creating a WBS?
    - A. Project team members
    - B. A preferred sequence of project activities
    - C. A project scope
    - D. Identified project deliverables
13. You are the IT project manager for a project to install a new mail server. Which of the following describes the best approach to creating the WBS?
    - A. Create a sample WBS and give it to the project team to complete.
    - B. Work with the project team to create a sample WBS and give it to management.
    - C. Work with the project team and the key stakeholders to create the WBS.
    - D. A project of this size does not need a WBS.
14. One of your primary concerns with the project scope is that the stakeholders may add requirements once the project is in motion. Why should the project scope be guarded against even simple additions?
    - A. It adds team members to the project.
    - B. It distracts the team members from the project.
    - C. Additions, even simple ones, can greatly impact the success of a project.
    - D. Additions, even simple ones, must be approved through the project sponsor.
15. What should signify the end of each phase?
    - A. A milestone that has been reached
    - B. A party for the project team
    - C. A date that has been established within the WBS
    - D. A definite deliverable result

Answers

1. **B.** A WBS is a decomposition of the project scope. It serves as input to five key processes within a project: cost estimating, cost budgeting, resource planning, risk management planning, and activity definition.
2. **C.** A work package is the smallest decomposed object in the WBS. Work packages can be scheduled, cost estimated, monitored, and controlled.
3. **C.** The 8/80 Rule is a guide that says a project activity should not take less than 8 hours or last more than 80 hours.
4. **C.** A WBS is deliverables-oriented decomposition of the project work. It is a process to ensure that all of the required deliverables are identified and broken down into manageable components so that resources and labor may be assigned to complete the project work.
5. **D.** The best choice is that you are utilizing an agile approach in the project so there is not a WBS. Agile projects have a prioritized product backlog and welcome changes throughout the project. Predictive projects use a WBS and are averse to changes. Agile projects don't have the same upfront planning that traditional waterfall projects rely on.
6. **C.** A requirements traceability matrix is a table that traces the individual project requirements to the actual creation of the requirements. It can include information about the requirement, such as the owner, the phase when the requirement is expected to be created, and other information.
7. **A.** Milestones are typically the end result of phases, and they show progress toward project completion. Once you have identified the project milestones, the project life cycle phases used in the WBS can become more clear and distinct.
8. **A.** The project charter is not part of the scope baseline. While you'll need a project charter to authorize the project and identify the project manager, it's not part of the scope baseline. The scope baseline includes the project scope statement, the WBS, and the WBS dictionary.
9. **C.** Of all the choices presented, only the interoperability of the software and existing hardware could be an assumption. The predetermined budget and schedule are constraints, as is the scalability requirement. Assumptions are things that are believed to be true though they may not have been proven to be true.
10. **A.** A code of accounts is the numbering sequence project managers use in the WBS to identify the various components of the project scope decomposition.
11. **D.** A valuable benefit of including stakeholders in the creation of the WBS is to promote the idea of shared ownership between the project team and the stakeholders.
12. **B.** The WBS is not concerned with the order of activities. Activity sequencing and scheduling, however, will be the process that helps the project manager determine the correct order and relationship of activities.

13. **C.** Creating a WBS is not a solo activity. The project manager should work with the project team and any key stakeholders to create a WBS.
14. **C.** Additions to the project scope can have huge impacts on the deliverables of a project. Often additions are tossed into the plans without adequate foresight or care, so their consequences can throw a perfect plan off balance. Do not change the scope of an existing project unless it is absolutely required.
15. **D.** Just as each project produces a definite deliverable, so should each phase. A milestone does not necessarily signify a phase has ended, because there can be multiple milestones within each phase. A party for the project team (while always an excellent idea) does not prove that a phase has officially ended. Dates, while targets for completion, do not signify the end of a phase—the deliverable proves the end of a project phase.

CHAPTER 5

Creating the Budget

This chapter covers the following topics:

- Creating a budget
- Bottom-up cost estimates
- Top-down cost estimates
- Budget at completion
- Zero-based budgeting
- Determining project expenses
- Tracking expenses

This chapter is about money. Have you ever noticed how people bristle when that word is mentioned? In some circles, it's not money; it's finances, working capital, currency, or funds. Whatever you want to call it, it's a large part of what you need to get your project done. Your project will need a budget to create the product or service your project customer expects. The material resources you'll need, such as software and hardware, cost money. The labor resources you'll need; the developers', database experts', and network engineers' time; as well as the brute force required to install the hardware and software also cost money.

Your project needs a budget to determine just how much money, er, capital, needs to be allotted, and when it needs to be available, so you can reach the project goal. Your project needs a plan to create estimates and predict the total cost of the project. Your project's budget needs proof of why it will cost the amount you say it will, input from vendors, quotes from suppliers, and estimates on work hours committed to the project. And your project needs a time-phased budget that ties the resources needed with the project schedule.

You'll need to estimate the project costs in order to get to the project budget. A budget may seem to be a necessary evil required by any project manager to get a project off the ground and into implementation. In reality, a budget is needed so that the project stakeholders can see how much it will cost to create the deliverable they desire. In addition, a budget is needed to confirm that the project manager truly knows what it is they need to deliver. A dreamy project goal is snapped into reality when management wants to hold you accountable for the cost of the project deliverables. With that in mind, let's get started.

VIDEO For a more detailed explanation, watch the *Estimating the Project Costs* video now.

Budget Basics

You need a budget to control and document project expenses—before the project work begins. When you are creating a feasibility plan, you'll no doubt include facts on the cost of the project and any ROI for the project. Now, once the project has been approved, or approved based on the financial obligations, you have to do a touch more research. Any bean counter in your organization wants to know what, exactly, your project will cost. As any project manager who has worked on IT implementations will tell you, "It's not as easy as it looks."

You also need a budget to get your arms around the scope of the project and what you can afford to include in your implementation. There will be instances when your budget won't be approved and you'll have to cut the "nice-to-have" features from the project scope or settle for fewer resources, trade-offs, and compromises to complete the project. In other scenarios, the project may have to be delayed until funds are available to continue. The worst-case scenario, of course, is that the project is approved but the funds to support the project are nonexistent.

A budget will serve as a financial guide to where the project is headed. Project managers who do their homework will have a clear vision of what the deliverables of the project will be and what it takes to reach those deliverables. Much of the information covered in this chapter is really defined in your project cost management plan. The project cost management plan, which I'll discuss in detail in Chapter 6, defines six things:

- **Control limits** These define the acceptable range of variances for the project costs for phases, vendors, and the entire execution.
- **Assignment of costs** The project manager needs to identify where the monies in the project will be spent and link the costs to the project deliverables.
- **Chart of accounts** This is a predefined table of costs and categories for project or organization use for commonly completed activities. For example, a programmer's time is $150 per hour regardless of which programmer is assigned to the project. Categories can include labor, supplies, cost of materials, and other known expenditures.
- **Project budget** This defines the process for creating, spending, and controlling the project budget.
- **Cost estimates** I'll discuss several approaches to cost estimates in this chapter.
- **Cost baseline** This is the agreed-upon level of costs tied to phases, milestones, and the project. As the project moves toward completion, the costs to reach certain milestones in execution should synch with the predicted costs of the cost baseline.

Phased Gate Estimating

As you begin to create a budget, you need to come up with a plan of attack. There are numerous ways to create a budget, some better than others. One approach IT project managers have a tendency to use is to write down a list of all the products that the organization needs to purchase to complete the project and add up the cost for each. At first glance, this seems like a viable solution; however, it opens the door for potentially overlooking important details, lack of true planning, and error. A better approach is to divide your project into phases and extract cost estimates for each phase of the project. This approach, called *phased gate estimating,* is ideal for large projects and describes the phases of the project, such as Ideation, Requirement gathering, Design, Development, Testing, and Deployment.

Phased gate estimating allows project managers to forecast the exact expenses for the pending phase of a project and provide more general estimates for phases downstream. Phased gate estimating is linked to step funding, where the project is funded in increments based on phases instead of the entire project being funded at once. Figure 5-1 is a depiction of phased gate estimating and how it's linked to step funding.

Phased gate estimating helps the project team get to work on immediate deliverables as they work toward milestones at the end of each phase. The immediate actions of a project should be foreseeable, as opposed to actions that will happen way off in the future. For example, you probably know what you're doing this weekend, but you don't know your plans for the weekend a year from now. Because IT changes so rapidly, accurate estimates are available for actions in the present and less so for actions in the future.

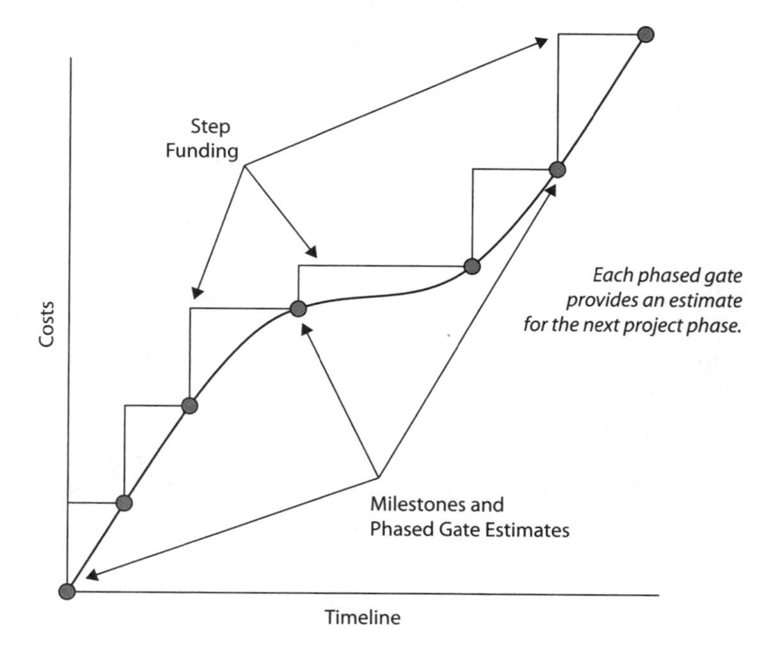

Figure 5-1 Phased gate estimates provide an accurate estimate for the near-time deliverables.

As discussed in Chapter 4, a key factor in any predictive project is the work breakdown structure, a deliverables-oriented decomposition of the project. From these lists of deliverables, the project manager can derive the activities required to deliver each component of the project. The major deliverables of the project, often associated with project milestones, are ideal for identifying the ends of phases within a project. For example, a project to create a new application will have some logical, visible milestones between its beginning and completion. A project manager using phased gate estimating can predict the cost of the project through the next foreseeable milestone.

Adaptive projects receive their budget from a steering committee or customer working with the product owner to predict the likely cost of the project. In an adaptive project, the budget is fixed and the scope may be trimmed based on performance. Recall that in adaptive project management, the items with the most business value are done first, and the items of lesser value are at the bottom of the product backlog. As the team completes the prioritized work, the budget is consumed; should the project run out of money before all of the items in the backlog are completed, it's not tragic, as the most important items are done. From there, additional projects may be launched to finish the product backlog items, the project may be considered done, or more funds may go into the current project to finalize the backlogged items.

When a project calls for phased gate estimating, the WBS will reflect the approach as well. A software development project has some obvious phases, just as a hardware rollout project will. A WBS in these instances can reflect the deliverables within the immediate phase with a nod to downstream phases that will come later in the project.

Determine the Estimate Type

As a project manager, you need to be familiar with three different categories of estimates. These estimates will dictate how much detail the project manager, product owner, or business analyst will need to provide in order to create an accurate estimate.

- **Rough order of magnitude** This estimate is "rough" and is used during the initiating processes and in top-down estimates. This estimate is based on high-level requirements. The range of variance for the estimate can be –25 percent to +75 percent.
- **Budget estimate** This estimate is also somewhat broad and is used early in the planning processes and also in top-down estimates, usually when you have an approved project scope. The range of variance for the estimate can be –10 percent to +25 percent.
- **Definitive estimate** This estimate is one of the most accurate. It is used late in the planning processes and is associated with bottom-up estimates (discussed next). The range of variance for the estimate can be –5 percent to +10 percent.

The percentages associated with these estimate types are pretty standard—but it's not unusual for an organization to develop a range of variances specific to that organization. For example, an IT integrator who performs the same type of projects for customers over and over will have a good understanding of what it takes financially to complete a typical project. The more familiar you are with your project work, the easier it becomes to predict project costs. When you're faced with a project scope that's significantly different from your experience, the range of variance can, and should, increase to reflect the unknowns in the project.

EXAM TIP The CompTIA Project+ exam won't quiz about the range of variance for each estimate type. Instead, know that the rough order of magnitude estimate is the least accurate but fast; the budget estimate is fairly accurate, based on the project scope, and is used for top-down estimates; and the definitive estimate, based on the WBS, is the most accurate but takes the longest to create.

Implementing Bottom-Up Cost Estimates

IT project managers love estimates; accountants don't. One of the toughest parts of your job as an IT project manager is to accurately predict the expenses your project will generate. As an IT professional, you know this is true, because there is so much in IT that fluctuates: RAM, new versions of software, the size of hard drives, the speed of processors, and just about any other facet of the IT world. And don't forget the cost of labor: it takes people to complete the project work, and there's often a ramp-up time as people join the project, learn the goals, and then really begin producing results. Everything becomes more efficient with technological advances.

The old adage "time is money" is never more true than when it comes to information technology. While the speed and prices of hardware and software may fluctuate, one of the largest expenses in an IT project is time. Why? Basically, if you, or your team, are not adequately prepared to implement the technology, the estimated time to install and roll out a plan can double or even triple. A project manager must take into account the learning curve to implement and manage the new technology. And there's the concept of knowledge work—the work that happens in your brain. It takes time to figure things out when doing software development, technical design, or other brain-centric work. You can't really predict how long it'll take someone to think their way to a solution.

Consider a project from management's perspective: time spent on a project is time away from core operations. The longer it takes to complete project work, the more the project costs in both labor on the unique endeavor and time lost on operational work that drives company income. Also consider the disruption that project work can have on the organization and how the disruption can take time away from operational work. It's imperative for a project manager to consider not just the project costs but also the indirect costs the project can have on the organization.

A project manager cannot always know their team's ability to implement a given technology. For example, suppose a project manager assigns Jacob to the development of an application. Jacob does have a proven track record with developing applications in Visual Basic; however, this application will have hooks into a SQL database. If Jacob does not have a clear understanding of the procedures to communicate with the SQL database, his reported estimated time might well be lower than the actual time he'll use to create the application. Worse still, if Jacob doesn't understand SQL at all, he'll need additional weeks to ramp up on the technology to make his application design flesh in with the existing SQL database. Jacob's weeks of training may incur additional expenses from your project budget and delay other workers and tasks that require Jacob to complete his portion of the project first.

The cost of team development needs to be included in your project budget, both from training and learning-curve perspectives. In other words, if you have a QA tester who will be using new software for error detection, not only do you have to figure in the cost of training that person, but you also have to remember that productivity on that piece of testing software will be about 60 percent of capacity in week one versus 90+ percent of capacity in week ten. If the project team lacks the skills to deliver, they must be trained. Lack of knowledge to do the project work guarantees project failure. It's no great discovery that so much of the knowledge surrounding information technology is disposable, although it's necessary for the imminent project. Consider all the old and discarded information you and your project team have learned about old versions of Linux, Windows 7, and even outdated hardware. At the time, the information was of incredible value; as technology changed, however, the information's value waned. The value of the training and knowledge to complete the project is what's important, not its value years from now.

Another fluctuating expense is hardware. Generally hardware is set at a fixed price and decreases in cost as newer, faster, better hardware becomes available. However, there are times when demand outweighs supply and the hardware costs increase. Also, as laptops, desktops, and servers drop in price, the demand for parts from manufacturers increases; this can cause hardware prices to remain steady while the hardware itself is significantly back-ordered. This, of course, throws your entire implementation plan askew. In addition, consider the unpredictable impact of supply chain issues, such as computer processors, in the past few years.

To avoid these pitfalls, a project manager should implement bottom-up cost estimates in a predictive project. *Bottom-up cost estimating* is the process of creating a detailed estimate for each work component (labor and materials) and accounting for each varying cost burden. The bottom-up cost estimates are based on the WBS and the WBS dictionary, as these documents define each element of the project deliverables. As Figure 5-2 illustrates, a project can be divided into phases, and then each phase can be assigned a cost value based on the work packages within the phase.

For example, an application development project can be divided into three phases in the WBS. Within each phase, the work to complete the phase relies on time, software, and hardware. The project manager can assign each of these factors a monetary value to complete the total phase of the project.

In other words, the project manager is starting from the bottom of the project—the genesis—and working toward the project deliverables. Each component of the project has a monetary requirement assigned to it, to ultimately predict the final cost.

Figure 5-2
A WBS can help estimate costs based on project phases.

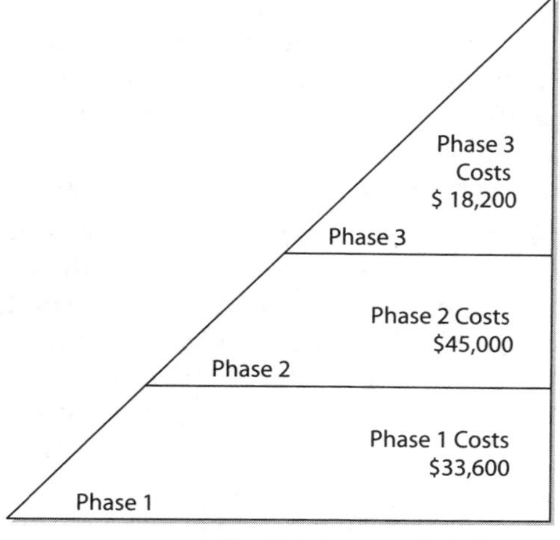

When you begin to create your budget in a predictive project, here are some issues to consider:

- *Create the WBS and the WBS dictionary.* These documents are needed to create the cost estimate for the project, as they show what deliverables are created in the project. You'll need the work packages to determine the resources that you'll need to create the deliverables. Resources include people, materials, and facilities.
- *Divide your project into phases.* By segmenting the entire project into phases, you'll find it easier to identify milestones and assign the amount of work and materials required to complete each phase. Once you break the project into phases, assigning dollars to each phase will be more manageable than assigning one lump sum to an entire project.
- *Address the integration phase.* By prepping the production environment for the onset of the implementation, budgetary concerns can address downtime, lag, required work hours, and the time the project manager requires to oversee that the tasks are being completed to keep the project on schedule.
- *Consider the fully burdened workload required to complete each phase of the project.* A *fully burdened workload* is the amount of work, in hours, required by the staff to complete each phase of the project. Part of the budget must include the work hours necessary to complete any given phase. Team members should have a dollar amount assigned to their work hours to predict the true expense of the implementation. (In some instances, it may be, unfortunately, beyond your control to predict the hourly rate of workers because of your organization's human resources department policy.) Additionally, when some work is outsourced, the hourly rate may include overhead, general and administrative expenses, a risk load factor, and a profit load. As a project manager, you should be aware of what ancillary, or additional, costs go into a true project cost.

- *Consider the costs for any specialized services.* Will you be using subject matter experts? Will the project include training for the implementation team? Will the project include a pilot team of ordinary users? Will you use testing services, security audits, or other firms to complete the project? Any of these special services are easy to overlook when you're calculating a project's budget, but they will come back to haunt you if you don't plan for their expenses before the project begins.
- *Consider the costs for equipment.* Of course, if you are purchasing new hardware, this is easy to account for. However, consider the value of leasing versus purchasing new hardware. Consider the impact of equipment dedicated to the project and any production machines that may be affected by the project's implementation, such as test servers, workstations reserved for testing, application development machines, the percentage of processor utilization, memory usage, and bandwidth impact.
- *Consider production costs.* Any project will have fringe costs such as expenses for photocopying, creating rollout manuals and user manuals, designing and developing web pages, and development.
- *Consider quality requirements.* The project needs to account for the level of testing required. How much regression testing, integration testing, and so forth should be included to meet the customer's quality standards?

  You'll also plan for any rework and testing as a result of quality control activities. When you or your team finds an error in the product, you'll need time, and likely money, to fix the problem before the product is released to production.
- *Consider risk.* Just as with the budget, risks are vague in the initial stages of a project. As the planning evolves, so does information on risks and risk management. You will need money for rework, risk mitigation, schedule delays, and workarounds. I'll talk more about risk planning in Chapter 6.
- *Consider reserve amounts.* All projects run into challenges. Smart project managers plan for the unknowns and for uncertainty. One way to plan for those things we can't know for certain is to keep a reserve amount to handle unforeseen circumstances. This is similar to the personal savings account we keep for emergencies, except this reserve amount is for our project. The amount may or may not be under your control, but it is useful to understand the concept and how to plan for it.

EXAM TIP The work breakdown structure is referenced frequently in a predictive project. If you're stuck on an exam question and one of the choices is the WBS, lean toward the WBS. Also know that the product backlog is a good tool to create a bottom-up estimate for the project, as it contains all of the project requirements and can be estimated for time and costs.

Once you've considered these different aspects of your project implementation, you're ready to begin calculating expenses. After you've broken down your project plan into phases, create an evolution of expenses for each phase. For example, in phase one of a project, consider the expenses required to complete this stage:

- Hardware to be purchased
- Software to be purchased
- Licensing issues
- Consultants
- Internal developers' time
- Percentage of time required by each team member to complete this phase of the project
- Risk and reserve funds
- Other expenses pertinent to your project

In the first phase of the project, you can complete the expenses required and then use the same template to move on to the second phase, the third phase, and so on, to create a table of expenses for each phase of the project.

Allowance for Change

When using bottom-up cost estimates, you need to calculate some allowance for change. When calculating time and costs for expenses, a project manager can create an average estimate for each phase of the project by factoring best- and worst-case scenarios into components that may fluctuate on price. An average estimate is called a three-point, or triangular, estimate. This approach uses the average of pessimistic, most likely, and optimistic estimates. Here's an example of an implementation phase for a new server-based application:

Component	Pessimistic	Most Likely	Optimistic	Three-Point Average
Server	$9,500	$8,000	$7,000	$8,166.67
Application	$5,500	$4,000	$3,600	$4,366.67
Licensing	$6,500	$6,000	$4,500	$5,666.67
Development	$10,200	$9,000	$8,500	$9,233.33
Testing	$7,500	$5,000	$4,500	$5,666.67
Documentation	$7,000	$6,000	$5,500	$6,166.67
Training	$9,500	$8,000	$7,500	$8,333.33

By including the best- and worst-case scenarios in your bottom-up cost estimates, you are factoring in an allowance up to the maximum amount, but predicting an average amount. Figure 5-3 depicts a simple predicted average for a project's expense. Most expenses within your project can follow this formula. This estimating approach is called the three-point estimate because you're using the worst-case, most likely, and optimistic approaches to create an average cost for the project expense. You can also use this approach in time estimating.

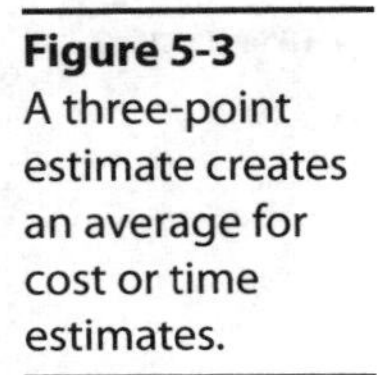

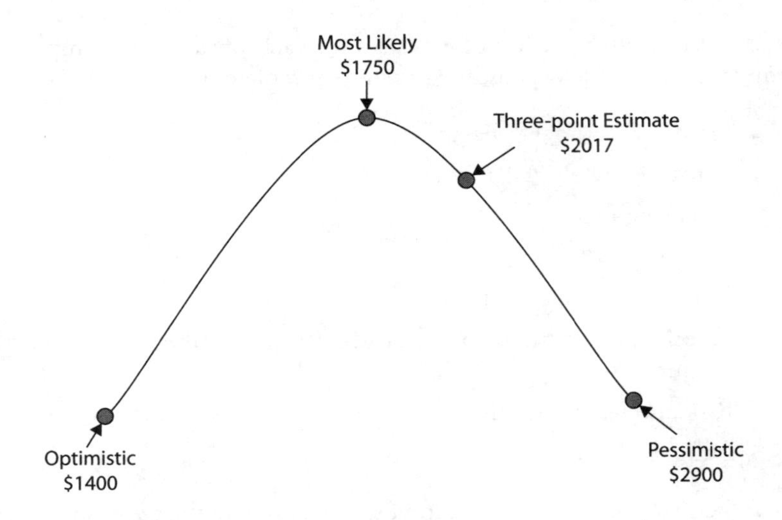

Figure 5-3 A three-point estimate creates an average for cost or time estimates.

Some elements of your estimates will not come close to the worst-case scenario, or even the average cost. Others will no doubt reach the worst-case scenario and perhaps even pass it. How do you determine the amount of time and the price value associated with each component? Here are factors that you should call upon to estimate your budget:

- **Prior experience** If you've worked with similar projects in the past, call upon your experience to predict how similar phases of work will fit within the scope of this project.
- **Historical information** Similar projects may have historical information that helps guide your current project's budget. In addition, are there mentors or other project managers you can call on for advice? Ask others how long certain elements took when they implemented similar projects within the organization or in their work history. Project team members may have experience with key areas of your plan, so their input is needed.
- **Fixed quotes** Vendors should be able to offer a fixed quote or a not-to-exceed (NTE) price on a deliverable. Typically, a fixed quote is for a product rather than a service and is valid 30 days from the time of the quote.
- **Standard costs** Your budget department may have preassigned standard costs for labor to perform tasks such as programming lines of code, installing hardware, or adding new servers. The cost of these activities may be found in organization-wide charts of accounts that represent types of work and their associated costs. A chart of accounts includes predetermined fees for standard work your organization does, regardless of materials or labor. For example, a network drop could have a chart of accounts fee of $175 per drop, regardless of the time or materials used in the network drop. This preassignment of values helps you easily estimate costs for a project, without having to justify each deliverable as a line item.

We'll talk more about time estimating in Chapter 6, but you should be keenly aware that time and money are interrelated. Time *is* money. In some organizations, the cost of the employee completing the work is not seen as a cost attributed to a project. In other organizations, however, the employee's time is billed to the project's customer. For example, an IT project to create a sales automation program may bill the sales department for the application developer's time. While the cost of the developer may not reflect the hourly rate of the employee, dollars are shifted from the sales department's budget to the IT department to account for the developer's time.

Tolerance for Budget Variance

As the cost of hardware, software, and services can fluctuate, project managers and management must agree on a tolerance level for the project's budget to be plus or minus a percentage of the predicted costs. Depending on your project and its budget, this may be only 1 to 2 percent or as large as 10 percent. Any variance in your project's budget can be unsettling, as it may reflect a lack of planning. Typically, management is more eager to deal with budgets under the predicted total costs than those that are over.

NOTE Projects that finish significantly under budget are not reasons to celebrate; this often indicates a lack of proper planning for project costs.

To circumvent any disagreements, management and the project manager must agree on the range of variance for a project. Don't use the range of variance as an additional cushion for your purchases now—you may need that percentage later in the project. In some companies, a variance in the budget can reflect the monetary rewards assigned to a project's success.

Creating a PERT Estimate

While finding the best- and worst-case scenarios is a quick and easy way to arrive at an average cost, you can use a slightly more sophisticated method called the *Program Evaluation and Review Technique,* also known as *PERT.* PERT is ideal for time and cost estimates to complete activities. PERT uses a weighted average to predict how long the activity may take. You'd say that as, "pessimistic plus the optimistic, plus four times the most likely, divided by six." It's divided by six because of one count for pessimistic, one count for optimistic, and four counts for most likely. The following table uses the same values as the three-point estimate but with PERT's formula instead (this formula is also included on the Estimating Worksheet that is part of this book's online resources; see Appendix C):

Component	Pessimistic	Most Likely	Optimistic	PERT Result
Server	$9,500	$8,000	$7,000	$8,083.33
Application	$5,500	$4,000	$3,600	$4,183.33
Licensing	$6,500	$6,000	$4,500	$5,833.33
Development	$10,200	$9,000	$8,500	$9,116.67
Testing	$7,500	$5,000	$4,500	$5,333.33
Documentation	$7,000	$6,000	$5,500	$6,083.33
Training	$9,500	$8,000	$7,500	$8,166.67

Using Top-Down Estimating

Top-down estimating allows a project manager to take a very similar project's budget, work some financial math magic, and arrive at a reasonable budget for the current project. Top-down estimates are often used by organizations that complete IT projects for other companies. Consider IT integrators who install servers, network cable, and network equipment. They'll have similar projects they can refer to when predicting the cost of current projects.

Within an organization, IT project managers also have projects that are similar to other projects they've completed in the past. Consider a project to roll out a new operating system using a disk-imaging server. If the project manager has rolled out other operating systems in the past using the disk-imaging server, they'll have a pretty good idea of how the current project will go. This historical information, also known as artifacts, on proven, completed applications allows the project manager to avoid spending time doing a bottom-up estimate; they can work from prior successful projects instead.

The problem with top-down estimates in the IT world, however, is that most IT projects have never been done before. Specifically, because IT changes so quickly and each environment is generally customized, top-down estimates are not as reliable or usable as bottom-up estimates.

Using Analogous Cost Estimating

If you find that you're launching projects that are similar to past accomplishments, analogous estimating may be your best bet. *Analogous estimating* relies on historical information to predict the cost of the current project. It is a type of top-down estimating. The process of analogous estimating takes the actual cost of a historical project as a basis for the current project. The cost of the historical project is applied to the cost of the current project with respect to the scope of the current project, its size, and other known variables.

This estimating approach takes less time to complete than other estimating models, but it is also less accurate. This top-down approach is good for fast estimates to get a general idea of what the project may cost.

Here's an example of analogous estimating: You completed the design and installation of an application for the sales department to track incoming phone calls from clients. Your IT help desk now wants you to create an application to track phone calls from internal users. The project deliverables are technically different, but both have fundamental characteristics that can guide you to create a reasonably reliable project cost estimate.

EXAM TIP You can do analogous estimating only if you have historical information. If your organization has never done this type of project work before, then you can't create an analogous estimate because there's nothing to create an analogy with.

Using Parametric Modeling

Another approach to top-down estimating is parametric modeling. *Parametric modeling* uses a mathematical model based on known parameters to predict the cost of a project. The parameters in the model can vary based on the type of work being completed. A parameter can be cost per cubic yard, cost per unit, and so on. For example, it costs $149 per software license and you need 200 licenses. Some quick math tells you that the total cost for the software is $29,800. A complex parameter can be cost per unit with adjustment factors based on the conditions of the project, such as delivery dates, penalties or bonuses, or additional materials used. Furthermore, the adjustment factors may have additional terms depending on project criteria.

For example, if you're managing an application development project, you may create a cost estimate based on the number of years of experience the application developer has with a given software language. Binita may have eight years of experience, while Sam only has two years of experience. Sam doesn't cost as much as Binita because Sam is considered less experienced than Binita. Sam can still get the work done; it just may take slightly longer than it would if Binita did the work.

When you think of parametric modeling, a parameter is generally used: cost per unit installed, cost per machine delivered, and so on. This approach doesn't always lend itself to IT projects because of the variables within the technology. Consider function point analysis: lines of code are not always reflective of the productivity, the number of servers, or even the number of programmers assigned to an activity.

Budget at Completion

The *budget at completion (BAC)* is the sum of the budgets for every phase of your predictive project. This is the estimated grand total of your project. If a project manager breaks down a project into phases—and they should—then each phase can be reflected with a dollar amount that needs to be allotted to that phase. The benefit of this approach is that an organization does not need to allot all of the BAC at the project's inception, but rather the initial amount required to set the project in motion, and another amount as each phase is completed.

The primary advantage of this approach is that an entity can continue to use the capital earmarked for the project until the next phase of the project is ready to proceed. A secondary advantage of the BAC is that it allows everyone involved in the project decisions to examine the costs of each phase of the project and then the project's grand total. So rather than seeing "Server upgrade costs: $25,128," management sees this:

Phase 1	**Start Date**	**Costs**
Server 1	November 3	$4,578
Server 2	November 3	$4,578
	Phase 1 Total	**$9,156**

(*continued*)

Phase 2	**Start Date**	**Costs**
Initiate clustering servers	November 10	$6,526
Install switch	November 12	$1,592
	Phase 2 Total	**$8,118**
Phase 3	**Start Date**	**Costs**
Add RAID 5 tower	November 17	$7,854
Test and document	November 19	$0
	Phase 3 Total	**$7,854**
Phase 4	**Start Date**	**Costs**
Migrate data from old servers to new (performed at night)	November 21	0
Put servers into production	November 22	0
	Phase 4 Total	**0**
	Budget at Completion	**$25,128**

As you can see, this approach to budgeting allows all parties to get a sense of what each phase will cost, when the monies will need to be allocated (in advance of the implementation date, of course), and the total cost of the project. This cash flow approach to project management creates cooperation among the project manager, the project customer, and management. The project manager should include phases that do not require any outlay of cash. In some situations, you may be required to add the number of hours estimated to complete each phase of the project to factor in the cost of an employee's or a consultant's time. The preceding sample shows only the hardware expense.

Zero-Based Budgeting

Another concept you'll likely encounter is zero-based budgeting. *Zero-based budgeting* means that the budget for a department or program to be created must always start at zero, rather than a dollar amount from a similar project, and then the new expenses are factored in. This long-winded approach generally is required each fiscal year. As Figure 5-4 depicts, zero-based budgeting requires a zero balance at the genesis. In other words, you can't take last year's budget for all projects in the IT department, add 20 percent to it, and claim that this new number is this year's upgraded budget. When you have to use zero-based budgeting for a project, you're required to reflect the true costs of each project deliverable with evidence of the costs. This can alleviate fluctuations in costs from past projects to the current project.

While this approach may seem similar to bottom-up estimating, it's often used for a series of projects, an entire department, or a long-term project that may last several years.

The biggest complaint IT project managers have with zero-based budgeting is that it feels as though you're doing your work twice. In reality, it forces you to ensure that the costs of goods and services have not changed—and if they have, the budget reflects

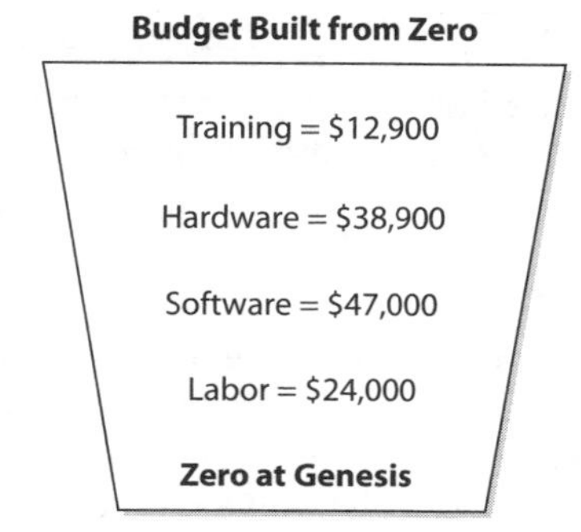

Figure 5-4 Zero-based budgeting requires a zero balance at the genesis.

the change in costs. Zero-based budgeting creates a sense of accountability for the project manager with regard to getting an accurate cost of the services and hardware to be purchased.

Some IT project managers will, however, rely on similar budgets and fudge their way through a new budget. Don't take this route! Why? Why not just take last year's figures, check out any major changes, and go with the number predicted? Well, it could cost the organization money and cost you your project and your job.

Imagine that you take last year's budget for server upgrades, add 20 percent to the budget, and claim it as this year's project budget. When it comes time to actually purchase the hardware, what will happen if the cost of the hardware from last year has increased due to supply and demand? Or what if the servers you used last year are no longer available and the next step requires purchasing a server that costs 30 percent more than a similar server last year? You'll have much explaining to do.

When you are asked to use zero-based budgeting, use it. Even if the project is identical to a previous project, investigate the costs of goods and services required to complete the project and report them accurately. It's not always fun, but that's why it's called work.

Determining Project Expenses

On the surface, it's easy to predict what a project will cost. Take the hardware required, add it up, and there's the amount needed, right? We all know it's not that easy. There are other factors involved in predicting the cost of a project. For starters, organizations have to examine if the capital expenses (CapEx) versus the operational expenses (OpEx) the project will affect the overall finances of the organization. Capital expenses are often related to projects, as these expenses are big purchases that will be implemented into the organization and used for many years. Operational expenses include expenses such as payroll, facilities, software management, and the general cost of doing business. Projects are affected by both types of expenses.

When predicting the project expenses, a project manager has to look at employees, the combination of employees working together or alone, hardware expenses, the determined scope of the project, and the necessary hardware to implement the plan. The total of these variables makes for long planning, calculating, and educated guesses as to the expense of a project. Careful planning and experience are the two best ingredients for cooking up an accurate budget.

EXAM TIP Operational expenses, such as rent, payroll, and keeping the lights on, are sometimes called the cost of doing business and are ongoing. Capital expenses are singular and are not ongoing.

The Cost of Goods

If you wanted to purchase one solid state drive (SSD), it would be easy to determine the cost of that one device. However, if you need to purchase two clustered servers, with giant amounts of RAM, terabytes of RAID drives, multiple NICs, eight processors, and more power than ever before, the calculation would be a little tougher. You could leave it all up to the manufacturer or your favorite salesperson, but would you get your dollars' worth?

Would it be better to assemble the servers in-house? Would it be better to have the manufacturer assemble the board and NICs, and then have your IT department add the RAM and the cluster RAID later? What about installing any operating systems through the manufacturer? Is your staff prepared and knowledgeable enough to assemble everything on their own? And is it even worth the time to assemble such a clustered server onsite? How much time will it take? These types of build-or-buy decisions should be based on the WBS. Each deliverable should be analyzed to assess whether it should be made or bought, or whether there is an option to build or buy that requires further investigation.

The decision to build or buy a product is a fundamental aspect of management. In some conditions, it is more cost effective to buy, while in others, it makes more sense to build an in-house solution. The build-or-buy analysis should happen in the initial scope definition to determine if the entire project should be completed in-house or procured. As the project evolves, additional build-or-buy decisions are needed.

The initial costs of the solution for the in-house or procured product must be considered, but so, too, must the ongoing expenses of the solutions. For example, a company may elect to lease a piece of equipment. The ongoing expenses of leasing the piece of equipment should be weighed against the expected ongoing expenses of purchasing the equipment and the monthly costs to maintain, insure, and manage the equipment.

For example, Figure 5-5 shows the mathematical approach to determining whether it is better to create a software program in-house or buy one from a software company. The in-house solution will cost your company $25,000 to create your own software package and, based on historical information, another $2,500 per month to maintain the software.

The software company has a solution that will cost your company $17,000 to purchase, but the software company requires a maintenance plan for each software program installed, which will cost your company $2,700 per month. The difference between making the software and buying the software is $8000. The difference between supporting the software the organization has made and allowing the external company to support their software is only $200 per month.

The $200 per month is divided into the difference between creating the software internally and buying the software, which is $8,000 divided by $200, or 40 months. If the software is to be replaced within 40 months, the company should buy the software. If the software that will be created will not be replaced within 40 months, the company should build the software.

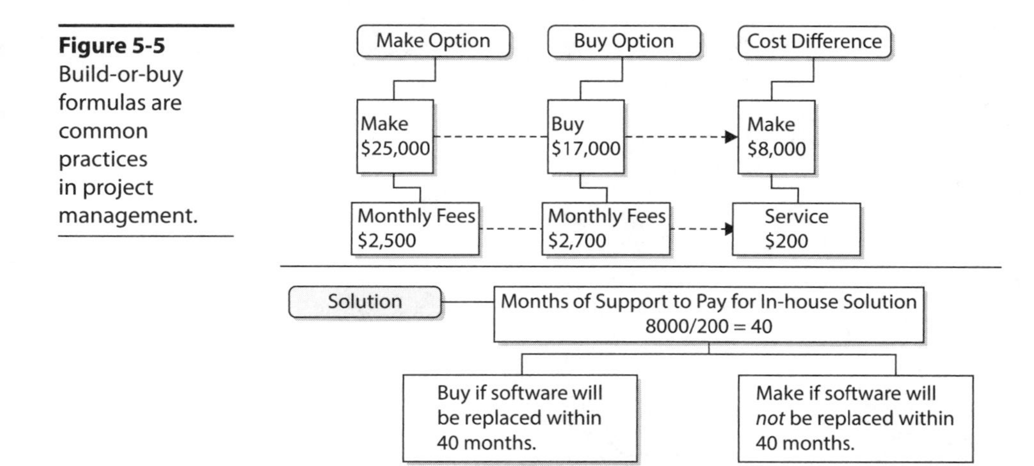

Figure 5-5 Build-or-buy formulas are common practices in project management.

There are multiple reasons an organization may choose to build rather than buy. A project team can build or buy as much as it needs to complete the project scope. Here are some common examples of reasons to build or buy:

Reasons to Build	Reasons to Buy
Less costly	Less costly
Use in-house skills	In-house skills aren't available or don't exist
Control of work	Small volume of work
Control of intellectual property	More efficient
Learn new skills	Transfer risks
Available staff	Available vendor
Focus on core project work	Allows project to focus on other work items

As you can guess, or maybe you've experienced, there are lots of avenues to consider when purchasing hardware that will need to be assembled and configured. In some instances, off-the-shelf hardware will be an appropriate solution for a project, while in other instances assembling the hardware onsite will be more cost effective. How can you know the difference? Figure 5-6 shows that the cost of hardware assembled by the manufacturer should not be greater than the amount of time it takes to assemble the hardware in-house.

You need to consider other factors when allowing hardware to be assembled through the manufacturer versus piecing the hardware together in-house:

- *How long will the assembly take?* If you or a staff member has experience assembling hardware components, you can make an accurate prediction as to the length of the assembly process. From that information, you can calculate the cost of the assembly process. This dollar amount, assigned to the assembly process, can help you determine if it is more cost effective to assemble the hardware in-house or to allow the vendor to assemble the hardware.

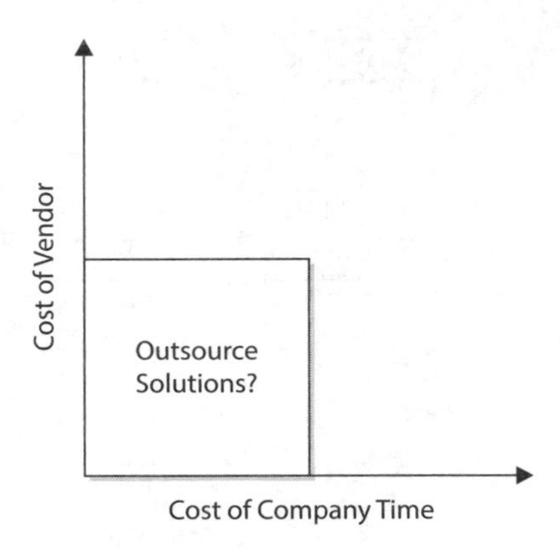

Figure 5-6 The vendor's costs should not outweigh the value of the cost of internal resources.

- *What other tasks can the technician do?* Consider the technician's time, the cost of the time, and the other responsibilities the technician could handle on the project. It may be more valuable to the project if the vendor assembles the hardware and the technician moves on to other aspects of the project.
- *Will the vendor guarantee the work?* If the vendor is to assemble the hardware, that vendor should guarantee their work. Incorrectly configured hardware by the vendor could bring your project to a grinding halt. A vendor that is assembling the hardware you are purchasing is going to charge you adequately for the time and materials it takes to build the component according to your specifications. The vendor's contract should include a guarantee that the hardware will arrive in working order and will work in your environment.
- *Is in-house assembly worth the headache?* The headache factor sometimes outweighs the money saved by doing the work in-house. In some instances, especially when the savings from doing the work in-house are nominal, it is more effective to allow the vendor to assemble the hardware. Let the vendor deal with installing the RAM, processors, and BIOS upgrades and configuration. Often it's not worth the headache to do the work in-house.
- *Can something new be learned?* Sometimes by creating the product in-house your team has an opportunity to learn a new competency that can be leveraged for other projects. This assumes, of course, that you have the time to learn, experiment, and try different approaches in the creation of the product.

Outsourcing

One of the most popular trends in information technology is to outsource practically any project. On some levels, this is both cost effective and extremely productive. In organizations that don't have a full-time IT manager, it is ideal to outsource the simplest of IT problems to a team or consultant. When you consider the cost of hiring a full-time IT professional at $85,000 per year base salary versus outsourcing to an integrator to service the computers and printers for $45,000, it's an easy decision. However, this opportunity, as with any industry, has created some less-than-desirable businesses. It's incredibly easy for anyone to market themselves as an IT expert, land a few accounts, and take advantage of otherwise unknowing clients. Not that this would happen to you.

When outsourcing a project or considering outsourcing a project, ask the following questions:

- *Is it cost effective?* Consider the time, learning curve, implementation process, and dollar amount associated with each variable and compare that to the figures from the vendor proposing to do the work.
- *Is it productive?* Again, consider the time of doing the work internally versus hiring an outside agency to complete the tasks. You should consider not only the dollar amount but also the time involved to complete the project.
- *Is the vendor reputable?* Ask the vendor for references of similar work they have done before. Ask them for industry credentials from Microsoft, Oracle, Project Management Institute (PMI), CompTIA, and others.
- *Is this an HR decision?* Outsourcing a technology project may not even be the project manager's decision. HR and management may have created contracts and agreements with staffing agents to complete the project work while you, the internal project manager, are to manage the external workers.
- *Consider culture differences.* Internal resources are familiar with the politics, priorities, and procedures within your organization. A vendor may have a completely different set of priorities or a different definition of quality or immediate deadlines.

Outsourcing is not always the best solution, but sometimes it's the most appealing to management. This is because the cost considerations, the internal learning curve, and other projects that may be on the horizon could conflict with the outsourced job. If you decide to consider outsourcing a project, get a fixed cost from the vendor—especially when proposing a budget to management. You may need to work with the vendor, or several vendors, to negotiate a fair cost for services and manage the purchase of the hardware separately to get a better sum price. Many vendors will give you a break on the price if you buy the hardware and the implementation through the same source. I'll discuss the gritty business of procurement in Chapter 6.

Estimating Work Hours

What's the most expensive element in any project? If you said time, you are correct! Time is the one component of a project that is the most difficult to predict, the hardest to manage, and the easiest to lose control of.

Think about your own day as an IT professional. How many times have you set out to complete a task—for example, something as simple as troubleshooting a printer for a particular user—only to be summoned for more tasks along the way? You go to the printer to make certain it's turned on, check the power, pop open the printer, and check the toner. While you're there, two folks begin asking you questions about how to create a macro for column numbering. Now your phone chirps with a text message reporting that the SQL server is running out of disk space and the transaction log needs to be cleared.

The printer looks fine, but the user still can't print. You get to their desk only to discover that their neighbor reports that their mouse won't work. Get the picture? Or is it too close to reality? It's just one thing after another all day long—and that user still can't print.

As an IT project manager, your time is very valuable and has to be guarded from interruptions by users, texts, and yet more users. I can hear you now, "Yeah, sure." Seriously, think about the percentage of your day that is committed to putting out fires in proportion to the percentage of your day that can be dedicated to a project. Now think of the people on your project team and the same interruptions and activities that may delay them from completing their project tasks.

While you, the IT project manager, may not be the individual performing each step of the project's implementation, you do have to be available to work with your team, resolve issues pertinent to the project's success, and have time to track and report the status of the project. In some organizations, the project manager may have to wear several hats, supporting the users, working on each phase of the project, and tracking the project status. In others, the project manager may have the luxury (or headache) of delegating the phases to individuals and managing several projects at once. In either situation, your ability to manage your time, and the time of your team, is crucial to the success of the project.

When you are budgeting a project, you can also use PERT or the three-point estimate approach for predicting team members' time and costs. Most project managers have a range of variance assigned to labor costs. For example, the cost of labor will be $4,000 +/– $400. In the following table, examine how much team members' average hourly rates cost the company from best- to worst-case scenarios to do a given task using a three-point estimate.

Team Member	Hourly Project Cost	Pessimistic Time	Most Likely Time	Optimistic Time	Three-Point Time Estimate	Three-Point Cost Estimate
Sally	$32	30	25	20	25	$800
Fred	$35	60	50	45	52	$1,808
Julieta	$40	55	44	36	45	$1,800
Selim	$20	80	65	55	67	$1,333
Holly	$15	40	35	30	35	$525

As the table illustrates, you can fairly accurately predict the cost associated with each team member's time by using the individual's hourly project cost rate; the time you predict it will take the team member to finish the task; and the best, worst, and average scenarios. This worksheet has been created for you in an Excel document called Estimating Worksheet that is part of this book's online resources (see Appendix C). You will be using the worksheet in an exercise at the end of this chapter.

Another advantage of this worksheet is that it can help you determine which tasks should be assigned to which team members. For example, you may not want to assign Julieta, who has an hourly rate of $40, to pulling cable—a mundane and tiresome chore.

A bigger bang for your budget dollars would be to assign this task to Holly or Selim. If you could, you may assign the task to both Holly and Selim, who have a combined hourly wage of $35. This would put two workers on the task and would cost less per hour than Julieta's hourly rate. In addition, two people could, in this instance of pulling cable, finish the job in nearly half the time, or better, that it would take one individual.

Of course, you'll have to consider a few things when assigning resources to tasks:

- *Consider skill sets.* Can the resources you'd like to assign actually complete the project work? Just because someone is available doesn't mean they're able and willing to complete the assignment.
- *Consider productivity.* Can a higher-paid resource complete the job more quickly and more cost effectively than a lower-paid, less-experienced resource?
- *Consider the law of diminishing returns.* Just because you can add more resources to a particular task doesn't mean the task time can be exponentially reduced. For example, adding two people to pulling network cable may ensure the activity is completed more quickly, but assigning four people to the same job doesn't mean it'll get done four times as fast.

Tracking Budgetary Expenses

It is very easy for expenses to spiral out of control. Imagine that you are buying a new server. You're talking with your favorite vendor and he's showing you that for a few hundred dollars more you can have eight processors instead of four. And you say, "Might as well." Then the vendor shows how for a couple hundred more you can add 800GB more storage. And you say, "Might as well." Then the vendor shows how for just a little more you can really up the RAM. Again you say, "Might as well."

"Might as well" are some dangerous words when it comes to shopping, aren't they? It's so easy to tack on some bells and whistles for just a few dollars more. Before you know it, those few dollars more have stretched your budget so thin you'll either have to ask for more funds or skimp on other areas of the project. And it's just not shopping that can ruin your budget. It's also personnel, human error, lack of planning, hidden costs, and general lack of research.

Not only do you need a detailed budget prior to making any purchases, but you also need a detailed method to track expenses as they are incurred. This is called working toward your BAC. By documenting each purchase as it's made, you can check the purchase price against your initial budget to confirm that what you planned for is what's actually implemented.

Runaway Projects

A *runaway project,* as its name implies, starts out well; gains speed, momentum, and scope; and then causes runaways with your budget, your work hours, and possibly your reputation or career. The biggest element of a runaway project is the budget. Project managers often try to throw money at a problem rather than completing root cause analysis. Too often in project management, there is an attitude of solving problems by spending more money.

Runaway projects happen for several reasons:

- **Lack of planning** Failure to plan for all aspects of the project can cause projects to fail in the beginning, not the end.
- **Lack of vision** Failure to create a definite purpose for the project can cause a project to fail.
- **Scope creep** Management and departments continue to add details and extras to an existing project scope without following the change control processes. Recall that the project scope is all of the required work—and only the required work. Scope creep is also called project poison, because it robs the project of time and monies for things that are in scope.
- **Lack of leadership** Without leadership, the project is bound to wander aimlessly and incur additional expenses.
- **Lack of a change control system (CCS)** A CCS is a formal process to evaluate, approve, or decline proposed changes and additions to the project scope.

You can prevent runaway projects by creating a definite, nearly unmovable plan for the project's implementation, budget, and scope. Any additional attributes of the project that are not key to its success should be set aside, regardless of the requestor. In all projects, however, there needs to be a process that will allow changes to enter the project plan. Chapter 10 will discuss this change management in great detail.

TIP Begin project cost monitoring early in the project. Early fluctuations in the project costs is often a good indicator of future project performance.

Here is an example of what appears to be a simple change to a project's scope: You are managing a project that will create an application with hooks into a SQL or Oracle database. The application will allow salespeople to place an order, check that order against warehouse inventory, and predict a ship date for the customer based on inventory or production.

The original plan of the application called only for tight coupling of the application and the database. (Tight coupling means the application has to be connected to the database to run.) Now, several weeks into development, management asks that you change the application to allow loose coupling. (Loose coupling allows the application to run without being directly connected to the database.) Can you see the problem now? Several weeks of development have been centric to tight coupling; now what appears to be a simple change does not reflect the work hours invested in the original application.

In this scenario, management is suddenly adamant about the loose coupling because it enables the salespeople to take their laptops into the field, take orders and store them locally, and then, once they are connected to the network in the office, actually synchronize the orders with the warehouse. The project manager must first meet with management and discuss the change and explain to management how the request will increase the scope of the project. When the scope of the project increases, additional funds will be required, in most instances.

Next, the project manager will have to meet with the developers and discuss the new application plans with them. The developers will, no doubt, curse management, slam their keyboards a few times, drink some sugary soda, and then start working the new plan into their project. Because of lack of planning, the project scope has increased, time has been wasted, dollars have been spent, and morale has suffered.

Keeping Track of Expenses

Before the project actually begins, you'll need to work within your organizational policies on how project expenses will be tracked and monitored. In some organizations, budgetary concerns are handled by management with some input from the project manager. In other organizations, the project manager is responsible for the day-to-day accounting of the project budget. There are multiple tools available to help you track the project expenses, but whichever one you use to keep track of your project expenditures, you'll need to include some basic elements:

- **Work hours** Work time is one of the most expensive elements to any project, so you should have a plan for team members to report their hours working on a given project. If you are working with vendors or consultants who will be billing by the hour, create a method for them to report their hours as well. You may need to create a formula to reflect overtime and weekend pay if that is applicable to your organization. Functional managers of your project team members will also want some accountability of their employees' time on your project.
- **Burn rate** How quickly the project consumes the project budget in relation to the work actually completed is called the *burn rate*. The faster you "burn" through the project budget, the more likely it is that the project will be over budget. To calculate the burn rate, you'll use the formula 1 / cost performance index (CPI). I'll explain CPI in Chapter 9, but for now just know it represents how well the project costs are performing. A burn rate greater than 1 is not good, as it's nearly impossible to recover the costs in future project activities.
- **Procured goods** Keep track of all hardware, tools, software, cables, and any other items that are purchased directly for your project. Your accounting software should have a method for entering any of these items. Also include petty cash items such as pizza dinners and miscellaneous items your team needs.
- **Software licensing** If your IT project includes software-licensing fees, be sure to document them. In some organizations, the IT department may pay for the initial licensing of the software, but as the software is released throughout the company, other departments have to pay to use the software from their budgets. An IT project manager should know how these fees are handled and from whose budgets these funds will flow.

- **Workstations and servers** If your IT project includes workstations and servers as part of the plan, document the purchase price and installation date of the computers. Obviously, in some plans the implementation of the workstations or servers may in itself be the project. The reason to document the actual expense of the computers is so that if they are recycled into other servers or workstations for future projects, you can reflect the original paid price of the PCs and then diminish the value of the computers in the new project. You likely won't have to get into the details of single-line deductions versus double-declining deductions for tax purposes, but your organization's accounting department may query your decisions and choice to recycle hardware.
- **Actual variances** Throughout your project, you may have small variances between what was estimated and the actual cost of the deliverable. For example, you may order supplies for the project at $440 and the actual invoice is $480. While it's only a $40 variance, it's still a variance that's going to add up and count against your budget at completion.

You can use a spreadsheet like the following example to keep track of budgetary expenses. This spreadsheet is for the first of three phases for a software upgrade. The actual Excel spreadsheet, named Budget Worksheet, is available as part of this book's online resources (see Appendix C). In this spreadsheet, you'll see a predicted cost baseline and the actual costs. When you track the actual costs, you can create a burn rate for the project. As mentioned earlier, a burn rate shows how quickly the project is "burning" through the allotted budget in relation to the progress the project is making. Be wary of burn rates; while they seem like a good idea, many projects have up-front expenditures such as purchasing equipment, hardware and software, and other resources.

Each project will, of course, have different needs for computing the expenses committed to that project. This example shows work hours, hardware and software purchased, and any incidentals. The formulas reflect a running total of each week of the project and a total for the project's expense at the phase the project is currently in. You will get a chance to practice creating a budget spreadsheet in an upcoming exercise.

Phase One						
Budget for phase	$160,000.00		Amount spent to date	$159,897.89		
			Variance	$102.11		
Work Hours	**Hourly Rate**	**Week 1 Hours**	**Week 2 Hours**	**Week 3 Hours**	**Hours to Date**	**Cost to Date**
Steve	$21.63	37.0	30.0	39.0	106.0	$2,292.78
Sally	$30.53	27.0	25.0	26.0	78.0	$2,381.34
Jane	$32.81	38.0	37.5	29.0	104.5	$3,428.65
John	$32.31	29.0	40.0	37.0	106.0	$3,424.86
Fred	$30.38	35.0	40.0	26.0	101.0	$3,068.38
Totals	**$147.66**	**166.0**	**172.5**	**157.0**	**495.5**	**$14,596.01**

Purchases	Cost	Number of Units	Totals
Server 1	$7,854	2	$15,708
Application	$89	950	$84,550
Licenses	$45	950	$42,750
Total			**$143,008**
Incidentals	**Cost**	**Number of Units**	**Totals**
Network card	$21	2	$42
Sound card	$45	4	$180
Mouse	$37	2	$74
Video card	$69	3	$207
RAM	$268	5	$1,340
Team dinner	$150	3	$450
Total			**$2,293**

As you can see, this project has ended phase 1 with a surplus of $102.11—an excellent reflection of planning and predicting by the project manager. While a surplus of this little amount is acceptable, a surplus of 10 percent or more of the predicted project phase budget is not a reason to celebrate.

Some IT project managers congratulate themselves for coming in under budget. However, there are several problems with large budget surpluses. The first problem is that it reflects poor planning on behalf of the IT project manager. An accurate plan will keep any surplus within 3 to 5 percent of the original budget, including the agreed-upon range of variance for the project. The second problem with surpluses is that they create an attitude of spending. Organizations with surpluses do not feel obligated to return the funds, but rather feel obligated to spend them to justify their original budget and to ensure that their budgets will be as fat on the next project. Gold plating happens when a project manager bypasses the change control processes and adds features to a project's deliverable only to consume the entire project budget. Poor planning is not a reason to celebrate.

CompTIA Project+ Exam Highlight: Managing Project Costs

You'll be faced with plenty of questions about cost management, creating cost estimates, and managing budgets on the CompTIA Project+ exam. You'll need to be familiar with the terms and processes covered in this chapter and how these processes relate to other areas of project management, such as risk, quality, and resource management.

1.2 Compare and contrast Agile vs. Waterfall concepts Adaptive and predictive projects, which CompTIA calls agile and waterfall projects, respectively, approach cost management differently. While predictive projects aim to predict the costs of all items in the project through intense planning, adaptive projects focus on delivering value as quickly as possible. Adaptive projects begin with a budget for the project based on the project vision and the product backlog. The team creates the items with the most business value first and then items with less value. Predictive projects create items based on phases and sequencing of activities to reach the end of the project, not to reach the business value. Neither approach is wrong or better than the other, but each approach has merits and advantages.

1.11 Explain important project procurement and vendor selection concepts If a project is utilizing a vendor, there will be contractual obligations, cost of the procured resource, and payment systems to follow. It is not unusual for a project to hire contract labor through a vendor to help the organization complete the current project and maintain operational activities. In addition to contract labor, a project may need to purchase or lease physical resources such as hardware, equipment, or space for the project team to work. In the next chapter I'll delve into more specifics on procurement management activities, but obviously procurement is related to cost management in any project.

2.1 Explain the value of artifacts in the discovery/concept preparation phase for a project When it comes to determining the cost of the project, the organization will determine if the project is related to an operational expense or a capital expense. Operational expenses (OpEx) are the costs of maintaining the operations, such as payroll, facilities, and keeping the lights on. Capital expenses (CapEx) are often larger purchases, like equipment, custom software, or new facilities, that will be used by the organization for years to come. Operational expenses are ongoing, while capital expenses are a single purchase of the implementation.

2.3 Given a scenario, perform activities during the project planning phase In a predictive project, cost planning is one of the most important planning activities. The cost management plan defines how the costs of the project will be estimated, budgeted, and controlled. In this chapter, I detailed several cost-estimating techniques that a project manager can use in different scenarios in the project. Recall that bottom-up estimating creates a cost estimate for each element of the WBS, while top-down estimating uses a similar process to predict the current project costs. Parametric estimating is an estimating approach that uses a parameter, such as cost per unit, to predict the project costs. Analogous estimating creates an analogy between the current project and similar, but completed, projects.

You also learned about three-point estimates and PERT estimates. Three-point estimates are really just an average of the optimistic, most likely, and pessimistic cost estimates. PERT, the Program Evaluation and Review Technique, uses the formula (Optimistic + [4 × Most Likely] + Pessimistic) / 6 to find the cost estimate. You can use PERT and three-point estimates for both costs and time estimates.

Often the project budget is consumed by the purchase of goods and services. The procurement management plan defines the complete procurement process, but I did touch on some key procurement information in this chapter. You learned about the need for purchasing goods and services and when it's appropriate, best for the project, and cost effective. You also learned about the buy-versus-build scenario. Remember that to find the most effective build-or-buy decision, you'll consider more than just the price. For the CompTIA Project+ exam, however, you may be faced with a decision to build or buy based on price only. To make the decision, find the difference between the build and buy costs. Then you'll find the difference between the monthly fees for buy and build. Finally, you'll divide the difference of the buy and build initial costs by the monthly fee difference. Sometimes it's more cost effective to let the vendor provide the solution, whereas other times you can recoup your out-of-pocket costs and build the solution in-house.

Chapter Review

Technology is an investment, not an expense. One of your roles as a project manager is to safeguard the investment dollars and ensure that the project is implemented successfully and within budget. This includes the planning, testing, integration, and, ultimately, the implementation phases. In some instances you will be forced to alter the plan, which will most likely alter the budget. Of course, adaptive projects expect the requirements to change and they adapt quickly to customer requirements.

An effective project manager can work with bottom-up cost estimates to accurately predict each phase of a project and what expenses will be associated with each phase. Typically, zero-based budgeting will determine estimates for IT projects, and this will require that you and your project team research the true costs of each component of the project to ascertain an accurate price for the product implementation. In some instances, best- and worst-case scenarios should be used so that you can predict an average amount of time, cost, and dollars needed to implement the technology.

Finally, you will need a good flow of communication between vendors, team members, and consultants to keep an accurate record of time invested, dollars committed, and incidental expenses incurred in a project. By tracking budgetary expenses, you can see weekly, or even daily, expenses incurred throughout a project. This will also allow you to see a running total of a project's phase and predict any overrun or the possibility of a budget surplus.

Exercises

These exercises allow you to apply the knowledge you have learned in this chapter and are followed by possible solutions.

Exercise 5-1: Complete an Estimating Worksheet

In this exercise, you will complete the Estimating Worksheet to predict and calculate the cost of each team member. Microsoft Excel is required to use the formulas to predict the cost of each task automatically.

Scenario: You are the IT project manager for Harding Enterprises. The project you are managing is an installation of new network cable, network cards, servers, and workstations throughout the entire company. In this first part of the budget planning exercises, you need to calculate the hourly rate of each worker.

Follow these steps to complete Exercise 5-1:

1. In this book's online resources, you'll find a Microsoft Excel file named Estimating Worksheet. (If you need assistance accessing this file, please refer to Appendix C.)
2. The Excel document has several spreadsheets. The first worksheet is titled "Time cost analysis." Navigate to this worksheet if you're not there already.
3. Hover your mouse over the comment marker in cell A1 and read the comments. Click in cell A3, enter **Rick Gordon**, and then press TAB to move to cell B3.
4. Hover your mouse over cell B1 to read the comment. In cell B3, enter Rick Gordon's yearly salary, **73500**, and press TAB to move to cell C3.
5. Rick Gordon's hourly rate is calculated for you based on his annual salary, divided by 52 weeks, and then divided again by 40 hours. Press TAB again to move to cell D3.
6. For this first task, enter **15** for the optimistic value time. Press TAB to move to cell E3.
7. Enter **24** for the most likely value and press TAB to move to cell F3.
8. For the worst time, enter **37** and press TAB. The time and cost for an activity is calculated. Press TAB to move on to cell G3.
9. Notice that cell G3 has already calculated the average time for Rick Gordon, and the average cost for Rick to complete the assigned task is displayed in cell H3.
10. Complete the remainder of the spreadsheet with the following information:

Team Member	Yearly Salary	Optimistic	Most Likely	Pessimistic
Samantha Murray	67500	20	25	34
Bradley Kiser	43200	25	32	42
Harriet Sutherland	37600	17	25	36
Fred Stephens	57600	40	55	75
Ruth Carze	67000	25	45	57

11. Based on your entries, answer the following questions:
    - **A.** What is the three-point estimate cost of Samantha Murray's time on the assigned task?
    - **B.** What is the cost of Bradley Kiser's time if he takes the pessimistic amount of predicted time?
    - **C.** What is the cost of Harriet Sutherland's time if she beats the best time estimate by two hours?
    - **D.** What is the average cost of Fred Stephens' time?
12. Review your work and then keep the document open, as you'll use your answers and entries for Exercise 5-2. You can save the spreadsheet to your hard disk if you would like to review your work and complete Exercise 5-2 later.

Exercise 5-2: Explore Project Cost Estimating

In this exercise, you will explore PERT and the three-point estimates in the Estimating Worksheet you used in Exercise 5-1.

In the Estimating Worksheet that you completed in Exercise 5-1, navigate to the worksheet called "3 Point Estimate." This worksheet is based on the entries you made in Exercise 5-1. You will also use the worksheets in this file called "PERT" and "PERT versus 3 Point."

1. What is the three-point estimate result for Ruth Carze on this entry?
2. Navigate to the worksheet titled "PERT." This worksheet is based on the values you've entered on the "Time cost analysis" worksheet.
3. What is the PERT result for Ruth Carze?
4. Return to the "Time cost analysis" worksheet and change the optimistic values for Ruth Carze to **20**.
5. Compare the three-point result and the PERT result for Ruth Carze with this new optimistic value.
6. Navigate to the worksheet titled "PERT versus 3 Point."
7. On this worksheet there are two charts that will compare the results of the three-point estimate and PERT. Hover your mouse over the first bar that represents Paul Samms in the 3 Point Estimate chart. What is the value that's displayed?
8. Now hover your mouse over the first bar that represents Paul Samms in the PERT Estimate chart. What is the value that's displayed?
9. Why is there a difference between the PERT value and the three-point value?

Exercise Solutions

The following offer possible solutions for the chapter exercises.

Exercise 5-1: Complete an Estimating Worksheet

In this exercise, you completed the Estimating Worksheet to predict and calculate the cost of each team member. The answers to the questions posed in step 11 are as follows:

A. The three-point cost estimate of Samantha Murray's time on the assigned task is $855.

B. The pessimistic cost of Bradley Kiser's time is derived by multiplying his hourly rate of $20.77 by 42 hours, which equals $872.34.

C. The cost of Harriet Sutherland's time if she beats the best time estimate by two hours is derived by multiplying Harriet Sutherland's cost per hour time, which is $18.08, by 15 hours, which equals $271.20.

D. The average cost of Fred Stephens' time is the three-point estimate of $1,569.

Exercise 5-2: Explore Project Cost Estimating

In this exercise you explored PERT and the three-point estimates in the Estimating Worksheet you used in Exercise 5-1. The Completed Estimating Worksheet in the book's online resources is the same file you'll use to review your answers in this exercise solution. The exercise steps are repeated here with the answers embedded:

1. What is the three-point estimate result for Ruth Carze on this entry?
   Your answer: The three-point estimate result is 40.67 hours (or 41 hours if you're rounding up).
2. Navigate to the worksheet titled "PERT." This worksheet is based on the values you've entered on the "Time cost analysis" worksheet.
3. What is the PERT result for Ruth Carze?
   Your answer: The PERT result is 42.83.
4. Return to the "Time cost analysis" worksheet and change the optimistic values for Ruth Carze to **20**.
5. Compare the three-point result and the PERT result for Ruth Carze with this new optimistic value.
6. Navigate to the worksheet titled "PERT versus 3 Point."
7. On this worksheet are two charts that will compare the results of the three-point estimate and PERT. Hover your mouse over the first bar that represents Paul Samms in the "3 Point Estimate" chart. What is the value that's displayed?
   Your answer: The value is 30.
8. Now hover your mouse over the first bar that represents Paul Samms in the "PERT Estimate" chart. What is the value that's displayed?
   Your answer: The value is 29.83.
9. Why is there a difference between the PERT value and the three-point value?
   Your answer: PERT uses a weighted average toward the most likely estimate, while the three-point estimate is an average of the optimistic, pessimistic, and most likely estimates.

Questions

1. You are the project manager for your organization. Management has asked that you create a detailed cost estimate for a new solution they'd like you to implement. What type of project estimating must account for every expense within a project before the work begins?

   A. Bottom-up estimating

   B. Top-down estimating

   C. Zero-based budgeting

   D. Parametric estimating

2. You are the project manager of the JHN Project. You have estimated the project will cost $129 for each unit installed. There are 1,200 units on this project. What type of estimate is this?
   A. Bottom-up
   B. Top-down
   C. Analogous
   D. Parametric
3. Harry is the project manager for his organization. Midori, his manager, has asked that Harry create a bottom-up cost estimate for a new project. What is bottom-up cost estimating?
   A. Using last year's budget plus 20 percent to equal the current year's budget
   B. Using this year's budget with a 20 percent plus or minus shift in the bottom line
   C. The process of working toward a zero balance as the bottom line in a budget
   D. The process of creating a detailed estimate for each work package in a WBS
4. You are the project manager for your organization. Your current project is similar to a project you finished recently for your organization. You decide to base the current project cost estimate on the previous project. What type of cost estimating are you using in this instance?
   A. Analogous
   B. Parametric
   C. Definitive
   D. Rough order of magnitude
5. What should a project manager do to predict the total cost of an IT implementation project accurately?
   A. List all of the expenses and add them up using best- and worst-case scenarios for each expense.
   B. List all of the expenses, including labor, and add them up using an average-case scenario for each expense.
   C. Divide the project into phases and assign a dollar amount to each phase.
   D. Divide the project into phases and estimate a dollar amount for each milestone within a phase.

6. What is a fully burdened workload?
   A. It is when an employee has reached their maximum number of hours allotted for any given project.
   B. It is when a consultant has reached their maximum number of hours allotted for billable time for a project or task within the project.
   C. It is the prediction of the number of hours required by staff to complete each phase of the project.
   D. It is the record of the number of hours required by staff to complete each phase of the project.
7. Why should an IT project manager use most likely, optimistic, and pessimistic scenarios when calculating the time required for a task?
   A. Some staff members will take longer than other staff members to do the same type of work.
   B. Each staff member will have a dollar amount assigned to the work hour. The optimistic and pessimistic scenarios can predict which staff member is the most valuable.
   C. The most likely, optimistic, and pessimistic scenarios allow an IT project manager to predict the average time expense required to complete a task.
   D. The most likely, optimistic, and pessimistic scenarios allow an IT project manager to predict the average amount of labor required to complete a task.
8. Reynaldo is the project manager for his organization and is completing a project for a customer. He and his project team have completed the project scope, and they could turn the project deliverables over to the customer and close the project, but they still have $12,500 in the project budget. Reynaldo decides to add some features to the software that would make the software better and to consume the remaining budget. This scenario is an example of which term?
   A. Scope creep
   B. Value-added change control
   C. Configuration management
   D. Gold plating
9. You are the project manager for your organization, and you'll need to procure servers and workstations from the vendor. You'd also like the vendor to install the operating system on each of the workstations based on an image you've created. In your procurement process, you've specified that the vendor must provide a fixed quote for the work. What is a primary advantage of an IT project manager requiring a vendor to deliver a fixed quote?
   A. It locks the vendor into the project.
   B. It prevents the vendor from adding features to the implementation.
   C. It allows the project manager to use the quote for up to one year.
   D. It allows the project manager to incorporate the quote into a proposed budget.

10. You are the project manager of an adaptive project. Your customer has requested several changes to the product backlog that they say are of top priority. You and the product manager agree and the product backlog is adjusted accordingly. Now, however, the project has insufficient funds to do all of the requirements in the product backlog. What should you do next?

 A. Refuse the changes to the product backlog that are causing the funds to be depleted.

 B. Ask the customer for additional funds.

 C. Submit a change request for additional labor to work on the new project requirements.

 D. Nothing. The most valuable items are completed first.

11. You are the project manager for your organization. Management has asked that you use PERT for estimating the time it takes to complete the project. Using PERT, what time would you record for an activity that has an optimistic time of 25 hours, a most likely time of 44 hours, and a pessimistic time of 76 hours?

 A. 44 hours

 B. 46 hours

 C. 48 hours

 D. 76 hours

12. What is the difference between a three-point estimate and a PERT estimate?

 A. PERT uses a weighted average of the most likely, whereas the three-point estimate is really just an average of all three estimates.

 B. PERT uses an average model to predict the most likely duration for time or costs, whereas a three-point estimate uses a straight average for the duration or costs.

 C. PERT and three-point estimates are actually the same model.

 D. PERT is used for time and cost estimates, whereas three-point estimates are used only for time.

13. You are the project manager for your organization and are trying to determine if you should buy or build a solution. You've determined that your project team could build the solution for $23,500 and the monthly maintenance on the solution would be $1,200. A vendor reports that they could build the solution for $17,000 but their monthly maintenance fee would be $1,800. Should you buy or build this solution?

 A. Build the solution if you'll keep it for less than 11 months.

 B. Buy the solution if you'll keep it for more than 11 months.

 C. Buy the solution if you'll keep it for less than 11 months.

 D. Build the solution if you'll replace it in less than 11 months.

14. You are the project manager of a project to install 8,500 workstations in an organization. You have estimated that the optimistic cost of the labor for each workstation will be $99, the most likely cost will be $149, and the worst-case cost for each workstation will be $225. What cost will you record for your cost estimate per workstation?

    A. $149

    B. $225

    C. $158

    D. It depends on the person doing the installation

15. Which one of the following is an example of an estimate using a parametric model?

    A. The current project is similar to the past project, so the project manager will base the current estimate on the previous project.

    B. Each item in the WBS must be cost-estimated to create the project cost estimate.

    C. There are 7,100 network drops, and each drop will cost $175.

    D. The organization is charged a set fee for all the connections to a server.

Answers

1. **A.** Bottom-up estimating requires the project manager to account for all expenses within the project to arrive at a grand total for the project.

2. **D.** This is an example of a parametric estimate. The units will cost $129 each; this is the parameter. As there are 1,200 units on the project, the estimate is calculated by multiplying the parameter of $129 by the total number of units needed, 1,200, for an estimate of $154,800.

3. **D.** Bottom-up estimating is a process that requires the project manager to create a detailed estimate for each work package in the WBS. Every work package can be cost estimated, and it's the sum of the work packages' cost estimates that help determine the bottom-up estimate.

4. **A.** This is an example of analogous estimating. By creating an analogy between the current project and the previous project, the project manager can predict current costs. To use an analogous estimate, you must have accurate historical information.

5. **C.** The project manager should not create one grand total for a project. In order for the project manager to see a true picture of the work, they should segment the project into phases and assign each phase a dollar amount based on the work to be completed within it.

6. **C.** A fully burdened workload is the prediction of the actual hours required by the team members to complete a given project. This process allows the project manager to predict the financial obligations corresponding to time and create a sense of urgency as to when each task must be completed.

7. **C.** Using the most likely, optimistic, and pessimistic scenarios allows a project manager to predict the average amount of time the team member requires to complete a task. The project manager uses this value to assign a dollar amount to the work to be completed.
8. **D.** This is an example of gold plating. Gold plating happens when the project manager or team adds deliverables that were not in the project scope in an effort to consume all of the project budget.
9. **D.** A fixed quote allows the project manager to use that dollar amount in a budget to predict the funds required to complete a project. It can also be used to determine which vendor will actually be awarded the job based on the price and hours to complete the work.
10. **D.** The product backlog is always prioritized so that the items with the most importance and the most business value are completed first in the project. Requirements that are of less value shift downward in the product backlog and may fall out of the possibilities for completion. These additional items may be moved into a new project, such as a future release.
11. **B.** The Program Evaluation Review Technique (PERT) is a time-estimating formula that accounts for the optimistic, pessimistic, and most likely estimates. The formula is P + O + (4ML) / 6. In this example, the closest result is 46 hours.
12. **A.** PERT uses a weighted average to predict the cost or time of an activity. It's considered weighted because the formula is P + O + (4ML) / 6. The three-point estimate is just an average of the pessimistic, optimistic, and most likely estimates.
13. **C.** To determine whether to build or buy, first find the difference between the in-house solution and the vendor's solution. In this example the difference is $6,500 more to build in-house. Next, find the difference of the monthly support for each solution, which is $600 more a month for vendor support. Divide $6,500 by $600, which tells you how long you'll need to use the in-house solution to equate to the cost of the vendor's solution. In this example, it's nearly 11 months, which means you should buy the solution if you'll keep it for less than 11 months (or build the solution if you'll keep it for more than 11 months).
14. **C.** If you use the three-point estimate approach in this example, you will calculate the average of the pessimistic, optimistic, and most likely estimates and record the cost estimate as $158 per workstation.
15. **C.** A parametric model is a value that can serve as a parameter to determine the project costs. In this example, the 7,100 network drops at $175 each is an example of a parametric model.

CHAPTER 6

Building the Project Plan

This chapter covers the following topics:

- Creating the project management plan
- Building the project schedule
- Designing a project network diagram
- Planning for project costs
- Writing the staffing management plan
- Assuring and controlling project quality
- Communicating with project stakeholders
- Analyzing project risks
- Implementing the procurement management plan

Remember the story of Noah and the ark? When did he build that ship? Aha! He built it *before* the rain started. That's the same idea with all of this planning you're doing for each of your projects. By effectively planning, analyzing, and examining your plan from different perspectives, you increase your chances of completing your project on time and on budget.

At this point of your project, you've already done a great deal of work. Look back at all you've accomplished: you've created a vision, researched the technology, partnered with management, created a budget, made a WBS, and assembled your team. Phew! That's a ton of progress, and the actual implementation of the project work hasn't started yet. Don't get discouraged—these activities that you've been doing are the building blocks of a strong foundation for the success of your project. Without all of those activities, your project would be doomed.

This chapter focuses on bringing your plan together. If you've done all of the preliminary work described in the earlier chapters, your plan is already coming together—which is grand. However, there are still a few more things that require your attention before you set your plan into motion. Particularly, you need to give some thought to the project schedule, the risk, and the procurement activities defined in your plan and determine if everything is reasonable. This is crucial for any project. What looks good on paper doesn't always work well in execution.

Building the Project Plans

Depending on the size of your project, your project plans will vary. The project plan is not one big plan, but rather a collection of subsidiary plans that detail how different conditions, scenarios, and actions will be managed. It is a formal document that is reviewed and, hopefully, approved by management. The project plan is not a novel that tells the story of how the project will move along, but rather a guide that allows for changes to the project plan as more details become available.

In a predictive project, rather than adaptive projects, the project plan may evolve, but there are some elements within the project that generally do not change—or are protected from change. Of course, the foundation of the project is the project scope. Recall that the project scope is all of the required work—and only the required work—to complete the project objectives. The project scope statement defines what the project will and won't accomplish. Once the project scope statement has been agreed upon, your change control system protects it.

Other elements of the project plan that should be immune from change are the project charter and the performance baselines. The project charter authorizes the project. It is a formal document that allows the project manager to manage the project work, resources, and schedule to deliver on the project scope. Performance baselines are time, cost, and scope objectives that the project manager must meet within the project delivery. These baselines rarely change unless an approved change happens in the project, and then these baselines are updated to reflect the approved change. In other words, you're supposed to have enough time and budget to meet the requirements of the project scope.

Addressing Project Factors

As you begin to plan out the project in both predictive and agile approaches, you need to address the project factors that will affect how the project operates. These factors are the environmental, social, and governance conditions and can have great influence over the project approach. You may need to work with your project sponsor or product owner to address these concerns. You don't want to ignore these factors; if you don't have a clear plan of attack and strategy on how these factors should be managed, they'll haunt your project throughout.

Environmental factors describe where the project work will take place. Consider how a project can affect the water, soil, air quality, and community and consider the ethical and regulatory requirements. For example, there are regulations on disposing computer equipment. You don't want a project to disrupt or spoil the environment for ethical reasons, but also consider the goodwill of the organization, and legal requirements you'll need to adhere to. Environmental conditions can also mean the organizational environment—such as how your project may disrupt the business, the day-to-day operations of the people working in the organization, and their attitude to the project and the change it ultimately brings to the organization. This is where politics can creep into a project, something that is inevitable in almost all projects, so you'll need a strategy to manage perceptions of your project and how it helps, or hinders, the people in the organization.

Social factors are directly related to the politics of the organization as well. It's no secret that there will be forces for and against your project. There will be power plays, pet requirements, and attitudes you'll have to contend with as your project marches towards its vision. Social factors also include how society, your organization's customers, and community view your project and what it aims to accomplish. Some projects target social good, such as setting up free Wi-Fi access in a park or an improvement in wastewater management, for example. It's always important to ask how the project is viewed by the positive and negative stakeholders and how you can improve upon their perception of the project.

Governance is about following the rules of the organization and regulations specific to your organization's industry. All organizations have their own way of conducting business, of getting things done. Consider how you request IT resources, how you hire contractors, or how your team schedules vacation time. There are rules or best practices within your organization to do these and other tasks that will be entirely different from how other organizations accomplish these same common activities. You want to know the rules, the governance, and work within those boundaries. Your project may also need to follow government regulations that are specific to your industry or the type of project work you're doing. These regulations are requirements, and you must stay in compliance or there will likely be fines, penalties, and even a pause in your project execution.

Planning for Security and Privacy

Project planning, especially in IT projects, also entails addressing the security and privacy requirements of the project. The project manager, product owner or project sponsor, key stakeholders, and the project team must consider the need for privacy and security of the data they'll be working with in the project. IT projects often encompass data that should remain private and secure, so you'll need a policy and strategy for all team members for accessing and working with such data in the project. Many organizations require operational security, such as background screening and security clearance requirements.

Your IT project plan will likely need to address the following privacy and security considerations:

- Data confidentiality and security
- Remote access
- Authentication protocols
- Regulatory requirements
- Physical security of equipment and technology
- Intellectual property and trade secrets

Security and safeguarding data and information is always an early consideration in project planning. You'll need to know your organization's security policies and restrictions in order to plan effectively. You may have to work not only with your IT department but also human resources and maybe even lawyers or a legal department to understand the security requirements and approaches you're allowed to take to remain in compliance. Throughout the project you'll need to examine the security approaches to ensure that the team remains in compliance.

Creating Project Plan Documents

When you and your project team create the elements of the project plan for a predictive project, you can start from scratch and build your plan, or you can rely on historical information to lend a hand. Many times project managers will find that their projects are similar to past projects they've completed. Rather than reinventing the project management wheel, they'll rely on past project plans to serve as templates for their current projects. There's nothing wrong with this approach at all—it's just working smart, not hard. Of course, when you use older plans as templates, you'll update the older plans to reflect your current project.

Regardless of which approach you take to building your project plan, there are some common project management documents you should include in your comprehensive plan:

- **Project charter** This document comes from someone in a supervisory position that is higher in the organizational chart than the immediate management of the project team. This document authorizes the project. All projects need a project charter.
- **Scope baseline** The project scope baseline is actually a combination of three documents: the project scope statement, the work breakdown structure (WBS), and the WBS dictionary. The scope baseline is used throughout the project when there are questions about the project requirements, execution concerns, risk management, cost estimating, issues, and other project management activities. You'll use the scope baseline as a guideline for all future project decisions.
- **Time and cost estimates for each work package** Recall that time and cost estimates reflect the labor and materials needed to deliver the project. This portion of the project plan will also detail how the estimate was derived, the degree of confidence in the estimate, and any assumptions associated with the estimates.
- **Performance measurement baselines** These baselines are boundaries or targets the project manager and the project team are expected to perform within. For example, the cost baseline may predict the amount of budget that should be spent by a given milestone, with an allowable variance. Time and cost are the most common baselines that you'll measure your project performance against.
- **Milestones and target dates for the milestones** Within your project, there should be easily identified milestones that signal you are moving toward project completion. Associated with these major milestones are some target dates that you and management agree on. This allows you and management to plan on resource use, consider adjunct processes within your business, and keep all stakeholders informed of where the project should be heading—and when.
- **Resource requirements** Resources are people, materials, tools, equipment, facilities, and services that you'll need to get your project done. There may be portions of your project plan that require procured resources or temporary specialized resources to complete. The required personnel, materials, and services should be identified, their availability determined, and their associated costs documented.

- **Issue log** Issues are risks that have occurred in the project. This doesn't mean that the risk was identified; an issue can happen because a risk was overlooked during planning. For example, if a team member quits the project, that's an issue; it was always a risk that people on the team could leave during the project execution. There will often be open issues and pending decisions as the plan is first created. This section of the plan identifies and documents the issues to be determined and allows the project to continue. Of course, the decisions and issues in this section of the project plan should be addressed accordingly, which may cause other areas of the project plan to be updated. In an adaptive project, issues are known as impediments or blockers and the project manager, coach, or scrum master is responsible for resolving those issues.
- **Assumptions log** An assumption is anything that you believe to be true but haven't yet proved to be true. In IT, you often have to make assumptions in your planning: the software and hardware will work together, the new hardware drivers will work with the existing operating system, the learning curve of the software isn't too drastic, and more. When you complete risk management planning, you'll test these assumptions to prove your theories true or false. Assumptions that prove false can become risks in the project.
- **Quality requirements** Quality is an esoteric concept—what's fast, good, and super to you may not be fast, good, or super to me. To have quality in an IT project, you must define the quality metrics, provide quantitative measurements that will equate to quality, and explain to the project stakeholders what the measurements mean and how they'll map to the requirements for quality acceptance within the project. Your organization may also subscribe to quality programs, such as Six Sigma or ISO programs. You'll need to include the project requirements for quality in these management-driven programs.
- **Calendars** Your project management plan needs two calendars: a resource calendar and a project calendar. The resource calendar defines when the project resources are needed, when the resources are available, and when the resources will be used in your project. Remember that resources are more than people—materials, facilities, and services are resources, too. The project calendar defines when the project work will take place. You'll define the working hours of the project, company holidays, and busy seasons when the project work may, or may not, take place.
- **Risk register** The risk register, sometimes called the risk log, is a document that defines each risk, its probability, impact, and overall risk score, and an issue owner. The risk register should also reference or document the defined risk response for each risk event. Some projects may also include a risk owner that owns the risk and is empowered to enact the planned risk response should a risk event come into fruition. All risks are updated, tracked, and documented if their status or outcome changes.

- **Stakeholder management strategy** Your project will likely have negative, neutral, and positive stakeholders. The stakeholder management strategy defines how you'll bolster support for your project, fend off the negative stakeholders, alleviate fears and concerns, and promote the support for your project. You'll use the stakeholder management strategy along with the project's communications management plan.
- **Supporting details** The supporting details are any relevant documentation that influenced your project decisions, any technical documentation, and any relevant standards the project will operate under.

These project documents will be unique for each project, and you might need to adjust some of the documents as the project moves forward. Changes to the project scope, for example, will cause changes to the cost and schedule baselines, but not to the project charter. You'll use these documents to help you refine your project plans, but also to help you in the project execution. The documents, plans, reports, and other communications that come out of project planning and project management are called *artifacts*. Project artifacts should be archived as part of the project. Typically, a project team member is responsible for these artifacts and their organization, location, and access.

Creating the Project Scope Management Plan

The *project scope management plan* defines how the project scope in a predictive project will be defined by the project management team in terms of the project requirements, how the scope will be monitored and controlled throughout the project, and how the scope will be verified by the project customers. The project scope is all of the required work for the project to be considered complete, and it's based on the requirements of the project stakeholders. This project plan focuses on the requirements and benefits of the project—things that are in scope. It also defines, in the project scope statement, the things that are considered out of scope.

This plan directs your creation of the project scope statement, the WBS, the WBS dictionary or the product backlog in an agile environment. Recall that these three elements comprise the project scope baseline. Once you've created the project scope statement and your stakeholders have approved the scope, you'll decompose the scope into the work packages. Each work package, as I'm sure you remember, can be estimated for time and cost within your project. Along with the WBS, you'll create a companion document—the WBS dictionary. This document defines what the work packages are, what resources are needed for the work packages, and the time and cost estimates for each work package.

The project scope management plan also defines how the scope might be allowed to change. If you've ever managed a project before, you know that stakeholders change their minds—often. The scope's change control system is defined in the project scope management plan—you don't want (or need) change requests fluttering into your inbox on a daily basis. In theory, if you've accurately collected the project requirements, there should be few to no change requests. In practice, however, stakeholders aren't as thorough as you are in collecting and offering requirements. Generally, change requests are needed for the project scope for four reasons:

- **Errors and omissions** Oops! The stakeholders or the project manager goofed and didn't accurately include all of the requirements in the project scope. When there are errors or omissions in the project scope, the scope must be revised to reflect the newly added requirements.
- **Value-added changes** Sometimes there is a great reason to add elements to the project scope. The additions can be profitable for the organization, make the deliverable better, or create benefits that are low in cost and high in value. Value-added changes are documented and folded into the project scope.
- **External events** There's a world outside of your project that doesn't care about your project. New releases from vendors, new laws and regulations, and competitors can all change your project scope. External events are a common element that can change the direction of your project scope.
- **Risk events** A risk is an uncertain event or condition that can have a positive or negative effect on your project. Some of your project work may be too risky, so it might be carved out of the project scope. Other events, such as a discount from a vendor if you add a few elements, is a good deal, so you might add the new elements to the project scope.

When a change request is introduced, the change must pass through the project's integrated change control system, which is part of the project's project management information system. Change requests must be documented and explained as to why the change is warranted. Undocumented change requests are not implemented in the project—get the change request in writing, define who started the change request, and explain why the change might be needed. Once the scope change request has been documented, it enters the scope change control system. Technically, as you can see in Figure 6-1, there are four change control systems: scope, schedule, costs, and contract.

In an adaptive project the product owner acts as the change control board. The product owner evaluates each change for its business value to the product and determines if it should be added to the product backlog. If the change is added to the product backlog, it is also prioritized in the backlog—it doesn't automatically go to the bottom or top of the list.

A scope change request passes through the *configuration management system,* which examines the scope change request's effect on the features and functions of the project's product. For example, a scope change request could be to add reporting features to a piece of software. The reporting features would affect the features and functions of the project's product, so the change would be evaluated, weighed for value, and then documented as part of the product scope, if approved.

EXAM TIP Adaptive projects welcome change and don't follow a change control process like waterfall projects do. Adaptive projects welcome and expect change, while predictive projects are averse to change.

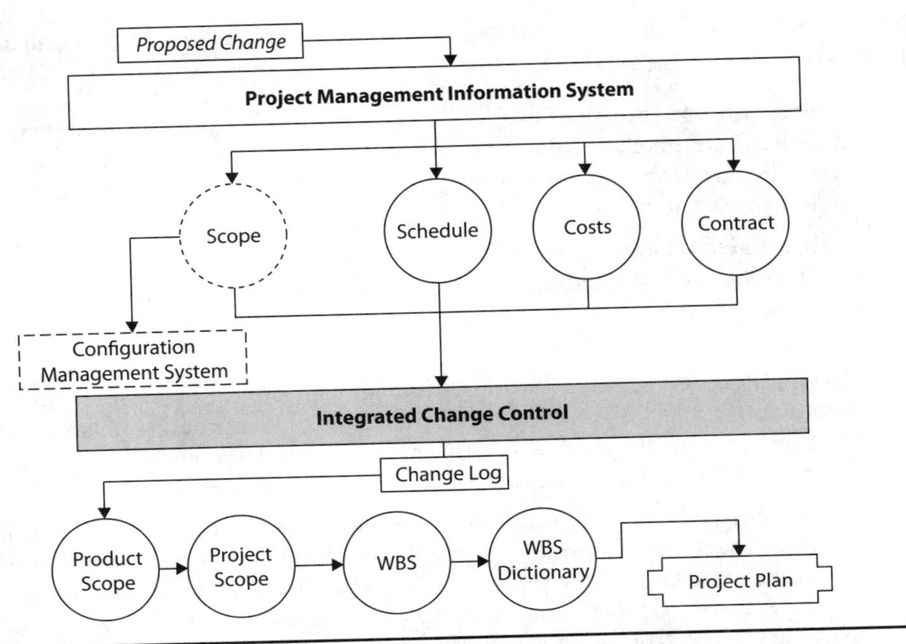

Figure 6-1 An approved change request has an effect on the entire project.

All change requests, scope or otherwise, pass through *integrated change control* to assess the effect of the change on the entire project. A scope change request could affect the project costs, schedule, quality, human resources, communications, risk, and even procurement. Integrated change control is the requirement that a proposed change request be examined for its true impact on the project. Once the change request is determined to be accepted or rejected, the result is communicated to the appropriate stakeholders and documented in the change log. Yes, even changes that aren't approved go into the change log as a record of what happened to the change and why it was rejected.

A change to the project scope can affect the product scope statement, the project scope, the WBS, and the WBS dictionary; it can even have ripples into the project plan. All of these documents and affected plans need to be updated to reflect the approved change. When a change enters the project scope and it's approved, then you'll also need the time and monies to reflect the approved change. Approved changes to the project's scope and budget will cause the corresponding baselines to be updated, too.

Changes that happen early in the project usually don't cost as much as changes that are requested later in the project. It's easier to change the project plan and its documents than to change the physical work that has already been completed in the project. The physical work of the project and the results of project execution are also affected by this scope management plan. The work results must be examined by the project stakeholders to verify that the results are acceptable to them. This examination is called *scope verification,* and the project's scope management plan defines when and how the verification will happen.

Scope verification usually happens at the end of project phases and always at the end of the project. The goal of scope verification is for the project customer to accept the deliverables that the project team has created. Scope verification is an inspection-driven process that the customers, key stakeholders, or project champions do with the project team and project manager. If there are mistakes or errors in the inspection, the project manager can create a "punch list" of things that must be corrected before the project deliverables are accepted by the project customer.

Defining the Project Schedule Management Plan

Your predictive project must have a definite set of deliverables that mark its end. Projects also require a finish date. Some projects' finish dates are a touch firmer than others—for example, building a football stadium before the season begins. Or consider a project that management says must be completed before a peak business period. Other projects, such as the release of a new e-mail program within an organization, can tend to go on forever and evolve into runaway projects.

Runaway projects stem from a loosely guarded project scope, poor planning, and lack of research. Of course, the longer a project takes to produce its deliverables, the more the project will cost. In addition, to make it personal, a missed deadline can impact bonuses, incentives, and raises for project managers and team members. The best way to reach a target date for completion is to plan, plan, plan. And then analyze the plan. And then adjust and readjust the plan until it is acceptable and the team is ready to implement the technology. The *project schedule management plan* defines several things for the project manager and the project team:

- What project activities need to be completed
- An estimate of how long the project activities will take to complete
- When the project activities will need to happen and in what order
- What resources are needed for the project activities and when the resources will be needed
- How the project schedule can be adjusted, compressed, analyzed, or manipulated for the best possible outcome for the project
- How the schedule will be monitored and controlled
- A cadence to the project communications, regularly scheduled events, and communications

A project schedule should be a reflection of the WBS, the accumulation of all of the work packages within the project, and the assignment of resources for each task. Most new project managers work around specific target dates for milestones, phases, and a completed project. This makes the most sense, right? IT professionals are used to working off a specific calendar for so much of their lives that this next concept can be a little confusing at first: if possible, do not schedule project tasks to happen on a specific date.

Monday	Tuesday	Wednesday	Thursday	Friday
3 Task 22: 3 Days	4	5	6	7
10	11 or	12 Task 22: 3 Days	13	14
17	18	19	20	21

Figure 6-2 Assign tasks to be completed in units of time rather than by specific dates.

Project managers should not work around specific dates when creating the project plan, but instead should initially work around units of time—for example, one day, two weeks, three months, and so on. Rather than saying a specific work unit will take place next Thursday and Friday, it's better to say that a specific task will take two days to complete. Why? Isn't next Thursday and Friday the same as two days? Yes, and no. Assigning two days to complete a task rather than two *specific* days allows you to move the task around within your project plan. Figure 6-2 demonstrates the concept of working in units of time rather than specific dates. This little trick allows for a process you'll learn more about later in this chapter: project compression.

Working with units of time rather than specific dates for each of the tasks within your project plan allows you to tally your plan to a specific amount of time—regardless of when the actual project is implemented. For example, if tasks within your project are all assigned a deadline based on the project start date of July 9 and end date of November 2, each task is very time constrained and date specific. However, this same project takes 90 workdays (depending on the year and allowing for weekends). When you assign tasks units of time within the span of 90 workdays, regardless of when the actual start date commences, the project can shift 90 days into the future from the start date.

Often, however, project schedules and deadlines are determined before the project even begins. We've all been there, right? We're handed a project to create an application that must be delivered by an unrealistic date. There hasn't been real reflection on the needed time to create a quality application by the given date. In these instances, the project manager still needs to address the project, the work decomposition, and the assignment of resources to complete the work, just as with a project where the end date is not known.

EXAM TIP Predictive projects utilize a work breakdown structure to show the components of the project scope. Adaptive projects utilize a product backlog to show all of the project requirements and features.

During project planning, you'll create the WBS based on the project scope. Once you've created the WBS, you can enter the activity list into Microsoft Project or your favorite project management information system (PMIS) software. Once you enter the

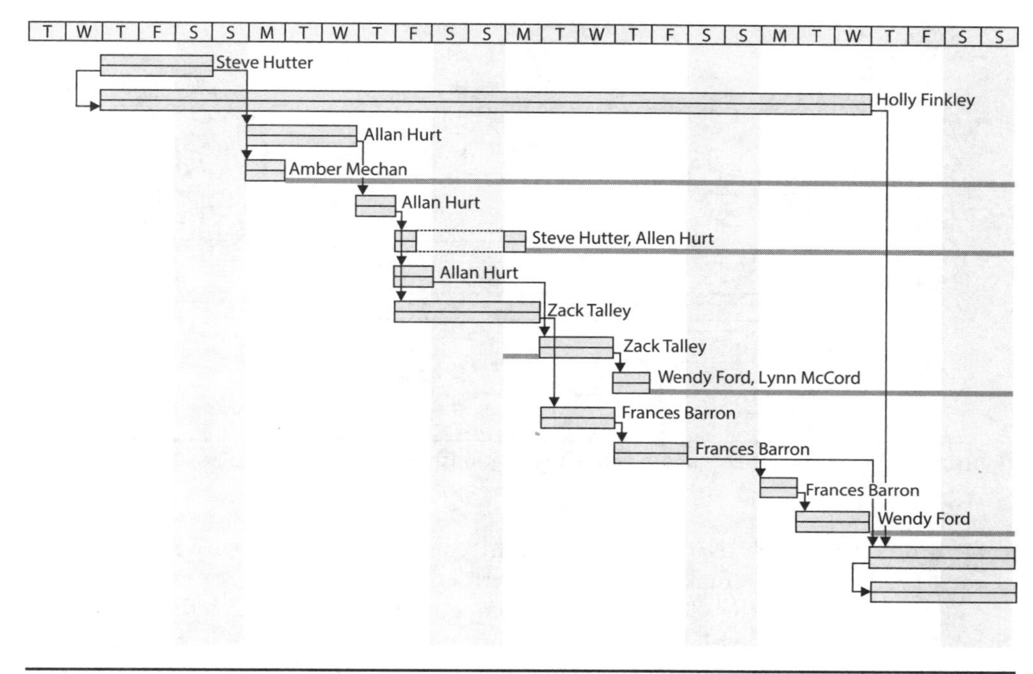

Figure 6-3 A Gantt chart maps activities to a project calendar.

activities into Microsoft Project, you can create a Gantt chart like the example shown in Figure 6-3. It shows a mapping of each of the units of work required to complete each phase of the project.

The Gantt chart is ideal for simple, short-term projects. It is a timeline of the events with consideration given to tasks that can be completed concurrently within a project's life span. Traditional Gantt charts have some drawbacks:

- Gantt charts do not display detailed information on each work unit. (Microsoft Project does allow project managers to add task information and notes within a Gantt chart on each task.)
- Gantt charts display only the order of tasks.
- Gantt charts do not clearly reflect the order of tasks in multiple phases.
- Gantt charts do not reflect the shortest path to completion.
- Gantt charts do not reflect the best usage of resources.

To address these issues, project managers can use a project network diagram (PND). PNDs are fluid mappings of the work to be completed. (Figure 6-4 shows a sample of a portion of a network diagram.) Incidentally, the terms "PND," "project network diagram," and "network diagram" all refer to the same workflow structure—don't let the different names confuse you. Such diagrams enable the project manager and the project

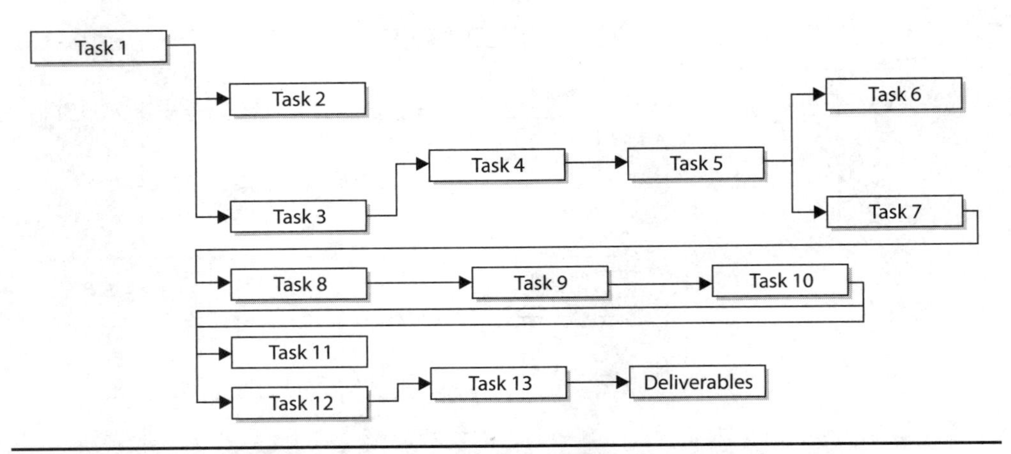

Figure 6-4 Network diagrams demonstrate the relationships between tasks.

team to tinker with the relationships between tasks and create alternative solutions to increase productivity, profitability, and the control of a project.

A PND visualizes the flow of work from conception to completion. Network diagrams provide detailed information on work units and enable project managers to analyze tasks, resources, and the allotted time for each task. You can use a PND to determine the flow of work to predict the earliest completion date. Network diagrams are ideal for these situations:

- **Detailed project planning** In large projects that may span several months, or even years, a network diagram is essential, as it can correlate each task in relation to the project scope. Through a network diagram, the project manager, management, and the project team can see the entire project plan from a high-level view and then zoom in on a specific portion of the project plan.
- **Implementation tracking** As tasks are completed on time, or over time, the number of time units used can accurately display the impact on dependent tasks within the project. If you use software to track the project implementation, the reflection of the impact is automated for you. Imagine a task that has four dependencies and is two weeks late in completion. The failure of the task to be completed on time now pushes the dependent tasks back by two weeks. A network diagram can illustrate this impact and allow the project manager to react to the changes by adjusting resources or other dependent tasks.
- **Contingency plans** Network diagrams enable a project manager to play out "what if?" scenarios with any work unit within the project plan. A project manager can adjust units of time to see the impact of the work units on the entire project. For example, it may be obvious to see an impact on dependent tasks when a work unit is two weeks late, but what about units that are completed early? Imagine that pay incentives are based on project completion dates—a series of work units that each have one day shaved off of their target completion times may have positive impacts on all future tasks.

- **Resource control** A network diagram shows the flow of work and the impact of the finished tasks on the rest of the project. By using the Gantt chart's assigned resources to a unit of work, a project manager can add or remove resources to a task to complete it faster or delay the completion. Resources can be both workers and physical objects such as routers, faster computers, and leased equipment.

Using the Precedence Diagramming Method

The most common method of creating a network diagram is the *precedence diagramming method (PDM)*. PDM requires the project manager to evaluate each project activity and determine the order in which the activities should be completed. You'll determine which activities are *successors* and which are *predecessors*. In other words, each activity in your project has some activities that must come before it in a particular order and activities that will come after the current activity, also in a particular order. Once this information is obtained, you can begin to snap the pieces of the PDM puzzle together. You must give careful consideration to the placement of each activity, as all activities are connected even if they are scheduled to run concurrently.

Each unit of work in a network diagram using PDM is represented by a rectangle called an *activity node*. Predecessors are linked to successors by lines and the flow of the work is always left to right, as Figure 6-5 demonstrates. One work unit can be both a successor and a predecessor. For example, the task to install and configure a web server may first require the installation of an operating system; only after the web server task is completed can the web pages the web server will host be added.

To begin creating the network diagram using PDM, a project manager and project team have to determine the order of tasks to be completed. Basically, a project manager asks, "What tasks must be completed before the next tasks can begin?" The activity list, which you derived from your WBS work packages, is what you are sequencing. You do not want to include activities that are too short in length to be useful. Based on your project, you may have work units that last days or weeks—generally not hours.

As you add tasks to the network diagram, draw a line between tasks to connect them in the successor/predecessor relationship. As expected, the network diagram is read left to right, top to bottom. Most tasks within the diagram will have a successor and predecessor—except for the first and last tasks in the project. Once the diagram has been completed with successor and predecessor tasks, it is considered a connected network. A project manager, or anyone involved in the project, can trace any activity through the network.

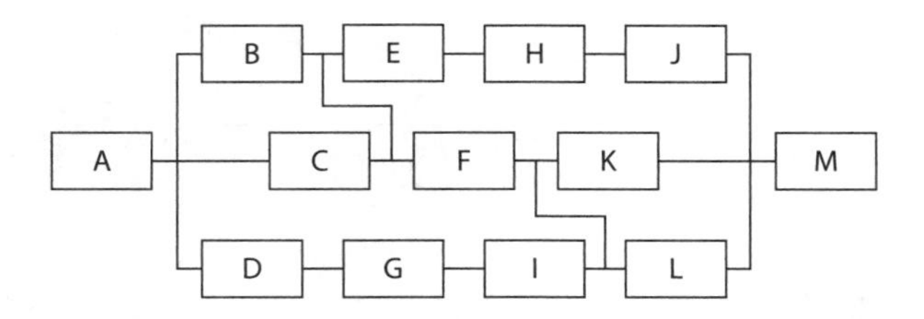

Figure 6-5 Network diagrams flow from left to right and connect predecessor and successor tasks.

The relationship between activities is what allows a project to move forward or to wait for other tasks to complete. An accurate description between tasks is required for the project manager to analyze and adjust each task in the network diagram. Figure 6-6 shows the dependency types you may have between tasks.

There are four types of dependencies:

- **Finish-to-start (FS)** This relationship is the most common. It simply requires the predecessor task to complete before the successor task can begin. An example is installing the network cards before connecting PCs to the Internet.
- **Start-to-start (SS)** These tasks are usually closely related in nature and should be started, but not necessarily completed, at the same time. An example is planning for the physical implementation of a network and determining each network's IP addressing configurations. All tasks are closely related and can be done in tandem.
- **Finish-to-finish (FF)** These tasks require that the predecessor task and the successor task be completed at nearly the same time. In this relationship the successor tasks cannot be completed before the predecessor task is completed. An example is rolling out a new software package and finishing the user training sessions. While users are in the new training session, the new software should be installed and configured on their workstations by the time the training session ends.
- **Start-to-finish (SF)** These rare tasks require that the predecessor not begin until the successor finishes. You won't find yourself using this activity all that often. It's often used with just-in-time scheduling for inventory and manufacturing instances. For example, if you're opening a coffee shop, you'll want your inventory of coffee to be on hand as close as possible to when the construction of your shop is done. The coffee has to be ordered to officially open the shop, and the construction has to be done before you can open the doors. You'll need to time the order of coffee with the completion of the construction in order to open your doors on time without your merchandise spoiling—or tying up your cash flow in inventory that's not being sold.

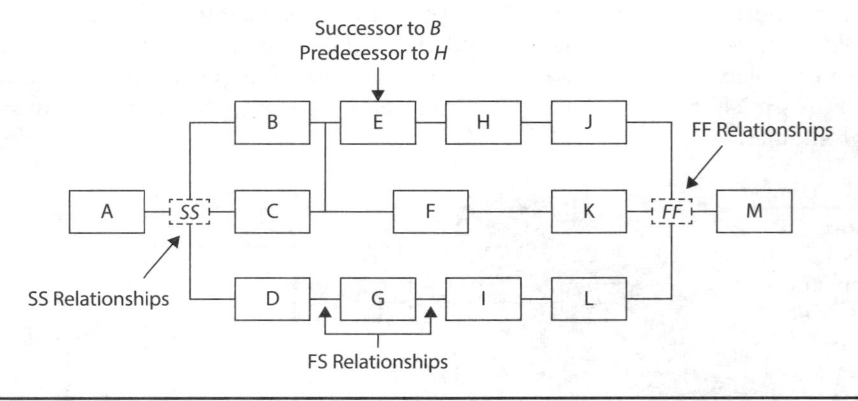

Figure 6-6 Dependencies describe the relationship between tasks.

NOTE Understanding the concept of the network diagrams helps with scheduling, but few project managers manually draw their network diagrams. Project management information systems, such as Microsoft Project, Basecamp, and other solutions, help sequence the flow of the work and can quickly expose bottlenecks, constraints, float, and other characteristics of your schedule.

Managing Logical Constraints

A *constraint* is a boundary or limit based on the project. You've dealt with constraints before: a preset budget for your project, an inflexible deadline, limited availability of computer hardware, or locating a resource with a specific skill. Constraints are any factors that can limit your options. The project manager and team must document constraints and their risks examined, and then the project manager must plan on how to meet the project objectives within the identified constraints.

When it comes to scheduling activities, you can also create constraints on the relationships you assign between your activities. For example, an FS relationship is constrained by the completion of the predecessor before the successor can begin. This is a natural constraint. This relationship between activities is sometimes called hard logic. *Hard logic* (aka *mandatory logic*) describes the matter-of-fact order of activities. For example, you must install the operating system before you install the application. On the other hand, *soft logic* is when the project manager decides to do tasks in a particular order based on experience, conditions in the project, time, or other reasons. This logic is also called *discretionary logic*. For example, it is a good "old school" practice to have completed all the coding before beginning the testing phase. It is not mandatory—you can unit-test certain modules that are complete before all the coding is done—but it is preferred to have all the coding complete before any testing begins.

In some conditions, you might want to schedule a task to wait before starting, and that's called lag time. *Lag time* means that you're adding time to the activity's start date. The opposite of this is called *lead time,* when you're moving the activity's start date backward to allow it to overlap with other activities. Lag time is always positive time or waiting time. Lead time is always negative time and often overlaps with other activities.

Often in project management, projects have preset deadlines that require project managers to work backward from the assigned completion date. The problem is that the person establishing the deadline may not realize the work required to complete a project by that given date. Unfortunately, this is often the way project management works: you're assigned a deadline and then you have to figure out how to complete the tasks by that date.

Whenever possible, avoid using specific dates for tasks unless it is absolutely required. With date constraints, you are signifying that a certain task must happen on a specific date, regardless of the completion of tasks before or after it. The best method of assigning tasks is to use a unit of time and then predict when the task may happen based on the best- and worst-case scenarios for the predecessor tasks' completion.

There are three types of date constraints:

- **No earlier than** This constraint specifies that a task may happen any time after a specific date, but not earlier than the given date.
- **No later than** This constraint is deadline oriented. The task must be completed by this date—or else.
- **On this date** This constraint is the most time oriented. There is no margin for adjustment, as the task must be completed on this date, no sooner or later.

You can set these constraints on a task by using your project management software.

Management constraints are dependency relationships imposed because of a decision by management—this includes the project manager. For example, suppose a project manager is overseeing the development of a web-based learning management system. The website will allow students to register for classes, check grades, and pay for their tuition, all online. The e-commerce portion of the project and the database development portion of the project are scheduled to happen concurrently. Because of the unique relationship between the two tasks, the project manager decides to rearrange the work schedule so that the database portion of the project must finish first and then the e-commerce portion of the project may begin. The project manager accomplishes this by changing the relationship between the tasks from start-to-start to finish-to-start. Now the database task must be completed before the development of the e-commerce portion. This is another example of soft logic.

EXAM TIP Constraints limit your options. Constraints are things like time, cost, and scope, but can also include quality expectations, security rules, and resource availability.

There are actually eight types of constraints that you may use in your project schedule management plan:

- **As soon as possible (ASAP)** When you specify a task constraint of ASAP, you schedule the associated task to occur as soon as it can. This is the default for all new tasks when assigning tasks from the start date. This constraint is flexible.
- **As late as possible (ALAP)** When you have a task with this flexible constraint, you schedule the task to occur as late as possible without delaying dependent tasks. This is the default for all new tasks when scheduling tasks from the end date.
- **Start no earlier than (SNET)** A task assigned the SNET constraint starts on or after a specific date. This constraint is semiflexible.
- **Start no later than (SNLT)** This semiflexible constraint requires that a task begin by a specific date at the latest.
- **Finish no earlier than (FNET)** This constraint requires that a task be completed on or after a specified date. This constraint is semiflexible.

- **Finish no later than (FNLT)** This semiflexible constraint requires that a task be completed on or before this date.
- **Must start on (MSO)** A task with this constraint must begin on a specific date. This constraint is inflexible.
- **Must finish on (MFO)** This inflexible constraint indicates a deadline-oriented task. The task must be completed by a specific date.

Managing Technical Constraints

Technical constraints stem from FS relationships. Most often within an IT project, tasks will be logically sequential to get from the start to the end. These constraints are the simplest and most likely ones you'll find in a project. The technical constraints you may encounter when building your network diagram fall into two major categories:

- **Discretionary constraints** These constraints allow the project manager to change the relationship between activities based on educated guesses. Imagine two tasks that are scheduled to run concurrently. Task A, the design on the web interface, must finish, however, before Task B, the development of the web application, is well under development. Because of the cost associated with the programmer, the project manager changes the relationship between the tasks from SS to FS. Now the first task must finish before the second task begins.
- **Resource constraints** A project manager may elect to schedule two tasks as FS rather than SS because of a limitation of a particular resource. For example, if you are managing a project that requires a C++ programmer for each task and you have only one programmer, then you will not be able to use SS relationships. The sequential tasks that require the programmer's talents dictate that the relationship between tasks be FS.

Managing Organizational Constraints

Within your organization, multiple projects may be loosely related. The completion of another project may be a key milestone for your own project to continue. Should another project within your organization be lagging, it can impact your own project's success. For example, suppose that you work for a manufacturing company that is upgrading its software to track the warehouse inventory. Your project is to develop a web application that allows clients to query for specific parts your company manufactures. The success of your project requires the warehouse inventory project to be completed before your project can end. You enter these relationships into your network diagram as FF, with the origin activity representing the foreign project.

Scheduling with Gates

We all know that gates are things that swing open and allow you to pass through. In project management, there are also gates that are part of the project governance. A *governance gate* governs how the project is allowed to move from one phase of the project into the next phase of the project. A governance gate defines the conditions that must exist for the project to move forward—for example, forms completed, budget reconciled, and stakeholder approval of the work completed so far.

Another common gate is tied to quality: you can't move on—that is, you can't pass through the quality gate—until certain conditions are met. A *quality gate* requires an inspection of the work your team has completed to confirm that it meets the quality requirements of your organization, PMO, customer, or even laws and regulations. If the quality is sufficient, then the quality gate will swing open and your project can move forward.

Consider, for example, a project that includes a phase to replace old network cabling with new cable in the plenum (the area above the drop ceilings). There are regulations, industry practices, and standards that all affect this project. The quality gates would require an inspection of the installed cable for accuracy, standard adherence, and compliance. If the cables pass the inspection, the project can continue onto its next gate; if not, the project is stalled until the problem is fixed. Once the problem is corrected, the gate swings open and the project can move forward.

Another gate, for projects in the public domain, is *legislative approval.* A government-sponsored project has different accountability than a project using private funds. Legislative approval would review the work and either cancel the project or continue to invest in the project based on its merit and success.

Building the Project Network Diagram

Because the project network diagram can be a long and detailed map of the project, you probably don't want to enter it into a computer on the first draft. One of the best methods of building and implementing the PND is on a whiteboard using sticky notes. A project manager, along with the project team, should begin by defining the origin work unit on a sticky note and then defining the project deliverables on another sticky note. On the left of the whiteboard, place the origin task, and on the right, place the deliverables. Now the project manager and the project team can use the activity list to identify the relationships between the units of work.

You and the team will continue to create the PND by adding activities in the order they should happen given upstream and downstream activities. This can be a long process, but it's necessary in order to complete the PND. Chances are, you'll be moving activities around and changing their relationships—that's why the whiteboard and sticky notes are so nice.

Once you have roughed out the network diagram, you'll refer to your time estimates for each activity. You can use the WBS, PERT, and supporting details to reiterate the amount of time allotted for each task. Once you've recorded the units of time, you may then begin to assign the resources to the tasks. Use sticky notes to move and strategize the relationships between the tasks, connecting each task with an arrow and identifying the relationship between the tasks to be implemented. You will have to consider the availability of the resources to determine if tasks can truly run concurrently within a network diagram. In other words, you can't assign a programmer to two activities that are supposed to happen at the same time.

After you have constructed the initial diagram, examine the activity lists and the WBS to determine if any tasks or project deliverables have been omitted. If you find omissions, update the WBS and task list to reflect the work and deliverables you've found. Examine the relationships directly between tasks but also the relationships of tasks upstream and downstream. Review these relationships to see if you can edit any of the tasks to save time or resources. If so, rearrange the necessary tasks to update the diagram. A balance of acceptable risk and predictable outcome is required to discern the type of relationships involving each task.

A project manager must also consider business cycles, holidays, and reasonable times for completing each task. For example, a company has sent new cellular modem cards to all of its employees working out of the office on laptops. Part of the deployment is a request that the users in the field connect to the corporate LAN as soon as they receive their cellular cards. There must be a reasonable amount of time allotted between the cards being shipped to the users in the field and the confirmation that the cards have been received.

Finally, once you've created the network diagram, break for a day or two to allow the team to ponder any additional tasks or other considerations in the workflow prior to implementing the plan. When you reconvene to finalize the network diagram, consider the amount of risk you've allowed into the project by asking these questions:

- Are there adequate resources to complete the project?
- Are the time estimations accurate?
- Are there too many concurrent tasks?
- Are resources spread too thin?
- Is this a proven plan?
- Is the plan realistic?

There are two other project network diagrams that you should be topically familiar with. The first is the *critical chain method (CCM).* CCM is similar to the *critical path method (CPM),* but it accounts for the availability of project resources, whereas CPM assumes the project resources are available all the time for the project work.

Another diagramming technique is *GERT,* the *Graphical Evaluation and Review Technique.* This diagramming approach has questions, branches, and loopbacks that allow you to move forward, backward, or sideways in the project according to the outcomes of the project work. For example, GERT may ask, "Are the results of network testing greater than 90 percent or less than 90 percent? If greater than 90 percent, go on to Activity D; if the results are less than 90 percent, return to Activity C."

EXAM TIP Adaptive projects don't have a project network diagram or GERT chart, but they do use a product roadmap to show when releases of the product will go into production. The product roadmap helps the product owner, the development team, the project manager role, and stakeholder define and analyze the project schedule.

Analyzing the Project Network Diagram

One of the most satisfying accomplishments in IT project management is to step back and, looking at the PND, follow the project conception through each task to the final deliverable of the project. Don't get too infatuated—this network diagram will likely change.

Now that the PND has been constructed, you can find the critical path. The *critical path* is the sequence of events that determines the project completion date. The critical path is the path with the longest duration from project start to project completion. For example, imagine that you have created and analyzed your network diagram. Most likely it depicts multiple paths from project start to completion. Usually, just one of the project paths will take longer than any of the other paths. This is the critical path. (I say "usually" it's just one because you could have two or more paths that tie for the longest paths in the project network diagram.) It's called the *critical* path because if any activities on it are delayed, the project completion date is also going to be late.

Calculating Project Float

Given you know that activities on the critical path cannot be delayed, what about activities that are not on the critical path? Can these activities be delayed? Yes, usually they can—but there is a limit to the amount of time an activity not on the critical path can be delayed. This limit is called *float*. (Sometimes this is called "slack," but it's the same business.) There are three different flavors of float:

- **Free float** The length of time a single activity can be delayed without delaying the early start of any successor activities
- **Total float** The length of time an activity can be delayed without delaying project completion
- **Project float** The length of time the project can be delayed without passing the customer's expected completion date

Most project managers allow their project management software to calculate the available float on each activity, but it's really not that hard to do manually. To find the float for each activity, first find the earliest possible start date and the earliest possible finish date for each activity by completing what's called the "forward pass." Once you've got this info, you do just the reverse through the "backward pass"—find the latest possible start and latest possible finish date for each activity. There are a few different methods for calculating project float. Here's one of the most common approaches.

Calculating float is one of the toughest topics in this book for most people to grasp. It's weird, it's confusing, and most project managers never do this stuff manually—they'll let Microsoft Project do it for them. I'll walk you through a scenario, but the process of calculating float is also a great example of how watching someone do this may be easier than reading about it. And for that reason, I've also created a video that demonstrates this entire process.

VIDEO For a more detailed explanation on calculating float, watch the *Finding Float* video now.

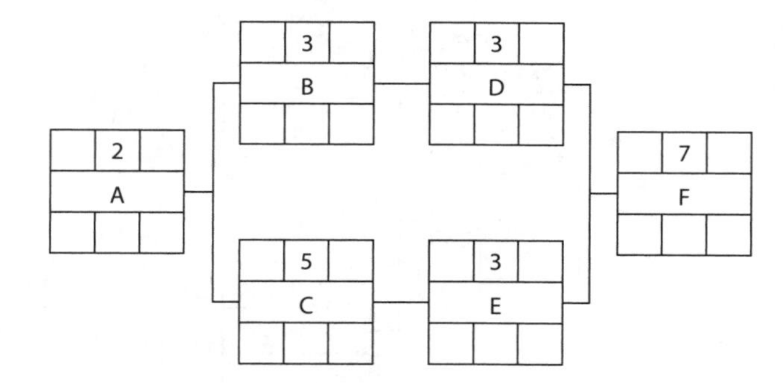

Figure 6-7 The longest path to completion is the critical path.

For this example, we'll be using the simple project network diagram shown in Figure 6-7. (You can print Figure 6-7 from the online resources that accompany this book if you'd like; it's in Adobe Acrobat PDF format. See Appendix C for accessing the online resources.) As you can see, there are two simple paths to completion: ABDF and ACEF. The number over each node represents the duration of the activity. If you add up the duration of each path, you'll find the critical path—the longest path to completion. In this example, it's ACEF because it takes 17 days, while ABDF takes only 15 days.

Now let's try the forward pass. (Again, there are different methods of finding float, so don't be concerned if you've been exposed to a different one.) Follow these steps:

1. Make the Early Start (ES) for Activity A one, because you'll start on Day one. Add the duration of the activity to the ES and you'll have three. Now this part trips up some folks: subtract one day from the value of the ES and the task duration to arrive at the Early Finish (EF) of the activity. The reason is that the duration of Activity A is only two days, not three, right? In other words, if you start on Day one, you should have two days of work to get to Day two. The EF for Activity A is two.
2. The next activities are Activity B and Activity C. The ES for both of these will be three. Why? Because Day three is the next day in the schedule, the earliest possible day to begin either activity.
3. Let's finish the ES for activities B, D, and F first. The EF for Activity B is the ES, plus the duration, minus one, for an ES of five. The ES for Activity D is six and the EF for Activity D is eight. The ES for Activity F is 9, and the EF for Activity F is 15.
4. Now let's do activities C, E, and F. The ES for Activity C is three, and the EF is seven. The ES for Activity E is eight, and the EF is ten. Activity F is the last activity in the project, so you can bet it will be on the critical path—with no float. The ES for Activity F is actually 11, because Activity F cannot begin until your project team completes Activity E. So, the EF for Activity F is actually 17. Figure 6-8 shows the project updated with all of the ES and EF dates.

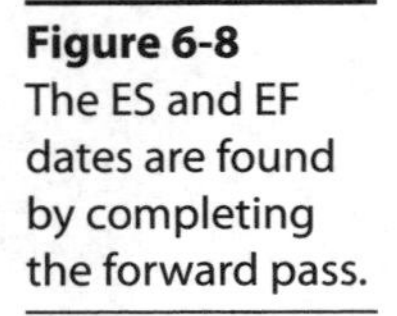

Figure 6-8 The ES and EF dates are found by completing the forward pass.

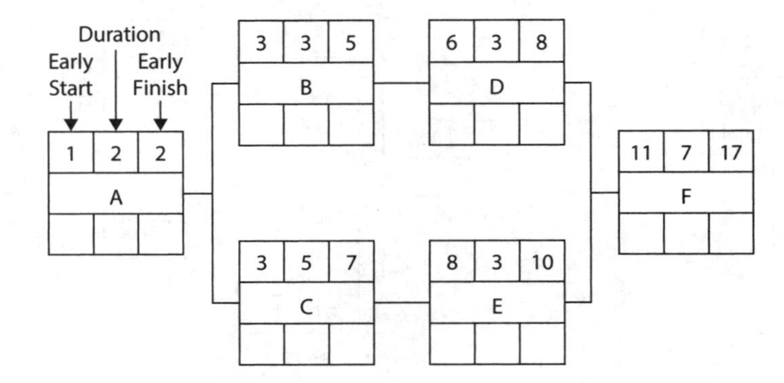

Now that the forward pass has been completed, it's time to do the backward pass. It's a cinch; just follow these steps:

1. Begin with the last activity in the network diagram, Activity F, which has an EF of 17. Make the Late Finish (LF) the same as the EF value: 17. The reason is that Day 17 is the latest day the project can finish without being late.
2. The Late Start (LS) for Activity F is the LF value, minus the duration of the activity, plus one. Yes, plus one. Because you're going backward in the network, you'll add one rather than subtract one. This accounts for the full day of work you have completed on the first activity and the last activity. So, Activity F has an LF value of 17, less the duration of 7, plus 1, which equals an LS of 11. It's no coincidence that the EF and the LF have the same value of 17. It's also no coincidence that the ES and LS have the same value of 11. This is because this activity is on the critical path.
3. Next let's do activities D, B, and A. The LF for Activity D is ten—one day prior to the ES of Activity F's LS. You get the LS for Activity D by subtracting the duration of the activity, plus one, which equals eight. The LF for Activity B is seven and the LS is five. The LS for Activity A is, well, it's the first activity in the project. Do you think it will have any float? Hey! You're right—it's on the critical path, so we can skip it for now.
4. Let's go back and complete the backward pass for E, C, and A. The LF for Activity E is ten, and the LS for Activity E is eight. The LF for Activity C is seven, while its LS is three. The LF for Activity A is two, and its LS is one. Figure 6-9 shows the completed backward pass.

To finalize the process of finding float, you'll subtract the LF from the EF and the ES from the LS on each activity. Wherever there's a zero, you have a task on the critical path; wherever there's a number, the activity has float. In this example, activities B and D have two days of float. Okay, technically they both don't have two days of float; there are two days of float on the whole project. Or you could say that Activity B and Activity D can each have one day of float, or either activity can have two days of float. However you slice it, if either activity goes two days beyond its expected completion time, this project is late.

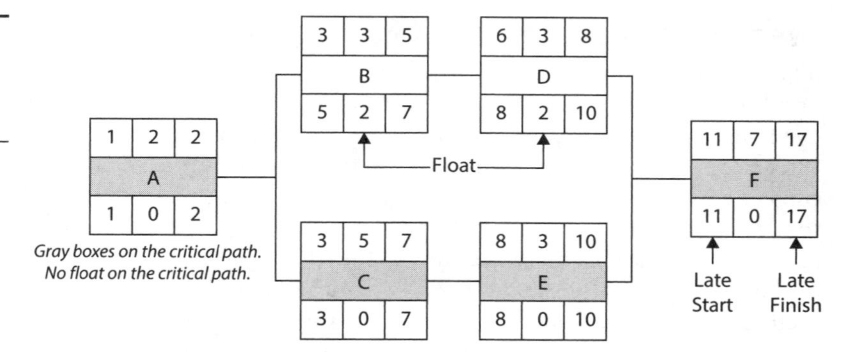

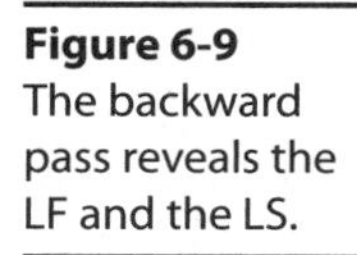
Figure 6-9 The backward pass reveals the LF and the LS.

Forecasting Adaptive Project Completion

Adaptive projects don't move through the same schedule approach that predictive projects do. Recall that adaptive projects work in iterations of two- to four-week time periods. The team selects the most important items from the top of the product backlog based on how much work they think they can complete in the iteration. The requirements from the product backlog are called *user stories* and are assigned *story points*. A user story is just a way of writing a requirement so everyone can understand it. User stories follow this formula:

> As a *role*
> I want this *feature*
> So I get this *value*

For example, "As a salesperson, I want to place orders through my phone app, so my customers can get their product delivered quickly." Based on how difficult the user story is to create, it's assigned a point value that is relative to the user stories. For example, an easy user story might have three points, but a difficult user story might have 13 points. The point value isn't hours of work, it's just relative sizing, like T-shirts get sized small, medium, large, and x-large.

Over time and with experience, the team will learn how many points they can actually complete in an iteration. For example, they may predict they can complete 36 stories points worth of work in four weeks. The number of points the team can complete may vary wildly at the beginning of the project as the team's efficiency settles, but when the number of story points accomplished settles, it's called the team's velocity. *Velocity* is how many points the team can reliably complete in an iteration. In this example, the velocity is 36 user story points. To predict how long the project will last, you'll examine how many user story points are in the product backlog and divide that number by the team's velocity.

For example, the product backlog has 128 user story points remaining and the team's velocity is 36. 128 divided by 36 is 3.55, which indicates four more iterations are needed to complete all of the user stories in the product backlog. If each iteration lasted four weeks, you could safely predict the project will take 16 weeks to complete based on the current velocity.

Creating the Project Cost Management Plan

The *project cost management plan* defines how the project will be estimated and budgeted and how the costs will be controlled. In Chapter 5, I detailed the costs of project management. This is a great example of how projects and project management planning are integrated—what you do in one area affects another. You need to plan specifically for project costs, but you need other project management plans in order to plan effectively for project costs. As more and more information becomes available in your project, you can plan in more and more detail.

The first portion of the project cost management plan should define how you'll complete project cost estimating. In your organization, you may have some specific rules about cost estimating, metrics you're required to use, and even a chart of accounts that defines standardized costs for certain types of project work. These organizational requirements that control how and what you plan are called *enterprise environmental factors.* Always follow the rules and policies of your organization when it comes to planning and estimating for costs.

As a reminder, to create accurate cost estimates, you'll need six things:

- Project scope baseline
- Project schedule
- Human resource plan
- Risk register
- Enterprise environmental factors
- Organizational process assets

You've seen all of these inputs to cost estimating before—except for the risk register and the human resource plan. Each of these inputs has some cost factor associated with it, which is why you'll need all of them to predict how much the project is going to cost the organization. I'll address the human resource plan and the risk register later in this chapter.

Cost budgeting, the second process that the cost management plan defines, is the aggregation of the costs of each work package in your project's WBS. It is the sum of all the costs the project will incur and is put into an authorized budget so that the project can execute the fully composed project management plan. The project's cost budget does not include any reserves for time or risks—these are special budgets for your project, not part of the project execution. The performance of your project is measured, often enough, against the approved project budget.

Your project's cost budget is largely built on the cost estimates of your project. That's why it's so important for you to document the basis of your cost estimates—how did you determine it would cost your project $24,987 for a particular server configuration? My point is that the project cost estimates are only as good as the supporting detail. If the supporting detail is fiction, then you can bet dollars to donuts the project's budget is going to be skewed as well.

One of the most important parts of the project cost management plan is how the project costs will be controlled and monitored. The project manager has the responsibility of tracking project costs for all areas of the project scope that have a cost factor. There must be cost reconciliation between what the project manager says the project will cost and what the project actually costs in execution. Monitoring and controlling of the project costs uses several approaches:

- Leading the team to do the work properly the first time to avoid cost overruns
- Thoroughly investigating the true costs of change requests
- Tracking costs to project deliverables against the total costs of the project by phase and overall project costs
- Measuring project performance against the costs of the project to reach said performance
- Stopping unapproved changes of all sorts from entering the project
- Making corrective actions to bring cost overruns back into alignment with the project cost baseline as much as possible

Basically, you've got to work with the project team and stakeholders to prevent waste by accurately planning the project work and then accurately doing the project work. When there are mistakes in the project, go to the problem immediately to try to find a solution for the problem. As a project manager, you're responsible for wastes, cost overruns, and deficits. The actions you take should always be linked to project performance.

Planning for Project Quality

Ask ten people what they think quality is, and you'll probably get ten different answers. *Quality,* when it comes to project management, is the entire project and how the project satisfies the stated and implied requirements of the project scope. Quality is creating for the project customer exactly what was promised in a way that's cost and time effective. Quality is more than just balancing the scope, time, and cost of a project, however—it's a fitness for use and a conformance to project requirements. In project management, you achieve quality by planning quality into the project—not inspecting the results of the sloppy work and then fixing problems later. It's always more cost effective to do the work properly the first time.

The *quality management plan* defines what quality is for your project, how you'll plan for quality, and then how you'll inspect the project work to ensure that quality exists within the project deliverables. The first step in planning for quality is to understand what the project customers want—this means you'll refer back to the scope baseline. By understanding what is expected of the project in exact terms, you can plan to achieve the expectations in exact deliverables. This is one of the reasons it's so important to get the customer stakeholder to approve the project scope statement and the WBS. You don't want to be planning to achieve quality for the wrong requirements.

When you begin to plan for project quality, you'll start with the requirements of the project, and then you'll need to address quality assurance. *Quality assurance (QA)* is an organization-wide quality policy or program that your organization subscribes to.

Quality assurance examines the quality requirements of the project, inspects the results of quality control, and ensures that the project is using the correct quality standards and terminology in the project. Your organization may use a home-grown quality policy that staff has written, or it may participate in a program such as Six Sigma or Total Quality Management. These quality programs usually have the same theme that W. Edwards Deming, the grandfather of quality, created in his Quality Circle: Plan, Do, Check, Act. My point is, your organization may already have a standardized approach to quality assurance that your project will need to follow. That's great!

EXAM TIP Quality assurance is planning to do the work correctly the first time. Quality control is inspecting the work to confirm that it conforms to the quality requirements. Remember, it's always more cost effective to do the work correctly the first time.

Your quality management plan also defines the *quality control technique* you'll use in your project. Quality control is the inspection of the work results that your project creates. By inspecting the work results, you'll know if the work has been done to plan, if the work adheres to the QA metrics, and whether the work needs to be corrected before the project moves forward or it's acceptable to you. Consider a project to install a network in a new office building. In this construction project, you might have 1,500 network connections to create throughout the building. For each network drop, you'd have a specific requirement of where the jack is to be positioned, how the network cable is punched down, and how the cables are secured between the network drop and the patch panel. The specifications of the work are the requirements, the planning of how you'll achieve the requirements is the quality assurance, and the inspection of the network is the quality control.

Along with your quality management plan, you may also create or adapt a process improvement plan. The *process improvement plan* is a project management plan that defines how the processes within the project can be analyzed for potential improvement. The process improvement plan is ideal for projects where there's a repeatable process, such as installing 1,500 network cables, or for project types that your organization may do over and over as part of your business. It's really a good idea to use a process improvement plan because it helps you, the project manager, really think about and understand the workflow of a project and how you might shave off some wasted time by getting rid of non–value-added activities.

You can use the process improvement plan to define and analyze several things about your project:

- **Process boundaries** You'll identify where a process begins and where a process stops. By understanding the process boundaries, you can document what conditions are needed for a particular process to begin and what conditions must be true for a process to stop.
- **Process configuration identification** This fancy-schmancy term just means that you identify all of the components within the process. You'll document how a process is completed, what the process interfaces are, and what each process in your workflow accomplishes.

- **Process metrics** If you don't measure, you can't improve, but in order to measure, you need to know what the metrics are for measurement. You can measure a process on speed, cost, efficiency, throughput, or whatever metric is most appropriate for the type of work the process is participating in.
- **Targets for improvement** Once you've completed process analysis, then you can begin to re-engineer the processes, set goals for process improvement, and try small changes to the process to measure what results the changes may bring.

I'll talk more about the execution of quality and the inspection of project deliverables in Chapter 11. It's perfectly acceptable for a project manager to use a previous project's quality management plan to help achieve quality in the current project. The application of quality control, however, will likely change for each project, as each project has different variables than the last.

Preparing for Managing a Project Team

I think you and I would agree that every organization in the world is unique when it comes to managing employees, the culture of the organization, and how much authority the project manager has over managing their project team. The enterprise environmental factors, the structure of the organization, the policies and guidelines of the organization, and the experience of the project manager all affect how much authority and decision-making ability the project manager has over the project team. As a project manager, you need to understand how you'll manage the team and still be within your "rights" as a project manager in your organization. You don't want to overstep your boundaries, but you'll need to manage your team effectively so that you can get them to complete their assignments as required.

Considering the conditions within your organization, you'll create two plans to address how you'll lead and manage your project team: the human resource plan and the staffing management plan. And, yes, as with other project management plans, you can adapt your current plan by drawing on plans from previous, similar projects. The first plan to discuss is the *human resource plan.* This plan defines the needed roles and responsibilities of the project team. A *role* is a definition of a type of team member who does a specific kind of actions—for example, network engineer, application developer, database administrator. Roles are generic descriptions of the project team members, not employees named specifically. Attached to the roles are the responsibilities or actions the roles are to fulfill on the project team.

NOTE Understand how much authority you have over the project team in your organization before assuming power you may not have.

One approach that's worked for me is that when I plan for the project work, I don't worry about who will be doing what. I focus on the role that will take care of the specific work. For example, you might assign the entire network cabling activities to the network engineer; later, once the initial planning is done, you'll determine how many specific people can fill the role of the network engineer and slice the activities across the people who can satisfy the one specific role in the project. This way, during planning the project team, you're not worried about getting a specific person, you're just identifying the skills the roles will need in order to accomplish the specific responsibilities.

The human resource plan also addresses the type of structure the project is operating in. Recall our conversation on functional, matrix, and projectized structures from Chapter 3? The type of structure you're in will help you as the project manager to determine who has to report to whom. In a functional structure, where the functional manager is in charge, the reporting structure is easy and shallow. Similarly, in a projectized structure, the team works on just one project at a time, and the project manager is in charge. In a matrix structure, however, your project team members have to report to you, their functional managers, and any other project managers that they work for. Communications, reporting, and resource management get messy in a matrix structure, so ample planning is needed.

The second plan, the *staffing management plan,* addresses how project team members will be brought on to and released from a project team. It's this plan that is so important in a matrix structure, as you'll need to coordinate timetables with other project managers and the project team members. This plan addresses the project calendar—when the project work is expected to take place. It also addresses the needed resource calendars—when the project resources are available for project work.

The staffing management plan also includes an assessment of what is needed for team development. This can include training, team activities, and expectations of team leadership roles for senior project team members. You can't expect the project team to complete the project without the appropriate skills to complete the project work. Training is an essential part of the team development, and it's also considered a cost of quality.

The staffing management plan can, and I believe should, include a definition of how a rewards and recognition system works. If your project can reward the project team for a job well done, then the project team should know the rules, the rewards, and how they can achieve those rewards. Rewards can be financial, time off, contributions to their employee review, or other incentives for each employee. What motivates one employee may not motivate someone else. As a general rule, you should shy away from "zero-sum rewards," which are rewards only one person can win, such as a reward for being employee of the month. As a side note, you will want to reward the project team with genuine praise when it's appropriate. This isn't something that's in your project plan, but when a project team member does a great job, you should acknowledge it. I'll talk more about team development in Chapter 7.

Writing the Project Communications Management Plan

Without a doubt, the best method to communicate with others is face-to-face communication. However, in most organizations, that's not always a possibility. So, as a project manager, you'll need to create a *communications management plan* to address other approaches for communicating and define what is and is not allowed, preferred, and required.

In order to create solid management alliances, you'll need to communicate. In order to communicate effectively, you'll need a communications management plan. Considering stakeholder analysis, the project manager and the project team can determine what communications are needed. There's no advantage to supplying stakeholders with information that is not needed or desired. Time spent creating and delivering unneeded information is a waste of resources.

The communications management plan helps to determine which person needs what information, when that person needs the information, the modality in which the information is expected, and which stakeholder will supply the information. The communications management plan identifies the appropriate communication technologies, types of forms, and types of meetings, and it creates a schedule of when communication is expected to occur.

A communications management plan organizes and documents the process, types, and expectations of communications. It provides the following:

- A system to gather, organize, store, and disseminate appropriate information to the appropriate people. The system includes procedures for correcting and updating incorrect information that may have been distributed.
- Specifics on how confidentially constraints of sensitive information within the project or organization will be addressed and documented, and the role of the individual(s) who may manage the confidential information. Project managers often have access to sensitive information, such as financial, human resources, and even classified government information. The plan must establish safeguards within the organization (or project) to ensure that confidential information is protected.
- Details on how needed information flows through the project to the correct individuals. The communication structure documents where the information will originate, to whom the information will be sent, and in what modality the information is acceptable.
- Information on how the information to be distributed should be organized, the level of expected detail for the types of communication, and the terminology expected within the communications.
- Schedules of when the various types of communication should occur. Some communications, such as status meetings, should happen on a regular schedule; other communications may be prompted by conditions within the project.
- Methods to retrieve information as needed.
- Instructions on how the communications management plan can be updated as the project progresses.

It's been said that 90 percent of a project manager's time is spent communicating. You'll be communicating with the project team, the project sponsor, business analysts, customers, vendors, and other stakeholders all the time. People will want information from you, and you'll want information from other people. One tool that you can add to the communications management plan is a *communications matrix,* shown in Figure 6-10. This matrix lists all of the stakeholders and maps out who needs to speak to whom. You can get fancy and create a legend beyond what I've done here, but basically it's a tool that can help you facilitate communication between certain stakeholders and the project.

	Steve	Holly	Sam	Joan	Ben
Steve		X	X		X
Holly	X			X	X
Sam	X	X			
Joan	X	X			X
Ben	X	X		X	

Figure 6-10 A communications matrix identifies which stakeholders must communicate.

Beyond the communications management plan, you'll also want to be aware of how you communicate. Fifty-five percent of all communication, for example, is nonverbal. When you consider how much IT project teams rely on e-mails, text messages, and web collaboration software, it's easy to see how communication can break down. Have you ever put a joke in an e-mail message, but the recipient didn't understand that you were joking? Your body language, facial expressions, and gestures all affect your message beyond the words you're communicating. That's why e-mails and even phone calls aren't always an effective mode of communication. In your communications management plan, you should plan for face-to-face meetings with your project team and other key stakeholders as often as possible. Face-to-face meetings are one of the best methods for accurate communication.

Your communications management plan should address how remote workers, sometimes called a virtual team, will communicate. What communication methods are allowed or required for these non–co-located members of the project team? Will you use collaborative software, instant messaging, web conferencing software, voice conferencing, video conferencing, or some other technical solution? You'll need to consider the pros and cons of communicating through any technology because it will likely affect the communications model and effectiveness of the communication. While static communications, such as e-mails, faxes, and printed materials, can be effective, they lack the interactive attribute of face-to-face communication.

Video conferencing, such as Zoom, allows stakeholders to see one another face-to-face through computer software. *Voice conferencing* is easy to do with nearly any phone system today, but you lose the effectiveness of the face-to-face communication video conferencing provides. Another possibility for your project might be to use social media for communication. Consider groups within Facebook, lists on Twitter, LinkedIn, and other social media. Project teams have leveraged these social media sites to communicate with one another.

You might also consider creating a Wiki page for stakeholders to access and review project information. Wiki pages are a repository that can give project background, project purpose, project sponsor information, the timeline, and other valuable information. Wiki pages, and even websites, can help you quickly share information on the public Internet or on your organization's private intranet.

Planning for Project Risk Management

Risks come in all shapes, forms, and sizes. Risks are uncertain conditions that can have a positive or negative effect on your project's ability to reach its goals and objectives. Risks are prevalent in IT projects; even the project itself is a risk, as any day a new technology may come along and outdate the technology your project centers on. Most risks, as luck would have it, are foreseeable events that you and the project team can identify and plan for. The *risk management plan* is a component of your overall project management plan that details how project risks will be identified, analyzed, responded to, and controlled. Risk management happens throughout the project, and you and the project team should always be on the lookout for new risk events.

You'll rely on several project components to help you plan for risk management:

- **Project scope baseline** The project scope, WBS, and WBS dictionary can help you identify risk events, plan for critical project deliverables, and keep an eye on where risks may be in the technical details of the project.
- **Cost management plan** There is a financial impact to most risk events, and you'll need to create a special reserve that's outside of the project budget just for the management of risk events. More on the risk reserve in a moment.
- **Schedule management plan** You'll also find that risks can affect the project schedule and not just the cost of the project. And some of your projects are extra sensitive to the project schedule, so you'll have reasons to query the schedule and look for events that can hinder the project progression.
- **Communications management plan** When risks are discovered, analyzed, or squashed, you'll need to communicate risk status and events to the appropriate people. Risk communication is an important part of the project manager's role.
- **Enterprise environmental factors** The rules and policies of your organization may require you to deal with a risk management department, follow particular risk analysis rules, or complete risk assessment forms. Always follow the rules of the organization.
- **Stakeholder register** This document will help you identify the stakeholders who need to be part of the risk management planning, the stakeholders affected by the identified risk events, and the stakeholders you'll need to communicate with about the risks in the project.
- **Organizational process assets** If you or someone in your organization has done a project like the one you're working on now, chances are you'll use historical project management plans and lessons learned documentation to help you manage the risks within your current project.

Planning for risk management relies heavily on these inputs, and you'll work with the project team, key stakeholders, subject matter experts, and management to help you plan how the risks in the project should be handled. You'll discover that some risk events are of little consequence and you can ignore them. Other risks are project-killers and you'll have to attack them any way you can to prevent them from coming to fruition. The key to risk management is to identify the risks as soon as possible in the project, analyze the risks, and then create a reasonable plan of attack.

Identifying the Project Risks

Risk identification is an iterative part of project planning. You and the project team are always on the lookout for project risks that can sneak into the project or that may have been overlooked. Risk identification must happen throughout the project, especially in an IT project, because as the project develops, new risks may develop. It would be nice if you could identify all the risks within a project before the execution begins, but the reality is, you'll probably find risks lurking in the project throughout the project life cycle. There are several tools you and the project team can use to identify risks within the project:

- **Review the project documents** A great place to start looking for project risks is to sit down with a mug of coffee and read all of your project documents. You can identify many risks just from what's in the project scope statement, the WBS, the resource requirements, supporting details of your project, your project management plans, contracts, and any other document that affects the decisions made in your project.
- **Brainstorming** You and your project team can rattle off as many risks as possible in a brainstorming session. In a brainstorming session, there's no set number of risks for a goal to identify; you're just encouraging the team to identify anything at all that might be a risk regardless of how silly or small the risk may seem. One technique that works for me is to examine the WBS as a group and create a similar structure, the risk breakdown structure, to group relative risks together.
- **Delphi Technique** This approach uses rounds of anonymous surveys to build consensus on risk events. Imagine that Frederico works for Bill, and both men were invited to a brainstorming session for Janelle's project. Frederico knows of a risk that could affect Janelle's project, but he doesn't want to say what it is because the risk event may make his boss, Bill, look bad. Frederico says nothing and Janelle's project crashes and burns. The Delphi Technique takes out the fear of repercussion, which Frederico was experiencing, by using anonymous surveys to identify risks. With this approach, Frederico can anonymously tell Janelle about the risk. Janelle then would compile all of the risk events and send them back to the survey participants for feedback. Now Bill can respond to the identified risk, if he needs to, and no one knows that Frederico was the identifier of the risk event. Depending on the response to the original survey results, Janelle can fire off another survey, or multiple surveys, until there's a consensus within the group on what risks are genuine and what risks are superficial.
- **Assumptions analysis** An assumptions log should be included as part of the project documents. All assumptions should be tested as much as possible to determine if the assumptions are valid or not. Assumptions can be risks if they're not tested and researched to determine how they'll affect the project.

An assumption that your software will work with the latest and greatest operating system can really wreck your project if the assumption proves false later in the project delivery.

- **Root cause analysis** This analysis approach examines an effect the project is experiencing and tries to determine why the effect is happening. You might plot out the result of the risks in a cause-and-effect diagram, sometimes called an Ishikawa diagram or a Fishbone diagram, as shown in Figure 6-11.
- **SWOT analysis** This approach examines the project in terms of its strengths, weaknesses, opportunities, and threats to test where the project could fail and where the project could improve. SWOT is a great technique to use on resources you've not used before or on requirements that you and your project team haven't tried before. If you're stumped in a brainstorming session, fire up SWOT to jump-start the meeting.

All of your identified risks, as your risk management plan will tell you, should go into a risk register. A *risk register*, also called a *risk log*, is a database of the identified risk events, what impact each event may have on the project objectives, and what effect each event will have on the project. The project manager and team will update the risk register throughout the project as new risk events are identified. You'll also record the outcome of the identified risk events in your project. A *risk report* is utilized to document the risk event, the response selected, and the outcome of the risk event.

You might encounter a risk/issues log in your organization. A *risk* is an uncertain event, while an *issue* is a certain event. Issues are generally risk events that have happened and now need to be managed. A risk/issues log documents the risks and issues within

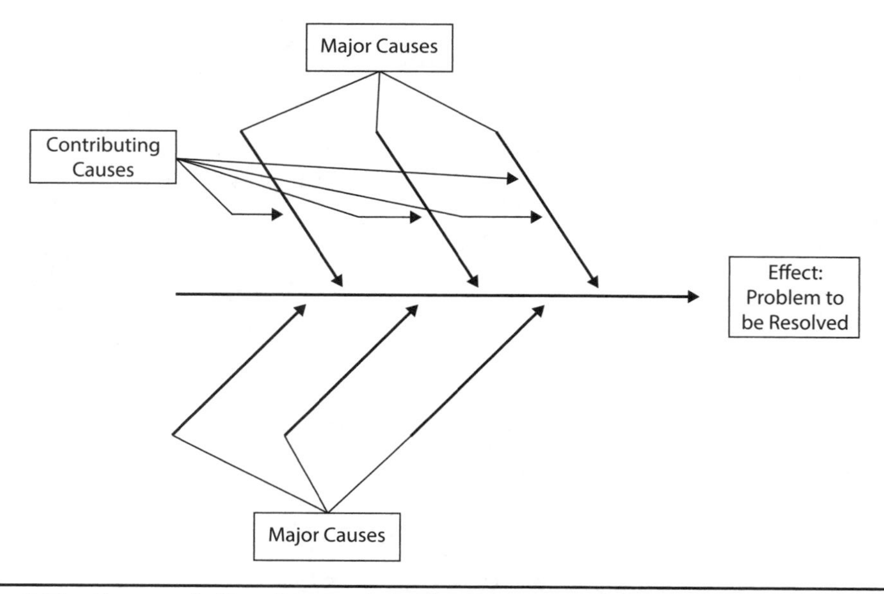

Figure 6-11 Cause-and-effect diagrams help identify causes and contributing causes.

the project; tracks their status, potential responses, and costs to the project budget and schedule; outlines plans for managing the risk or issue; and records the final outcome of the risk or issue. Note that keeping risks and issues separate in two logs, a risk log and an issues log, is not uncommon.

Analyzing Identified Risks

The risk management plan should detail how the identified risks will be analyzed. You should analyze the risks to determine how probable the risks are to occur and, if the risks do occur, what the impact of the risk events will be. The usual approach to risk analysis is first to perform qualitative risk analysis and then, for risks that qualify, to perform quantitative risk analysis. Your organization, as part of its enterprise environmental factors, may have risk policies already established on how you should analyze project risks. The risk management processes will also help you to rank and prioritize risk events based on when the risks may happen, their severity, costs, or other factors your organization deems important.

Qualitative risk analysis is a high-level, fast approach to risk analysis. It's somewhat subjective and doesn't offer much in-depth review and analysis of the risk events. When you perform qualitative risk analysis, you can base the predicted probability and impact on past experience, gut instinct, and other subjective inputs. You don't have much to go on as far as research during qualitative risk analysis—it's a fast approach to qualifying the identified risks for further analysis.

With qualitative risk analysis, you can use a risk matrix, sometimes called a probability-impact matrix, to score the risk events on an ordinal scale. Figure 6-12 shows a heuristic for how risks are ranked and responded to through qualitative risks.

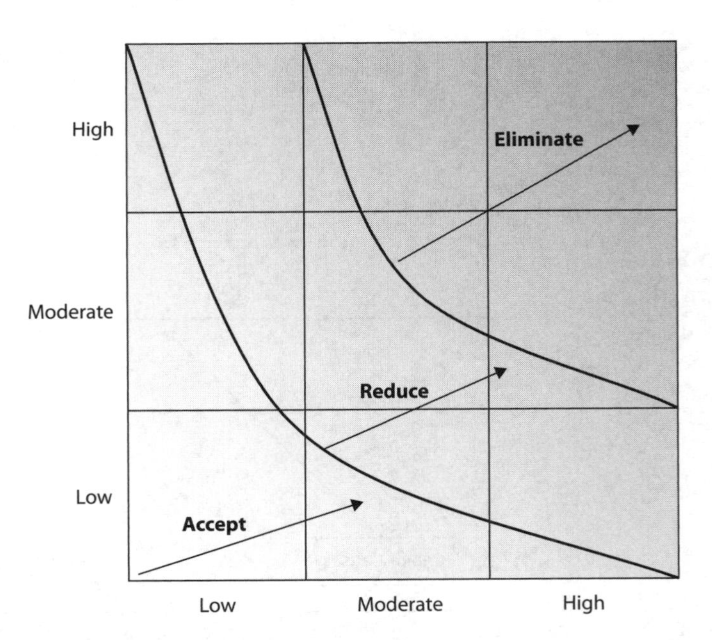

Figure 6-12 Qualitative risk analysis uses an ordinal scale to rank risks.

Because this is a subjective approach to risk analysis, the project manager should periodically review the low-level risks just to confirm they're not becoming more serious during the project execution.

Quantitative risk analysis takes the more serious risks from qualitative analysis and really researches the risk events to find their true probability and impact. Quantitative risk analysis takes time and often a budget, as you'll want to test the risk events in a controlled environment whenever possible. You want to determine which risk events are most likely to happen and what their related impact on the project will be. This approach allows you to create a risk ranking by impact and probability and by which risks are most imminent in the project.

During the quantitative risk analysis process, you and the project team are trying to quantify the risk event value. The impact, in quantitative analysis, is usually a dollar amount that shows the true impact on the project should the risk event happen. The last column in the risk-impact matrix is the risk event value, which you calculate by multiplying the probability of the risk by the impact for the risk event value score. The sum of the risk event value scores is the project's risk exposure.

Be aware that both positive and negative risk events are possible. Negative risk events are the risks that can hinder the project's success, and they're really the events most project managers focus on. Positive risk events are things that can happen to save your project time and money, and even create opportunities for your project.

Creating Risk Responses

Once you've identified the risks within your project, you can use any of seven risk responses for each risk. You should document the chosen response along with the risk event in your risk register. It's also a good idea to identify a trigger for when the risk response is appropriate. A *trigger* is a warning sign or condition that the risk event is coming true so it's time to fire the risk response to counteract the risk event. There are three risk responses for negative risk events:

- **Avoidance** This risk response seeks to avoid the risk event by creating workarounds, changing the project schedule, adjusting the project objectives, or taking other action to avoid the identified risk event.
- **Transference** This risk response transfers the risk to a third party, usually for a fee. Imagine a dangerous activity, such as working with electricity. Rather than owning the risk in-house, you could hire an electrician to own and manage the risk for you.
- **Mitigation** This risk response looks to spend extra time or monies to reduce the probability and/or impact of a risk event. You might spend more on a server with a fault-tolerant RAID system that's integrated rather than using a software RAID application among multiple hard drives. You've spent more for a faster, more reliable solution and reduced the probability of data loss.

There are also three risk responses for positive risk events you may find in your projects:

- **Exploitation** This positive risk response aims to take advantage of a positive risk. Imagine that your IT project creates a byproduct that could be sold to help offset the project costs. Or perhaps you and the project team could take advantage of a holiday weekend to work on the project uninterrupted.
- **Enhancement** This positive risk response tries to make the conditions just right for a positive risk to happen. You could save a tremendous amount of time and project costs if you were able to finish a particular milestone by a given date. To reach the milestone, you add extra resources to help the effort-driven work so that the team can complete the milestone by the specific date.
- **Sharing** This positive risk response allows your project team to partner or team with another entity to realize an opportunity that you may not have been able to realize on your own. Sharing examples are teaming agreements, joint ventures, partnerships, and special-purpose companies.

The seventh risk response is appropriate for both positive and negative risk events: *acceptance*. Smaller risks that have little probability and little impact are often accepted. It's also possible that low-probability risks with huge impacts could also be accepted: consider meteorites. Assumptions that may prove false, such as the weather, travel delays, and hardware interoperability, are accepted.

Creating the Stakeholder Management Plan

The *stakeholder management plan* addresses how you identify, engage, and maintain the engagement of your project stakeholders. This is a special plan that is closely related to the project's communications management plan. Recall that the communications management plan defines who needs what information, when the information is needed, and the expected modality of the communication. The stakeholder management plan describes the recipients of the communication and the expectations of how they'll communicate with the project manager and other stakeholders.

The stakeholder management plan is more than just communication for stakeholders; it also defines stakeholder identification and engagement. The stakeholder management plan defines four things for your project:

- How you will identify and documented stakeholders
- How you will manage stakeholders, based on their needs, expectations, and effects on the project success
- How you will manage stakeholder engagement throughout the project
- How you will monitor, maintain, and influence stakeholder engagement

EXAM TIP Stakeholder identification is an initiation process. You should complete stakeholder identification as early as possible in the project so you don't overlook key stakeholders. It's always challenging to gain stakeholder support if you didn't identify and involve them early in the project decisions.

Performing a Stakeholder Analysis

While all stakeholders are important, some stakeholders are more important than others. I know that sounds harsh, but you don't want to spend all your time communicating with and managing a stakeholder that has very little influence or power over your project. Stakeholder analysis is a method to discover, prioritize, and group stakeholders based on their power, interest, influence, and impact on the project success. There are three general steps to performing stakeholder analysis:

- **Broad stakeholder identification** Identify and document all of the stakeholders, identify their roles and responsibilities, and determine which stakeholders are decision makers who can affect the project outcomes, objectives, and approach.
- **Stakeholder analysis** Analyze the list of stakeholders for how they can affect the project by their decisions, power, and support of the project goals. You can then begin creating groups of stakeholders by their attitude toward the project.
- **Stakeholder engagement** Begin to plan how to address, engage, and influence these stakeholders to support the project goals, objectives, and approach to achieving the project goals.

Creating Stakeholder Classification Models

Once you have identified the groups, or classifications, of stakeholders, you can take your analysis and planning a bit deeper by creating a model to map out the stakeholders in your project. There are four common classification models you can use:

- **Power/interest grid** Maps stakeholders based on their power (authority) and their interest in the project, as shown in Figure 6-13
- **Power/influence grid** Maps stakeholders based on their authority and influence or involvement with the project
- **Influence/impact grid** Maps stakeholders based on their influence and the possibility that they could affect the project approach, goals, and objectives
- **Salience model** Classifies stakeholders based on their power, urgency for information, and legitimacy in the project

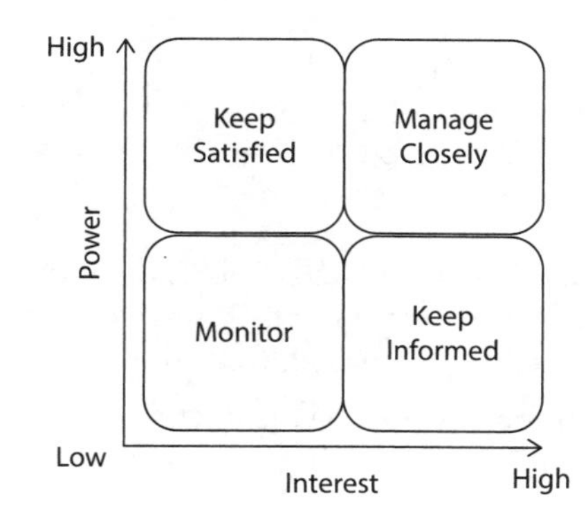

Figure 6-13 A power/interest grid maps out stakeholders' power in relation to the interest in a project.

These four models, and the entire stakeholder management plan, are usually not shared with everyone in the project, because they may include sensitive information. In other words, you don't want word getting back to Allen the CFO that he's a pain in the project with little legitimacy in the project planning. These models and classifications will help you better engage, and maintain engagement, of the most appropriate stakeholders.

Building the Project Procurement Management Plan

The final plan you'll need to create as part of your project management plan is the *procurement management plan.* This plan defines the process you'll need to complete in order to create contractual relationships with your vendors. As with most of these plans, your organization may have specific rules on how you, your procurement office, and the vendors all interact. As always, I'll define the generally accepted practices this plan defines, but out there in the real world, you should always follow the rules of your organization.

The procurement management plan defines four things for your project:

- The decision to procure for your project
- How the procurements are to take place according to your organization
- The administration of the procurement and the contracts the process will create
- The closing of contracts once both parties have fulfilled their obligation

Preparing to Purchase

The first step in creating this plan is to make certain you actually have a need to purchase something. Some things in your project will be easy to see you need to purchase them: materials, resources, facilities, equipment, software, hardware, and services. Other things, however, won't be so easy. You and the project team need to complete a build-or-buy analysis and weigh the benefits of purchasing products and services versus building them in-house. Not all decisions to procure are financially based, as sometimes there are other reasons, such as experience gained, available staffing, control of the intellectual rights, and other ancillary benefits beyond the cost of the product you're purchasing.

Once you've made a decision to procure, you should create a statement of work (SOW) document. The SOW defines what the vendor is to provide for the project, provides ample information so that the vendor may create an effective response for you, and includes enough information to understand what the expected results of the contractual relationship may be. The SOW is a document from you, the buyer, to the vendor. Along with the SOW, you'll need to create one of four exploratory documents, depending on what you want back from the vendor:

- **Request for information (RFI)** This document asks the vendor for more information about their products or services. It means that your organization may be interested in purchasing from the vendor in the future.
- **Request for quote (RFQ)** This document asks the vendor to provide just a price for the work described in the statement of work.

- **Invitation for bid (IFB)** This document is nearly the same as the RFQ; you want the vendor to provide a price only for the work described in the SOW. You might see this also called a request for bid (RFB).
- **Request for proposal (RFP)** This document asks the vendor to create a proposal for the work defined in the SOW. A proposal is not a quick thing to create; the vendor must examine the work, create ideas and solutions for you, and show you the details of the solution they'd create for you based on your SOW. An RFP is asking the vendor to share with you their approach for the work and a price to complete the work.

In some instances, depending on the size of the work to be contracted, you may invite all of the vendors to a bidders' conference. The bidders' conference is a meeting where all of the vendors that are interested in performing the work on your project can ask you questions about the SOW you've created. Often this meeting will lead to updates and revisions of the SOW that will be redistributed to the vendors. The bidders' conference is simply a meeting of all the vendors so that all participating vendors can receive the same information to create their bids, quotes, or proposals.

Once you've narrowed your search to two or three vendors, it's time to interview each one to determine who will be awarded the project. In the interview process, which the vendor will probably consider a sales call, remind yourself that this is a "first date"—it's a chance to find out more information about the vendor.

Document all parts of the meeting: How difficult was it to arrange a meeting time? How polite was the salesperson? Did the salesperson bring a technical consultant to the meeting? All of these little details will help you make an informed decision. In such meetings, pay attention to the following regarding vendors' representatives:

- Do they pay attention to details? Are they on time? Dressed professionally and appropriately for your business? Are their shoes shined and professional? How vendors pay attention to the details in their appearance and presentation to win your business will be an indicator of how they will treat you once they've won your business.
- How organized are their materials? When a salesperson opens their briefcase, can they quickly locate sales materials? Are the brochures and materials well prepared and neat, and not wrinkled or dog-eared? Again, this shows attention to detail, something every project requires from the start.
- What is their body language saying? Pay attention to how they are seated, where their hands are, and how animated their answers become. A salesperson should show genuine interest in your project and be excited to chat with you. If they seem bored now, they will likely be bored when you call to discuss concerns down the road.
- What does your gut say? Gut instinct is not used enough. The meeting with the salesperson should leave you with a confident, informed feeling. If your gut tells you something is wrong, then chances are something is. If you're not 100 percent certain, and you probably shouldn't be after one meeting, do more research or ask for another meeting with the project integrators.

Once you receive the quote, bids, or proposals, considering what you've asked for, you can select which vendor to purchase your products and services from. You can create screening systems, such as that the vendor must have a 120-day warranty on their product, to screen out vendors who don't qualify. You can also use a scoring model to choose the most desirable vendor. A scoring model identifies several categories of qualification, such as experience, cost, and certification. Each category is assigned a value of points, and then you'll measure each vendor and their proposals and award them points in each category. The vendor with the most points is awarded the contract.

Clinching the Deal

When you've just about made your decision, it's time to follow up with a phone call to a few of the references the vendor has provided. Most of these references will no doubt be excellent and prepped. Not that anything's wrong with that; everyone wants to put their best foot forward. Ask the vendor what type of work was performed for the client and when the work was done.

If the work the vendor completed for the client is not directly similar to your project, ask the vendor to provide you with another reference for whom the vendor has done similar work. In addition, the date of the work should be fairly recent—hopefully within the past six months.

Once you have made your decision to award the project to a particular vendor, ask that vendor to provide the scope of the project in writing, including the price, in the form of a contract. The vendor may, and should, have their own contract that they use whenever implementing technology. Review the vendor's contract and, if necessary, have your organization's legal department look it over and make any amendments or changes. Your organization may have, and prefer to use, its own contract for the agreement.

As painful as contracts are, they protect you and the vendor. The contract should require that the vendor guarantee their work for a specified amount of time. The technology to be implemented will determine the amount of time expressed in the warranty and the type of guarantee provided. Your organization may use a letter of intent and/or a memorandum of understanding. The letter of intent means that your organization intends to do business with the vendor; it's usually not binding, but it lets the vendor know they've won the contract. A memorandum of understanding precedes the actual contract written by the attorneys. This document spells out the agreement for both parties before the attorneys (and their fees) get involved and finalize the terms.

Many different types of contracts are available. The project work, the expected duration of the project, and the relationship between the buyer and seller will determine the contract type. Here's a quick overview of the common contract types and their attributes:

Contract Type	Acronym	Attribute	Risk Issues
Cost-plus-fixed-fee	CPFF	Actual costs plus profit margin for seller	Cost overruns represent risk to the buyer.
Cost-plus-percentage-of-cost	CPPC	Actual costs plus profit margin for seller	Cost overruns represent risk to the buyer. This is a dangerous contract type for the buyer.

Contract Type	Acronym	Attribute	Risk Issues
Cost-plus-incentive-fee	CPIF	Actual costs plus profit margin for seller	Cost overruns represent risk to the buyer.
Fixed-price	FP	Agreed price for contracted product; can include incentives for the seller	Seller assumes risk.
Lump-sum		Agreed price for contracted product; can include incentives for the seller	Seller assumes risk.
Firm-fixed-price	FFP	Agreed price for contracted product	Seller assumes risk.
Fixed-price-incentive-fee	FPIF	Agreed price for contracted product; can include incentives for the seller	Seller assumes risk.
Time-and-materials	T&M	Price assigned for the time and materials provided by the seller	Contracts without "not-to-exceed" (NTE) clauses can lead to cost overruns.
Unit-price		Price assigned for a measurable unit of product or time (for example, $130 for engineer's time on the project)	Risk varies with the product. Time represents the biggest risk if the amount needed is not specified in the contract.

Before any implementation begins, and once the contract details have been worked out, do some prep work. For example, if the project is an operating system upgrade on your servers, create a full backup or system image of your servers. If the technology is a new application to be developed with hooks into your database, assign the appropriate levels of access security to the database for the developers, but don't give the developers greater permission than what they need to accomplish their work. In other words, prepare for the worst-case scenario, but hope that you never have to use it.

Another part of the contract is keeping the deal. Both parties have responsibilities to live up to in the contract. The procurement management plan defines how the contract will be monitored and controlled. This means you'll plan for audits and inspections, and you'll expect feedback from people on the project who deal directly with the vendor's inputs and actions. You will also need to define in the procurement management plan how changes to the contract may happen. There is a contract change control system, and the terms of the contract should also define the process for adding changes to the contract. Usually, changes can be added directly to the contract for a fee, though some vendors may prefer a new contract or an addendum to an existing contract. It's not unusual to expect to pay a change request fee in addition to the fees to complete the actual change request.

Planning to Close the Contract

Your procurement management plan must also include details about how the contract will be officially closed. When the vendor reports that they've completed their procured work, the details of the contract may direct you on what to do next. In most instances, you'll inspect the results of the procured work to ensure that you've received all that was expected for the project and promised by the vendor. You should examine the financial details of the contract, confirm any invoices, and then finalize the payment for the vendor. Much of this work may be outside of the domain of the project manager, but you should work with your procurement office or project sponsor to ensure that the vendor is paid according to the terms of the contract.

In some instances, the vendor is no longer needed on the project. This situation could come about for a number of reasons: from poor performance to a change in project scope. When the vendor is no longer needed, a contract termination request is generated by the project manager or the contract authority in your organization. This request is subject to the terms of the contract. Your organization may be liable for early termination if the contract is to be terminated for any reason other than poor performance. It's never a pleasant experience to terminate a contract early, so it's important to follow the terms of the contract and document the experience.

It's possible that a contract termination, faulty work, or other disagreement between the vendor and the client can lead to a claim by one party or the other. The buyer in the contract sends the seller a cease-and-desist letter communicating their demand for the work to end. This letter defines why the contract is being terminated and how no additional work from the seller will be approved for payment.

The procurement management plan and the contract should detail how claims will be managed. Ideally, the vendor and the buyer settle the claim themselves without having to escalate the claim to litigation. Litigation, lawsuits, and attorneys can be one of the worst experiences for resolving a claim within a project. It's best to find a solution for the project without having to go to court to find a solution.

When a contract is closed, the vendor should receive a certificate of completion. This document verifies that the vendor has completed the contracted work and is free from the project with the exception of any warranty clause of the contract. All contracts should become part of the project's supporting detail, its archives, and then the organizational process assets. Your organization may also use a seller rating system that requires you to document your experience with the vendor once you've closed out the contract. The seller rating system allows other project managers and managers in your organization to query the system to check on vendors before they create contracts for new work and services.

CompTIA Project+ Exam Highlight: Planning Processes

As you might have expected, there are loads of CompTIA Project+ exam objectives covered in this chapter. The common theme you can use for your exam is that there are knowledge areas, and each knowledge area gets its own project management subsidiary plan in predictive projects: scope management plan, schedule management plan, cost management plan, quality management plan, human resource plan, communications management plan, risk management plan, stakeholder management plan, process improvement plan, and the procurement management plan—the final plan is the complete project management plan, and that's from project integration management. Adaptive projects plan, but not to the same depth as a predictive project. Let's look at the exam specifics covered in this chapter.

1.3 Given a scenario, apply the change control process throughout the project life cycle Changes are likely to happen within a project regardless of the project manager's best efforts to create an accurate project scope statement. Changes can result from errors and omissions, value-added objectives, external events, and changes from risk events. In a predictive project, when changes are proposed, they move through the project's change control system as a documented change request to be reviewed, evaluated, and approved or declined. Every change to project scope should also move through configuration management to determine the change's effect on the features and functions. You'll also need to use project integrated change control to determine the change's effect on the entire project. Adaptive projects welcome change, and the product owner approves and prioritizes the changes in the product backlog.

1.4 Given a scenario, perform risk management activities Risk is an uncertain event that can have a positive or negative effect on the project's ability to reach its objectives. The risk management plan defines how risks will be identified, how qualitative and quantitative analysis will happen, and how the risks will be monitored and controlled. The risk management plan also defines the scales of probability and impact and instructs the project manager and team on how the risk contingency reserve and risk responses should be created. The risk register records the status of both positive and negative risk events.

1.5 Given a scenario, perform issue management activities Issues are risk events that have happened. Issues are communicated, discussed, and recorded in the project's issue log for resolution. All issues are assigned an issue owner to manage the issue, and there is usually a deadline to resolve the issue in the project. Some issues are of greater importance than others based on the severity, project impact, and urgency for resolution. During project planning, the issue log and its rules are created and communicated for the project team. In an adaptive project, issues are known as impediments or blockers and the project manager, or scrum master, is responsible for resolving those issues.

1.6 Given a scenario, apply schedule development and management activities and techniques The project schedule is based on the WBS and the activity list it creates. The activities that you map out in a project network diagram will create the benefits and deliverables the project scope promises. The resource requirements are the people, materials, tools, and facilities that you'll need to complete the project scope. The project schedule uses constraints, hard and soft logic, and duration compression techniques to map the project work to the project calendar or milestone chart.

The exam objective includes PERT, the Program Evaluation and Review Technique, which is useful in time and cost estimating. It also includes use of a Gantt chart. A Gantt chart maps the project objectives to a network diagram, and then that network diagram is applied to a project calendar. The Gantt chart helps the project team visualize when the project work will take place with the consideration of holidays, weekends, and other pauses in the project execution. Finally, this objective includes the critical path method, float, and the examination of what project activities can be delayed given their predecessors and successor activities. You also learned about GERT, for projects that require branching and loopbacks, and the critical chain method, for considering the availability of project resources as indicated by the project network diagram.

1.7 Compare and contrast quality management concepts and performance management concepts Quality is achieved by defining the quality metrics, planning quality into the project, and then inspecting the project deliverables to ensure the existence of quality. Quality assurance is the plan to do the work properly the first time, while quality control is the inspection of the work to prove the results of quality. Quality expectations should be established as part of the project requirements so that the project manager and the project team know what they are supposed to be aiming toward. The quality management plan defines quality assurance and quality control for the project.

1.8 Compare and contrast communication management concepts Much of the project manager's time is spent communicating with project stakeholders—internal and external to the project. The project communications management plan defines who needs what information, when the information is needed, and the modality by which the information should be delivered to the stakeholder. The CompTIA Project+ exam will also test you on how best to communicate with the project team and other stakeholders. This includes the idea of ad hoc meetings versus planned meetings, face-to-face communications, communicating with virtual teams in your project, and even communication approaches such as social media.

1.11 Explain important project procurement and vendor selection concepts Procurement is a pretty common activity in most projects, and the procurement management plan directs these processes. In this chapter, we discussed the procurement planning process and the creation of exploratory documents such as the RFI, RFQ, IFB, and the RFP. You remember those documents, right? We also talked about the vendor response, the types of contracts that your project may use, and how an organization could use a memorandum of understanding prior to the actual contract being written. The contract that is signed by both parties may include a nondisclosure agreement, a statement of work for the contracted portion of the project, warranty information, and certainly payment terms, such as a purchase order.

2.3 Given a scenario, perform activities during the project planning phase This entire chapter is all about planning how the project should operate, the factors, regulations, and conditions that may affect the project, and how the project manager and team will perform. During project planning you'll address the project scope, schedule, cost, quality, human and physical resources, communication, risk, procurement, and stakeholder management. Planning is not a one-time event but happens throughout the project as the project conditions warrant. You'll shift from project execution, project monitoring and controlling, and project planning, as needed, throughout the project. There is no set rule or magic potion as to what you should do next, but each condition, organization, and project event should guide you through planning, execution, and control.

3.1 Given a scenario, use the appropriate tools throughout the project life cycle We discussed several tools that you can use during project planning to better plan and manage the project work. In this chapter, you learned about the Gantt chart, project network diagram, and milestone charts and how they can help you and stakeholders visualize the flow of the project work. This chapter also introduced the issue log, change log, risk register, and risk report to record and communicate risk and issues throughout the project.

4.2 Explain relevant information security concepts impacting project management concepts All projects must consider the security requirements of the project and the organization. Depending on the type of project work, the project management plan should address the physical security, operational security, digital security, data security, and the organizational security policy and restrictions. IT projects often interact with systems and data that not everyone should have access to view. You must plan for (and remain) in compliance with organizational security practices and requirements.

Chapter Review

The project management plan is really a collection of project management documents and subsidiary project plans. You'll need loads of documentation to help your project move from initiation to closing, and you can expect the plan to evolve as the project rolls along. Each plan in the project follows the project management life cycle: plan, execute, monitor and control, and then close that knowledge area. While each plan is separate, the project integration management ensures that all of the plans work together for the betterment of the project.

You could argue, and I think correctly, that much of the project plan begins with the project scope. You have to know what the project aims to accomplish before you can plan how to achieve your project objectives. One of the most important parts of project planning is to create the project's scope management plan and scope baseline. Recall that the scope baseline is really three separate documents: the project scope statement, the WBS, and the WBS dictionary. These three documents serve as a guideline for all future project decisions.

Once you've created the WBS or product backlog, you can define the activity list, and once you put the project activities in the correct sequence, they should happen in that order. By using the precedence diagramming method, you and your project team can map out the entire project from the start to the end. Once you've created the project network diagram, you can implement schedule compression, if needed, by designating tasks that may run concurrently or be completed faster with additional resources. This is also the fun process of finding the project's critical path, float, and opportunities to improve the project performance on schedule.

You'll also need to consider the cost of the resources your project needs. The cost management plan defines how costs will be estimated, budgeted, and controlled. Your organization may have an approach that you have to follow with cost management—though there are some common principles you'll use as a project manager. Associated with project costs is project risk. Risks can affect both project costs and the project schedule, so a risk management plan is needed to identify, analyze, and respond to risk events.

As a project manager, you will focus primarily on getting things done, but you'll need two important skills: managing the project team and communications. Your project team will look to you to provide the resources, training, and leadership so that they can complete their assignments. You'll need project communications planning to ensure that you're communicating the right information, to the right people, at the right time, and in the right manner. Your human resource plan and the project communications management plan are needed for these skills.

The final plan you'll create as part of your project management plan is the procurement management plan. This plan defines how the project may purchase the goods and services it needs to complete the project objectives. The procurement management plan directs the decision to purchase, the vendor selection, the contract types that you're allowed to use, and the administration of the contract once you've reached a deal with the vendor.

Creating the project plan is a type of alchemy. Adequate research, technical skills, and the ability to use logic and reason are required to create a solid, consistent plan for the project team, management, and the project manager to follow. Patience and vision are two attributes of a successful project manager. Just ask Noah.

Exercises

These exercises allow you to apply the knowledge you have learned in this chapter and are followed by possible solutions.

Exercise 6-1: Create a Project Network Diagram

In this exercise, you will create a project network diagram for a fictional company named Donaldson Investments and Holdings. You will be given the core information on the project and then create the diagram based on the information supplied. This is based on the WBS you examined in Chapter 5.

As you have learned, the creation of a PND can be a long and tedious process. If during this exercise you need some prompts to create the PND, you can refer to the Network Diagram Worksheet file in the accompanying online resources. (If you need assistance accessing this file, please refer to Appendix C.) The worksheet has key tasks completed to help you develop the PND.

The worksheet has key assignments completed to coach you through the creation of the diagram. If you would prefer not to use the worksheet to receive any coaching, you can create your own network diagram.

Scenario: Jennifer is the project manager for Donaldson Investment and Holdings. The project is an implementation of a new mail server and the mail clients on all of the workstations. Here is the project's task list. To organize your PND, begin by mapping out the network diagram by identifying the sequence of tasks and the relationship between tasks.

Task Name
Install Windows Server
Install Exchange Server
Configure Exchange Server
Link to other servers
Create test user accounts
Set mailbox rules
Create client installation packages
Create system computer policies
Test policies
Develop installation procedures
Test installation procedures
Create CD image for remote access users
Test CD image
Host pilot users class
Create class workbook for training
Host pilot users training class (groups of ten)
Send image to pilot users
Hold pilot users forum to discuss usage
Analyze pilot user feedback
Finalize and test install image
Hold user training classes
Roll out based on attendance

Exercise 6-2: Create a Risk Contingency Reserve

Another part of project planning is to evaluate the project risks and determine the risk contingency reserve. The following is a simple risk matrix where a few risks have been identified, their probability has been predicted, and their impact on the project has been assigned. Your job is to calculate the risk event value for each risk event and then to predict this project's risk contingency reserve. Remember that both positive and negative risks must be accounted for.

To complete this exercise, multiply the value of the probability by the impact of the risk for the risk event value. The sum of the risk event values will represent the project's risk exposure, and the inverse amount of the risk exposure is the needed contingency reserve.

Risk	Probability	Impact	Risk Event Value
Accuracy of requirements	0.20	–$10,000	
Resource availability	0.40	–$13,000	
Errors and omission in requirements gathering	0.10	–$19,000	
Delays from vendor	0.60	–$28,000	
Delays from client approval	0.15	–$8,000	
Discount from vendor	0.40	$12,000	
Data loss	0.10	–$35,000	
Network latency	0.25	–$8,000	

1. What is the project's risk exposure?
2. What is the needed contingency reserve?
3. What risks have a greater impact than the total contingency reserve?
4. What should you do with these risk events that have a higher impact than the contingency reserve?

Exercise Solutions

The following offer possible solutions for the chapter exercises.

Exercise 6-1: Create a Project Network Diagram

The network diagram solution can be found in the online resources. See the file named Completed Exercise One.

Exercise 6-2: Create a Risk Contingency Reserve

The following table is the completed risk contingency reserve. You can find this by multiplying the probability and the impact for the risk event value. The sum of the risk event value is inverted to reveal the risk contingency reserve needed for the project.

Risk	Probability	Impact	Risk Event Value
Accuracy of requirements	0.20	–$10,000	–$2,000
Resource availability	0.40	–$13,000	–$5,200
Errors and omission in requirements gathering	0.10	–$19,000	–$1,900
Delays from vendor	0.60	–$28,000	–$16,800
Delays from client approval	0.15	–$8,000	–$1,200
Discount from vendor	0.40	$12,000	$4,800
Data loss	0.10	–$35,000	–$3,500
Network latency	0.25	–$8,000	–$2000

1. The project's risk exposure is –$27,800.
2. The needed contingency reserve is $27,800.
3. Risks associated with delays from the vendor and data loss have a greater impact than the total contingency reserve. Delays from the vendor will cost the project $28,000, and data loss could cost the project $35,000.
4. The risks of vendor delays and data loss can be combated with mitigation to reduce the probability and/or impact of the events. If these events happen, they'll have a strong adverse effect on the project's ability to maintain costs. It's often more cost effective and time effective to spend more to alleviate these types of risks through mitigation than to allow the risks to exist within the project.

Questions

1. You are the project manager for your organization and are creating the project schedule. You know that a project schedule is composed of all the following components except for which one?
   - **A.** The network of all of the tasks within the project
   - **B.** The assignment of resources
   - **C.** The budget for the project
   - **D.** The reflection of the WBS

2. Juan is the project manager for his organization, and he's creating the project management plan. He is reviewing the project scope baseline with his project team so that they'll understand the requirements and direction of the project. Which of the following is the correct definition of the project scope baseline?
   A. It is all the work, and only the required work, to meet the project objectives.
   B. It is the combination of the project scope statement, the WBS, and the WBS dictionary.
   C. It is the combination of the project requirements, the project scope, and the WBS.
   D. It is the consideration of the project requirements, assumptions, and constraints that affect the project planning.
3. You are the project manager for your organization, and you're creating with your project team the project management plan. Which project management plan defines how the project deliverables will be inspected by the project team?
   A. Process improvement plan
   B. Scope management plan
   C. Risk management plan
   D. Quality management plan
4. Which project management plan defines integrated change control?
   A. Quality management plan
   B. Scope management plan
   C. Change control plan
   D. Risk management plan
5. You are the project manager for your organization. Management has approved the project scope, and they've asked you and your project team to begin creating a project network diagram. What is a project network diagram?
   A. An expansion of the WBS
   B. An expansion of the Gantt chart
   C. A sequential mapping of the project work
   D. A topology of a project phase
6. You are the project manager for your organization, and you're working with your project team to create the WBS. You and the project team are examining each deliverable of the project and are subdividing the deliverables into smaller elements. Once you've created the WBS, what document can you create next in your project planning?
   A. Activity list
   B. Project network diagram

C. Risk management plan

D. Schedule management plan

7. You and the project team have identified the critical path and are now reviewing the available float in your project network diagram. All of the following statements about float are incorrect except for which one?

A. Every project will have float.

B. Only complex projects will have float.

C. Float is the amount of time an activity can be delayed without increasing the project costs.

D. Float is the amount of time an activity can be delayed without causing the project to be late.

8. You are the project manager for your organization and have been assigned a new project team. Many of the project team members haven't worked with each other before, so you elect to do some team-building exercises to help the team learn to trust one another. What project management plan defines this process of team development?

A. Project management plan

B. Quality management plan

C. Team leadership plan

D. Project human resource plan

9. What law states that work will expand to fill the amount of time allotted to it?

A. Law of Diminishing Returns

B. Moore's Law

C. Parkinson's Law

D. Murphy's Law

10. You are the project manager of a web server and website upgrade. You have assigned one employee the task of creating the web pages and another employee the task of developing the web pages. The two can begin work on their assigned tasks at the same time. What type of relationship do these tasks have?

A. FS

B. SS

C. FF

D. SF

11. Philippe is the project manager of a network upgrade. All of the client workstations are to be replaced, and this task has been assigned to Steve, Haroun, and Beth. Once the physical workstations are in place, Matteo will release an automated script to deploy an operating system to each of the new workstations. What type of relationship best describes these two tasks?

A. FS

B. SS

C. FF

D. SF

12. You are the project manager of the NHQ Project and are working with your team to create risk responses. One risk, managing an electrical component in your project work, is deemed too dangerous to handle internally, so you and the project team elect to hire a vendor to manage this portion of your project. What type of risk response is this?

A. Sharing

B. Mitigation

C. Avoidance

D. Transference

13. Why should a project manager avoid assigning specific dates to tasks when at all possible during the creation of the network diagram?

A. Dates cannot change; tasks can.

B. Dates require the activity to happen at a specific time.

C. Tasks assigned to dates do not consider successor tasks.

D. Tasks assigned to dates do not consider both successor and predecessor tasks.

14. You are the project manager for your organization and are about to begin the procurement process. You have decided that you want the vendors to provide a fixed price for the materials your project will need. You create the SOW for the work. What other document should you send to the vendors?

A. RFP

B. RFI

C. IFB

D. PO

15. Why must lag times be scheduled between tasks in a project network diagram?

A. Lag times allow the team to take a break.

B. Lag times reflect instances when task overruns are anticipated.

C. Lag times reflect weekends and holidays.

D. Lag times allow other events to be completed before successor tasks can begin.

Answers

1. **C.** While the project budget is important, it is not part of the project schedule. A project schedule is the compilation of all of the tasks to be completed within a project and the assignment of resources, and it should reflect the entries of the WBS. If you are using Microsoft Project, then you can streamline your efforts, as Microsoft Project will allow you to create the network diagram and work toward the WBS.
2. **B.** The project scope baseline is composed of three documents: the project scope statement, the WBS, and the WBS dictionary. These three documents serve as the guideline for all future project decisions.
3. **D.** Quality control is the process that requires the project team to inspect the project work for the existence of quality. The quality management plan defines how quality control should be used, how errors are managed, and what techniques will be implemented.
4. **B.** The scope management plan defines how changes to the project scope may be allowed to enter the project. When a change is proposed, the project manager, the change control board, and even the project team will examine the change to see its effect on the entire project.
5. **C.** A project network diagram is a fluid mapping of the entire project, not just one phase.
6. **A.** Once you and your project team have created the WBS, you can then create the activity list. The activity list is based on the work packages; recall that work packages are the smallest item in the WBS.
7. **D.** Float is the amount of time an activity's completion can be delayed without delaying the project end date. Not every project will have float, though most projects do—including simple or complex projects.
8. **D.** Team development is part of the project human resource plan. It's essential for project team members to learn to trust one another, work well together, and become a cohesive team unit for the execution of the project.
9. **C.** When project team members and project managers add time to their time estimates, they'll likely succumb to Parkinson's Law: work expands to fill the time allotted to it.
10. **B.** These tasks have SS dependency, which means "start to start."
11. **A.** The tasks involving the replacement of workstations must be completed before the script can run; the dependency is FS, or "finish to start."
12. **D.** The best answer to this question is the transference risk response. You have transferred the risk of the dangerous electrical work to a vendor to manage for you. When you do transference, you're usually hiring someone and there's a fee associated with the risk response.

13. **D.** Tasks assigned to dates do not consider successor and predecessor tasks. This becomes a huge problem when upstream tasks are delayed by several days; the task assigned to a specific date does not change to reflect the changes of the tasks upstream. Whenever possible, do not assign tasks to a specific date. Examples of assigning tasks to specific dates can include when a particular resource is available or a consultant is scheduled to be present.
14. **C.** When you want the vendors to provide a fixed fee for their services or materials, you ask for a bid or quote. In this example, you'd send an IFB—an invitation for bid document—also called a request for bid (RFB).
15. **D.** Lag time is waiting time and can be used to allow other events to be completed before successor tasks can begin. For example, a dependent task is to mail a survey to all of the network users. Before the successor task, analyzing the user surveys, can happen, there must be time allotted for the users to respond to the survey. It's not an actual task, but it still requires times within the diagram.

CHAPTER 7

Organizing a Project Team

This chapter covers the following topics:

- Assessing internal skills
- Serving as a project coordinator
- Working with a project scheduler
- Managing team issues
- Defining agile roles and responsibilities
- Hosting agile project ceremonies
- Hiring a team

Think of your favorite caper movie. Remember how the heist team in the movie is assembled? Each member has a specialty and other necessary skills to get the job done. Notice how there's never an extra character walking around slurping coffee, dodging work, and whining about how tough their job is? Unfortunately, in the world of IT project management, you'll have to work with both types of characters—the specialists and the extras.

As a project manager, you will recruit the diehard dedicated workers who are genuinely interested in the success of the project. These team members are exciting to be around, because they love to learn, love technology, and work hard for the team and the success of the project. The other type of team members you'll encounter are nothing less than a pain in the, er, neck. These folks couldn't care less about the project, the success of the organization, or anyone else on the team. Their goal is to complete their required hours, draw a paycheck, and get on with their lives.

The reality is, however, most people want to do a good job. Most team members are generally interested in the success of the project. If you get stuck with one of the rotten apples, there are methods to work with them—and around them. This chapter will focus on how you, the project manager, can assemble a team that works well together. Your team may not be in any caper movies, but parts of the project can be just as exciting.

VIDEO For a more detailed explanation, watch the *Leading a Team* video now.

Assessing Internal Skills

Whether you get to handpick your project team or your team is assigned to you by management, you will still need to get a grasp on the experience levels of each team member. If you have an understanding of what your team members are capable of doing, the process of assigning tasks and creating the project plan will go much easier for you.

As a project manager, you must create a method to ascertain the skills of your team. It would, no doubt, be disastrous to your project if you assembled a team only to learn later that they were not qualified to do the work assigned to them. In some cases, this task will be easier to do than in others, especially if you've worked with the team members before, interviewed the team members, or completed a skills assessment worksheet.

Identifying Resource Requirements

Before you can start building your project team, you need to know what resources you'll need for the project to be successful. Recall that resources are materials, facilities, tools, equipment, and, especially in this chapter, people. Projects are completed by people, and you'll need the right people to do the right work to complete the project objectives. Depending on the nature of the project, you'll probably have a good idea what the project needs in order to be successful. You can, and should, use the project's scope baseline to help you and the project team identify the types of resources that you need on your project.

By referencing the scope baseline, and in particular the WBS or product backlog, you can identify the resources you'll need to complete the work packages within the project. By mapping the resources to the WBS or product backlog, you can create a resource breakdown structure. A resource breakdown structure illustrates what resources are needed to create corresponding deliverables. This can also help when you assign resources to project activities—there may be only so many resources available, so you'll need to create a project schedule that considers the availability of the project resources to create the identified project deliverables.

EXAM TIP Resources are people and materials, equipment, and facilities. People represent the labor needed to do the project work and are often the most expensive part of the project budget. Adaptive projects frown on calling teams resources—people are people, resources are things, though management commonly says human resources.

This examination of resource requirements can also help you discover a resource shortage. When reviewing the work, the interproject dependencies, and the calendars of the available resources, you may discover that you don't have the needed resources to complete the work as planned. Now you'll have to either reschedule the project work, acquire resources to alleviate the resource shortage, or even change your project plan based on the lack of resources available to complete the project.

Avoiding resource shortages, overallocation issues, and interproject contention is ideal, but it takes accurate mapping of who's doing what in the organization, and actual reporting of time spent on assignments. When you consider all the moving pieces within a project and within an organization, it's often a real challenge to avoid these issues and have the right resources you need when you need them. Communication, system-wide scheduling of resources, and reporting of issues or risks within projects that may affect resource scheduling are needed but not always easy to achieve or maintain.

Experience Is the Best Barometer

As you gain experience as a project manager, you will learn which people you'd like on your team—and which you wouldn't. Your first requirement for team members is the competency of the individuals to complete the work, but a close second is their attitude, willingness to work, and how well they can get along and work with others. If you are a consultant brought into the mix to manage an IT implementation, you'll have to learn about the team members, their goals, and their abilities.

You must use strategies to recruit and woo knowledgeable and hard-working team members onto your team. This means, of course, you'll have to do fact-finding missions to gain information on your recruits. As Figure 7-1 demonstrates, you have available to you many methods to assess internal skills—this is your resource pool.

Once you've started your fact-finding mission, rely on multiple methods to assess internal skills:

- **Prior projects** Obviously, if you've worked with your team members prior to this project, you'll have a good idea who's capable of what tasks. You'll also have a record, through historical information, of who's reliable, dependable, and thorough, and who has other traits of a good worker.
- **Organizational knowledge** You may not have worked directly with particular team members who have been assigned to your project, but you might have a good idea of their track record. Let's face the facts: in your organization, it's likely that there are people you haven't worked with, but you know the type of workers they are by their reputation, their ability to accomplish tasks, and what others say about them. Gossip is one thing, but proven success (or failure) is another. The best way to learn about someone, of course, is not through hearsay, but to work with them or speak directly with their manager or other project managers.

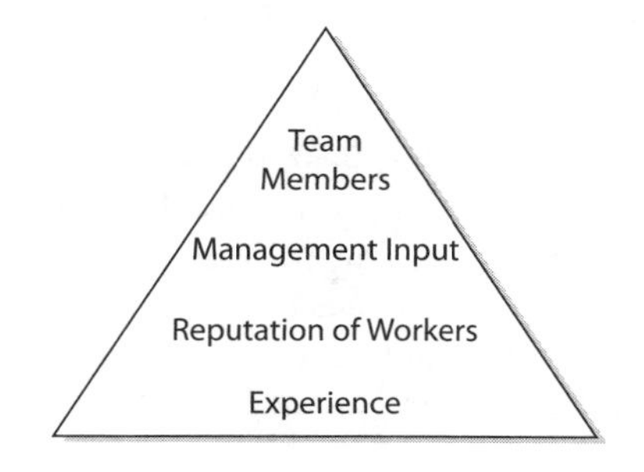

Figure 7-1 Assessing the internal resource pool can help with resource identification.

- **Recommendation of management** You may not have the luxury of selecting your team members like you're picking a kickball team at recess. You'll probably be able to recruit some members of your team, but not all of them. Functional or senior management will have an inside track on the abilities of employees and can, and will, recommend members for your project. Management will also be able to select individuals who can commit time to the project.
- **Recommendation of team members** Most likely, you will have other IT professionals within your organization whom you trust and confide in. These folks can help you by recommending other winners for your team. These individuals are likely in the trenches working side-by-side with other IT pros. Use their "scouting" to find excellent members to work on your project.

Résumés and Skill Assessments

Another source, if you have access to the document, is the résumé for each team member. A résumé can quickly sum up the skill set of a team member. You may want the project team to create quick résumés for you in order to learn about the experiences of individual members. Use caution with this approach, however. Résumé submission has the connotation of getting, or keeping, a job, and your team members may panic. If you want to use this method but are uneasy using the word "résumé," have the team members create a listing of the projects, their skills, and other past accomplishments. This will give you a way to quickly understand the collection of talent and then assign work to the team.

You might also rely on websites such as LinkedIn, where your colleagues have posted their résumés, skill sets, and credentials. Of course, use that approach with caution, because the individuals may have bloated their experience or not updated their credentials in some time. (And as I'm writing this, I'm assuming for future readers that LinkedIn still exists as I know it today.)

A collection of skills will also allow you to determine if you have the resources to complete the project. For example, if you're about to create a database that will span 18 states with multiple servers that will provide real-time transactions for clients, it'll be tough to do if none of your team members has worked with relational databases before.

TIP Many organizations have an internal database of employees and skill sets that can be searched. This helps managers and resources find one another. The challenge, of course, is that the database must be kept up to date based on new education and credentials the employees earn.

Create a Roles and Responsibilities Matrix

A *roles and responsibilities matrix* is a tool used to identify all of the roles within a project and the associated responsibilities to the project work. This matrix is an excellent way to identify the needed roles of the project participants, identify what actions they'll need to take in the project, and ultimately determine if you have all of the roles to complete the identified responsibilities. Here's a quick example of a matrix for a software rollout project.

	Project manager	Application developer	Network engineer	ZENworks expert
Create the application	A	C	P	
Test the application	A	P	P	
Package the application	R		R	P
Test the application release	R	R		C
Push the application to the workstations	A		P	C

Here's the legend for this matrix:

A = Approves
R = Reviews
P = Participant
C = Creator

The roles and responsibilities matrix can help the project manager identify the needed resources to complete the project work—and determine if the resources exist within the organization's resource pool. Later in the project, the project manager will use an even more precise matrix, the *responsibility assignment matrix (RAM),* to identify which tasks are assigned to which individuals. Another form of a responsibility assignment matrix is a RACI chart. A RACI chart is similar to the previous example matrix, but it uses the legend of Responsible, Accountable, Consult, and Inform. And that's why it's called a RACI chart—the first letter of each responsibility spells RACI.

Learning Is Hard Work

Within the IT employment world, a requirement for certification has become practically mandatory. Certifications such as the PMP, PMI-ACP, PRINCE2, AWS Credentials, Microsoft Certified Systems Engineer, and Oracle DBA—even industry certifications like CompTIA's Project+, A+, and Network+—are proof of your knowledge in a particular area of technology.

Individuals can earn these certifications based on training, experience, or a combination of both. Certifications are certainly a way to demonstrate that professionals have worked with the technology, understand the major concepts, and were able to pass the exam. Certifications do not, however, make the individual a master of all technologies.

Within your team, whether members have certifications or not, you'll need to assess if they need additional training to complete the project. Training is always seen as one of two things: an expense or an investment. Training is an expense if the experience does not increase the ability of the team to implement tasks. Training is an investment if the experience greatly increases the ability of the team to complete the project.

When searching for a training provider, consider these questions:

- What is the experience of the trainer?
- Can the trainer customize the class to your project?
- Would hiring a mentor be a better solution than classroom training?

- What materials are included with the class?
- What is the cost of the course?
- Is there an in-house training department that can deliver the training, provide assistance in developing the curriculum in-house, or assist in contracting with an outside trainer?
- Would it be more cost effective to host the training session in-house?
- Are there viable online solutions that can save time and provide value?

These questions will help you determine if training is right for your project team. In some instances, standard introductory courses are fine. Typically, the more customized the project, the more customized the class should be as well. Don't assume that just because a training center is the biggest, it's also the best. No matter how luxurious a training room, how delicious the cookies provided, or how slick the brochures are, the success of the class rests on the shoulders of the trainer. And online education platforms, such as Udemy (https://www.udemy.com), are becoming a quick way to ramp up on new technologies and approaches for the project team.

Creating a Team

You can't approach creating a team the way you would approach baking a cake or completing a paint-by-number picture. As you deal with multiple individuals, you'll discover their personalities, their ambitions, and their motivations. Being a project manager is as much about being a leader as it is about managing tasks, deadlines, and resources.

You will, through experience, learn how to recognize the leaders within the team. You'll have to look for the members who are willing to go the extra mile, who do what it takes to do a job right, and who are willing to help others excel. These attributes signal the type of members you want on your team. The easiest way to create teams with this type of worker? Set the example yourself.

Imagine yourself as a team member on your project. How would you like the project manager to act? Or call upon your own experience: What have previous project managers taught you by their actions? By setting the example of how your team should work, you're following ageless advice: leading by doing.

Defining Project Manager Power

Project managers have responsibility. And with that responsibility comes power. When it comes to the project team, you are seen as someone with some degree of power. Get used to it, but don't let it go to your head. While the project manager must have some degree of power to get the project work done, the level of degree is also likely relevant to the organizational structure you're working in. For example, recall that a functional organization gives the power to the functional manager, and the project manager may be known as just a project coordinator.

A project manager does, however, wield a certain amount of power in most organizations. The project team can see this power, correctly or incorrectly, based on their relationship with you. Their perception of your power—and how you use your project management power—will influence the project team and how they accomplish their project work.

There are five types of project manager powers:

- **Expert** The authority of the project manager comes from experience with the technology the project focuses on.
- **Reward/penalty** The project manager has the authority to give or withhold something of value to team members.
- **Formal** The project manager has been assigned by senior management and is in charge of the project. This is also known as *positional power*.
- **Coercive** The project manager has the authority to discipline the project team members. This is also known as *penalty power*. When the team is afraid of the project manager, it's coercive.
- **Referent** The project team personally knows the project manager. Referent can also mean the project manager refers to the person who assigned them to the position—for example, "The CEO assigned me to this position, so we'll do it this way." This power can also mean the project team wants to work on the project or with the project manager because of the high priority and impact of the project.

EXAM TIP None of these powers are better than the other powers, just different powers for different situations. Use the correct power when it's most appropriate to get progress and move the project forward.

Hello! My Name Is…

If your team works together on a regular basis, then chances are the team has already established camaraderie. The spirit of teamwork is not something that can be born overnight—or even in a matter of days. Camaraderie is created from experiences of the teammates. A team's successful installation of software, or even a failure, creates a sense of unity among the team.

It's mandatory on just about any project that team members work together. Here's where things get tricky. Among those team members, you've got ambition, jealousies, secret agendas, uncertainties, and anxiety pooling in and seeping through the workers on your project. One of your first goals will be to establish some order in the team and change the members' focus to the end result of the project. Personal ambitions are fine, but when they take precedence over the good of the project, they can have a detrimental effect on the success of a project.

By motivating your team to focus on the project deliverables, you can, like a magician, misdirect their attention from their own agendas to the project's success. You can spark the creation of a true team by demonstrating how the members are all in this together. How can you do this? How can you motivate your team and change the focus from self-centric to project-centric? Here are some methods:

- *Show the team members what's in it for them.* Remember the WIIFM principle—"What's In It For Me." Show your team members what they personally have to reap from the project. You may do this by telling them about monetary bonuses they'll receive. Maybe your team will get extra vacation days or promotions. At the very least, they'll be rewarded with adding this project to their list of accomplishments. Who knows? You'll have to find some way to make this project more personally important for each team member.
- *Show the team what this project means to the organization.* By demonstrating the impact that this implementation has for the entire organization, you can position the importance of the success (or failure) of the project squarely on the team's shoulders. This method gives the team a sense of ownership and a sense of responsibility.
- *Show the team why this is exciting.* IT project managers sometimes lose the sense of excitement wrapped up in technology. Show your team why this project is cool, exciting, and fun, and the implementation will hardly be like work. Remember, IT pros typically love technology—so let them have some fun! It is okay to have a good time and enjoy your work.
- *Show the team members their importance.* Team members need to know that their work is valued and appreciated. You can't fake this stuff. Develop a sense of caring, a sense of pride, and tell your team members when they do a good job. Let them own the technology, use the technology, and be proud of their work.

Where Do You Live?

In today's world, it's typical of a single project to span the globe. No doubt it's difficult for team members to feel like they are part of the same team when some team members are based in London and their counterparts are in Phoenix. Ideally, co-located teams communicate better, work together better, and have a stronger sense of ownership. Reality, however, proves that virtual teams exist in organizations, and the project manager must take extra measures to ensure that the project succeeds regardless of the geographical boundaries. When dealing with a virtual team, you will likely build your team around subteams. A *subteam* is simply a squadron of team members unique to one task within the project or within each geographical area.

For example, as depicted in Figure 7-2, a company is implementing Oracle servers throughout its enterprise. The company has 12 locations throughout the United States. Some of the same tasks that need to be accomplished in Ohio will also need to be performed in Texas.

Rather than having a six-member team fly around the globe to meet with other teams, the project manager implements 12 subteams, each with six members. Of the six members, one is the team leader for that location. The team members in each location report to their immediate team leader. All of the team leaders report to the project manager,

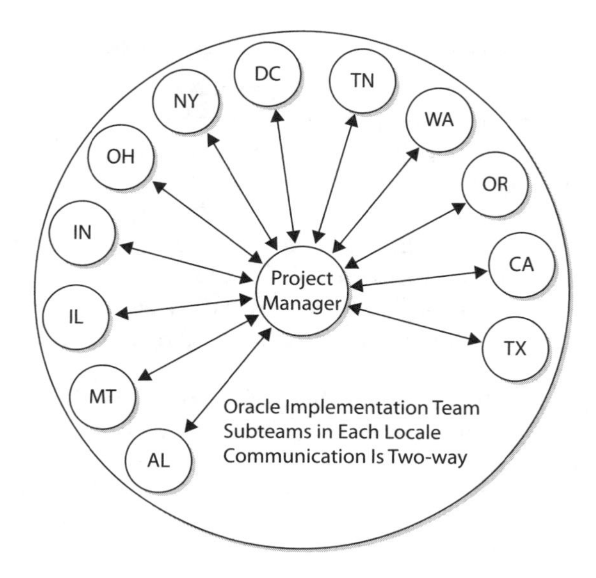

Figure 7-2 Virtual teams and subteams demand more attention to communication than co-located teams.

the 73rd member of the team. Implementation of the Oracle servers at each location will follow a standard procedure for the installation and configuration. The path to success should be the same at each location regardless of geography.

Certainly, not all projects will map out this smoothly. Team members at some sites may not have the technical know-how of those at others, and travel will be required. In other instances, some sites will require more configuration than others, or an increase in security, or other variances. The lesson to be learned is that when teams are dispersed, a chain of command must be established to create uniformity and smooth implementation. Unfortunately, the phrase "out of sight, out of mind" often proves true when dealing with dispersed project teams.

Finally, when working with multiple subteams, communication is paramount. Team leaders and the project manager should have regularly scheduled meetings either in person or through teleconferences or videoconferences. In addition, team leaders should have the ability to contact other team members around the globe. You'll need to plan for how to overcome communication challenges such as time zones, language barriers, and cultural differences.

EXAM TIP Of course, the pandemic has changed how project teams work. While the CompTIA Project+ exam may not quiz you on the impact of the pandemic, it'll certainly address remote or virtual teams. You might see virtual teams also referred to as *non-collocated teams*, which is a big way of saying we're not all in one location.

In some instances, team members from different teams will need to work together as well. For example, if the communication between two servers has to be configured, the teammates responsible for this step of the configuration will need to coordinate their configurations and installation with the teammates who have identical responsibilities in other locations.

Building Relationships

When an individual joins your team, you and that individual have a relationship: project manager to team member. Immediately the team member knows their role in the project as a team member, and you know your role in relation to the team member as the project manager.

What may not be known, however, is the relationships between team members. You may need to introduce each team member and explain why that person is on the team and what responsibilities that person has. Don't expect your team members to figure things out for themselves. In a large project, it would be ideal to have a directory of the team members, their contact information, and their arsenal of talents made available to the whole team.

On all projects, your team will have to work together very quickly. It's not a bad idea to bring the team together in some type of activity away from the workplace. With virtual teams, it's ideal to bring everyone into a central location for face-to-face meetings and team-building activities. The following are examples of team-building exercises:

- A bowling excursion
- A hike and overnight stay in the wilds, or some other outdoor activity
- A weekend resort meeting to learn about each other and discuss the project
- A trip to your local pool hall for an impromptu round of team pool

Serving as a Project Coordinator

If you're fairly new to project management or you're working in a functional organization, you may be given the role and title of project coordinator rather than project manager. The project coordinator is an individual who does just that: coordinates the project. This isn't a demeaning role, and I think it sounds better than the cursed "junior project manager" some organizations assign. The point isn't to get hung up on the title of project coordinator, but rather to embrace and complete the duties of the project coordinator effectively.

The project coordinator completes many of the project management duties, but with a bit more oversight from the project sponsor, an experienced project manager, or the functional manger. If you're assigned the role of the project coordinator, do the job well as you work to advance in your organization or career—it's a stepping stone to bigger and better things.

Exploring the Project Coordinator Roles and Responsibilities

The project coordinator, for the most part, does the same activities as the project manager, but with oversight from management. This oversight can come from a senior project manager, the project sponsor, or the functional manager. The project coordinator will work to identify the project requirements, write and decompose the project scope, and build out the activities with the project team. All of these activities, however, are usually done with more "check-ins" with the person overseeing the project management duties.

For example, the supervisor of the project coordinator would coach the project coordinator on the next activities to complete in their project management duties. The project coordinator would ask questions for clarity, listen to some advice, and then go on to complete the activities and report back to their supervisor. As the project coordinator becomes more and more confident and experienced, and the supervisor begins to trust and see the increasing competence of the project coordinator, there'll be less and less coaching from the supervisor, and the project coordinator will complete more and more project management duties unsupervised.

If the project involves external customers, the project coordinator may not be working with customers directly and alone, but rather in the presence of the supervisor or project sponsor. This is *not* meant to be an insult to the project coordinator; instead, it ensures that the project coordinator is providing correct information, managing the relationship properly, and not affecting the business of the organization in a negative manner. Like all things, as the project coordinator gains experience and the supervisor learns to put more trust in the project coordinator, more responsibilities will shift directly to the project coordinator.

Supporting the Project Manager

While most of our conversation about the project coordinator so far has described the project coordinator working under the guidance of a supervisor role, you may encounter another function of the project coordinator. In some instances, the project coordinator may not be actually managing the project, but working with a project manager as more of an assistant. The project coordinator in this role will follow the leadership of the project manager and complete duties assigned by the project manager.

This type of project coordinator will help the project manager with day-to-day project management activities. This can include tedious and less-than-glamorous things such as keeping minutes in a meeting, replying to e-mails on behalf of the project manager, and completing errands and tasks that help the project move along. As the project coordinator gains more experience and the project manager learns to trust and lean on the project coordinator, there may be more assignments and duties that are project management–centric, such as scheduling, leading team-building activities, writing reports, and planning.

Just because a project coordinator is seen as a helper to the project manager doesn't necessarily mean that this isn't an important position. Some projects can be large in scope and unwieldy to manage alone. A detailed and organized project coordinator can be a boon for the project manager and the project's success. Often the project coordinator and the project manager work as a team in the project, with the role of the project manager having a slightly senior role over that of the project coordinator. It's important not to get hung up on who's in charge, but to put the focus on getting the right things done to ensure that the project keeps moving toward its goal of completion.

Working with a Project Scheduler

A few years ago I was consulting as a project manager for a large insurance company. The organization assigned me 20–25 projects to manage at once. Of course, I was stunned, stupefied, and scratching my noggin. Twenty-five projects? How in the world can one project manager do that? Well, it turns out that these projects were all small-to-medium–scoped projects that each would take between one month and six months to complete. Second, the team assigned had done these types of projects over and over, so they knew what to expect in the workload—and, finally, I was assigned a project scheduler.

A project scheduler is the fabled, mythical creature of the project management landscape—managing a project with a full-time scheduler is like finding a pot of gold at the end of the rainbow. It is fantastic! The project scheduler is an expert at helping the project manager schedule project tasks, build the schedule, and ensure that enough resources are available to complete the work. The project scheduler is a wonderful ally to have in any project: a good project scheduler can automate the scheduling, track resources, compile reports, and take a load off the project manager's desk. A good scheduler with a great team is an absolute joy to work with.

Scheduling Project Activities

The project scheduler first needs something to schedule—that is, the project activities. This means you, the project manager, will first need to have worked through the scope definition, the WBS creation, and the activities list creation. The scheduler will then, with the project manager and project team, flesh out the project schedule based on the ordering of the events, constraints, and other details that affect the ordering of the activities. The project scheduler will most likely use advanced scheduling software (such as Primavera Professional Project Management or Microsoft Project) to schedule the work.

Using the scheduling software, the scheduler will follow some typical steps to build the schedule:

1. Define and enter the activities' duration estimates.
2. Sequence the activities in the order in which they should happen.
3. Identify the resources who will work on different activities.
4. Arrange the paths to completion in a project network diagram.
5. Apply any constraints such as vacations, holidays, and company business cycles.
6. Discuss risks that may affect the schedule.
7. Gain consensus from the project manager and the project team on the schedule.

Your organization may have additional steps and requirements for creating the project schedule, but these are the most common steps when working with a project scheduler. These steps likely won't be finalized in one meeting—in fact, the scheduler may collect information from the project team and project manager and then go off to work on the schedule. It takes concentration to build a comprehensive schedule, and there'll likely be trade-offs and discussions with the project manager, the project team, and maybe even management, before the schedule is approved.

Managing Resources with the Project Scheduler

Once the schedule has been created and approved, the scheduler doesn't go away. In fact, you'll probably see more and more of the scheduler as the project moves forward. In my consulting assignment, the scheduler came to our weekly team meetings, and we quickly buzzed through the status of each assignment, how many hours the team member had worked on the assignment, the percentage of the work completed, and how much longer the team may need on the task. This gave everyone a chance to report on their work, to anticipate any tasks that may be late, and to be held accountable to one another in the project reporting.

The scheduler can also help the project manager communicate the schedule to management and stakeholders. The scheduler can create reports that will help the project manager communicate several things to management:

- **Work completed in ratio to work remaining** Reports how many items or assignments have been completed, how many assignments remain to be completed, and, in some cases, how much money has been spent on labor to complete the work. This helps with predicting project completion.
- **Resource allocation** Reports how many hours of labor are being allocated to the project in a given time period, such as so far in the project or in the current week or month, and expectations for the remainder of the project.
- **Resource over allocation** Identifies any resources that are working more than a predetermined number of hours on the project in a given time period. For example, a team member can work no more than 40 hours per week in a project. If that person is schedule to work 47 hours this week in the project, they are overallocated 7 hours.
- **Resource shortage** Consider the activities that don't have resources assigned or available on the project team. Keep an eye towards tasks that have a resource shortage and plan for how the resources will be acquired for the activities. If the project is running against a deadline, consider adding resources to activities that can be reduced with additional effort.
- **Benched resources** Identifies resources that aren't being used in the project at any given time. This helps the project manager see opportunities, or risks, that need to be addressed.
- **Inter-project resource management** A scheduler can help identify who's being used on what projects in the organization and identify conflicts or dependencies between projects. This helps with better planning and improved communications among the project managers and the project teams.

The role of the project scheduler is to help the project manager better create the schedule, identify risks and opportunities in the project, collect activity completion updates, and create a more cohesive environment in the project. If you're lucky enough to have a scheduler on your project, don't get in their way; let them do their jobs, rely on their expertise, and know that they are available to help you manage a better project.

EXAM TIP Project schedulers are a stakeholder but aren't necessarily part of the project team. Other roles that help the project can include subject matter experts, business analysts, testers, quality assurance specialists, developers, engineers, consultants, and all sorts of people. Your project can get help from many different resources that are considered stakeholders to the project but aren't necessarily part of the project team.

Managing Team Issues

Without a doubt, people will have conflicts. Fortunately, in most offices, people are mature enough to bite their tongues, try to work peacefully, and, as a whole, strive to finish the project happily and effectively together. Conflict is natural and can be a healthy event to discover new ideas, better approaches, and solutions for the project, if the conflict is done with mutual respect.

Most disagreements in IT project management happen when two or more people feel very passionate about a particular IT topic. For example, one person believes a network should be built in a particular order, while another feels it should be constructed using a different approach. Or two developers on a project get upset with each other about the way an application is created. Generally, both parties in the argument are good people who just feel strongly about a certain work methodology.

There are, of course, a fair percentage of contrary and pessimistic people in the world. These people don't play well with others and are obnoxious at times. They don't care about other people's feelings, and much of the time they don't care about the success of your project.

Unfortunately, you will have to deal with disagreements, troublemakers, and obnoxious people to find a way to resolve differences and keep up the project's momentum.

Dealing with Team Disagreements

There will be instances when the project team, management, and other stakeholders disagree on the progress, decisions, and proposed solutions within the project. It's essential for the project manager to keep calm, lead, and direct the parties to a sensible solution that's best for the project. Here are seven reasons for conflict in order of most common to least common:

1. Schedules
2. Priorities
3. Resources
4. Technical beliefs
5. Administrative policies and procedures
6. Project costs
7. Personalities

So what's a project manager to do with all the potential for strife in a project? There are six different approaches to conflict resolution:

- **Problem solving** This approach confronts the problem head-on and is the preferred method of conflict resolution. You may see this approach as "confronting" rather than problem solving. Problem solving calls for additional research to find the best solution for the problem. Problem solving is a win-win solution that can be used if there is time to work through and resolve the issue. It works to build relationships and trust.
- **Forcing** The person with the power makes the decision. The decision made may not be the best decision for the project, but it's fast. As expected, this autocratic approach does little for team development and is a win-lose solution. Use it when the stakes are high and time is of the essence, or if inter-team relationships are not important.
- **Compromising** This approach requires both parties to give up something. The decision made is a blend of both sides of the argument. Because neither party really wins, it is considered a lose-lose solution. The project manager can use this approach when the relationships are equal and you can't "win." This approach can also be used to avoid a fight.
- **Avoiding** The person who can make the decision in the disagreement, such as the project manager, simply avoids the decision. This approach avoids the conflict, but it can cause resentment on the project team as people wait for the project manager to make a decision. The project manager could be avoiding the issue because it seems trivial, it could disrupt progress, or they are hoping the project will finish with a needed solution.
- **Smoothing** The project manager smooths out the conflict by minimizing the perceived size of the problem. It is a temporary solution but can calm team relations and boisterous discussions. Smoothing may be acceptable when time is of the essence or when any of the proposed solutions would work. This is considered a lose-lose situation, because no one really wins in the long term. The project manager can use smoothing to emphasize areas of agreement between the stakeholders in contention and minimize areas of conflict. Use it to maintain relationships and when the issue is not critical.
- **Withdrawing** This is the worst conflict resolution approach, because one side of the argument walks away from the problem—usually in disgust. The conflict is not resolved, and it is considered a yield-lose solution. The approach can be used, however, to provide a cooling-off period or when the issue is not critical.

Phases of Team Development

Teams develop over time—not instantaneously. As a project team comes together, there are likely people on the team who have worked together previously, just as there may be people on the project team who have never met. Because projects are temporary, the relationships among project team members are also often viewed as temporary. The project manager can see—and sometimes guide—the natural process of team development.

The goal of team development is not for everyone to like each other, have a good time, and create lifelong friendships. All of that is nice, but the real goal is to develop a team that can accurately and effectively complete the project scope. Within team development, the project team will pass through five stages:

- **Forming** In this stage, the project team comes together and learns about one another. Project team members begin to understand each other and find out who's who and what others are like.
- **Storming** This stage promises action. There's a struggle for project team control and momentum of who's going to lead the team. During this phase, people figure out the hierarchy of the team and the informal roles of team members.
- **Norming** Once control on the project team has been established, the project team's focus shifts toward the project work. This is where people learn to work together.
- **Performing** Team members have settled into their roles and can focus on completing the project work as a team. During this stage, a synergy is developed; this is the stage at which high-performance teams come into play.
- **Adjourning** The project team, like the project, is not a permanent fixture in the organization. At some point, the members of the team disperse to work on other projects and join different project teams.

Project Management Is Not a Democracy

Despite what some feel-good books and inspiring stories would like to have you believe, project management is not a democracy. Someone has to be in charge, and that someone is you, the project manager. The success of the project rests on your shoulders, and it is your job to work with your team members to motivate them to finish the project on schedule.

This does not mean that you have permission to grump around and boss any member of your team. It also does not mean that you should step in and break up any disagreements between team members. In fact, you should allow some discussion and some disagreement among the team members.

This is what teams have to do: they have to work things out on their own. Team members have to learn to work together, to give and take, to compromise. Step back and let the team first try to work through disagreements before you step in and settle issues. If you step into the mix too early, your team members will run to you with every problem.

Ultimately, you are in charge. If your team members cannot, or will not, work out a solution among themselves, you'll be forced to make a decision. When you find yourself in this situation, you can work through the problem using several recommended steps to conflict resolution:

1. *Pay attention.* Meet with both parties and explain the purpose of the meeting: to find a solution to the problem. If the parties are amicable, this meeting can happen with both parties present. If the team members detest each other, or the disagreement is a complaint against another team member, meet with each member in confidence to hear that person's side of the story.

2. *Listen.* Ask the team members to describe the problem. Allow each to speak their case fully without interrupting, and then ask questions to clarify any of the facts.
3. *Resolve.* Often, if the meeting takes place with both conflicting team members, a resolution will quickly boil to the surface. Chances are you won't even have to make a decision. People have a way of suddenly wanting to work together when a third party listens to their complaints. They both realize how foolish their actions have been, and one, or both, of the team members will cheer up and decide to work together.
4. *Wait.* If the resolution doesn't happen in your meeting, don't make an immediate decision. Tell the team members how important it is to you, and to the project, that they find a way to work together. Sometimes even this touch of direction will be enough for the team members to begin compromising. If they still won't budge, tell them you'll think it over and will make a decision within a day or two—if the decision can wait that long. By delaying an immediate decision, you allow the team members time to think about what has happened and give them another opportunity to resolve the problem.
5. *Act.* If the team members will not budge on their positions, then you will have to make a decision. And then stick to it. If necessary, gather any additional facts, research, and investigations. Drawing on your evidence, call the team members into a meeting again and acknowledge both of their positions on the problem. Then share with them, based on your findings, why you've made the decision that you have made. In your announcement, don't embarrass the team member who has been put out by your decision. If the losing team member wants to argue their point again, stop them. Don't be rude, but stop them. The team members have both been given the opportunity to plead their case, and once your decision has been made, your decision should be final.

Dealing with Personalities

In any organization, you'll find many different personality types, so it's likely that there are some people in your organization who just grate on your nerves like fingernails on a chalkboard. These individuals are always happy to share their discontent, their opinion, or their "unique point of view." Unfortunately, you will have to find a way to work with, or around, these people.

Here are some personality types you may encounter and how you can deal with them:

Personality	Attributes	Resolution
The Imaginary Leader	These individuals think they are managing the project this week and will be running the company next week. You know the type—always first to raise their hands in school and remind the teacher if they forgot to assign homework.	These people really do want to lead—they just don't know how! Give them an opportunity by allowing them to conduct an occasional team meeting or organize upcoming activities. If you can, try to show them how to lead with tact instead of with rudeness.

(*continued*)

Personality	Attributes	Resolution
The Mouse	These individuals are afraid of doing any activity on the project without explicit directions from you. They're so afraid they'll make a disastrous mistake that they require your guidance on each part of their work.	Encourage these types to take charge of their duties. Tell them that you have confidence in them to do the tasks that you've assigned to them. If they do make a mistake, work through it with them to build their confidence.
Your Favorite Aunt/Uncle	This person is the office clown—always playing gags, streaming toilet paper around someone's cubicle, telling jokes, and sharing stories around the office. Not only are these types of people great fun, but they're also great time wasters.	Often these folks don't have enough to do, and they assume everyone else has a similar workload. Give these people more assignments and they'll have less time to kill. If that doesn't work, politely share with them that their jovial activities are appreciated, but not always necessary.
The Cowboy	These people love excitement. They are happy to try anything out (like rebooting a server midmorning) just to see what happens. Their experience may be great, but their swagger, ten-gallon hat, and stunts aren't always well thought out.	To deal with the cowboy types, encourage their enthusiasm but discourage their ability to make on-the-spot decisions without thinking about the results of their actions. These individuals are generally smart and eager to help but need a touch more guidance from you.
The Prune	These sourpusses are as much fun as a pocket full of thumbtacks. They don't care about your project, they think the technology sucks, and they take their hourly breaks every 20 minutes.	Granted, these folks are hard to work with. They've got more problems personally than the project you are managing. You can start by befriending them and then sharing the value of their work on the project with their superiors. This transfers some responsibility of the work onto those prunes.

Leading People in Adaptive Projects

In an adaptive project you wouldn't say that you're managing people, but leading people. In adaptive projects, especially using the most-popular Scrum approach, project teams are self-led and self-managed. The team decides who'll do what and who's leading what portion of the project work, and they'll work together to resolve differences and find good solutions. In Scrum, there are three primary roles to know:

- **Scrum master** This is analogous to the project manager, but the scrum master role is to coach, mentor, and ensure that everyone is following the rules of scrum.

This role is different than a project manager's role in a waterfall project, as the scrum master isn't directing people what to do throughout the project, but coaching people on what to do. The scrum master is sometimes called the *servant leader*, as they'll get the team what they need to get the work done.

- **Product owner** This role is the business liaison between the team and the business people. This role is responsible for gathering, organizing, and documenting the project requirements. You may recall that the product owner is responsible for prioritizing the requirements, called *user stories*, in the product backlog.
- **Development team** The development team describes the core workers of the agile project. The development team is self-led and self-organized. They'll select chunks of work from the product backlog, determine who'll do what specific tasks to create the work, and support one another to get the work done. The development team is protected from interruptions by the scrum master, and they're free to innovate and experiment to get things done.

Ceremonies are events or meetings in an adaptive project, specifically a scrum project. Ceremonies are key events that are timeboxed and represent the iterative nature of adaptive projects. Each role has specific responsibilities in each ceremony.

- **Sprint planning** Sprint planning is the first ceremony of each sprint. During sprint planning the development team, product owner, and scrum master work together to select the most important items from the product backlog that the development team will create during the sprint. The development team will create a sprint backlog of tasks and they'll decide who'll do what tasks in the sprint.
- **Daily scrum** Every workday in the sprint the development team and the scrum master meet for 15 minutes to answer three questions of each person: What did you accomplish since our last meeting? What will you work on today? Are there any impediments or blockers preventing your progress? The daily scrum is sometimes called a stand-up meeting as the participants stand to keep conversations and the meeting to the 15-minute duration or less.
- **Sprint review** At the end of the sprint, the development team, product owner, scrum master, and key stakeholders meet for a demonstration by the development team. The sprint review is the development team demonstrating what they have accomplished during the last sprint. It's an opportunity for the team to show their work and for the stakeholders to offer any feedback, ask for changes, or approve the work completed so far in the project.
- **Sprint retrospective** This final ceremony of the sprint allows the development team, product owner, and scrum master to review what has and has not worked well in the project. It is not an opportunity to blame and argue with one another, but to have open, transparent discussions about what needs improvement in the project. This ceremony is led by the scrum master and the focus is on how the team can improve the project during the next sprint.

After the sprint retrospective, the team moves back to sprint planning and the entire cycle begins again. It is an iterative approach to getting work done, and the ceremonies help define the roles and responsibilities of each team member. For new agile teams, the agile framework can be a little confusing, so the scrum master ensures that everyone follows the rules and guidelines, but after a few sprints of project work, the development and stakeholders will grasp the rules of agile and things will begin to run much more smoothly.

Use Experience

The final method for resolving disputes among team members may be the most effective: use experience. When team members approach you with a problem that they just can't seem to work out between themselves, you have to listen to both sides of the situation.

If you have experience with the problem, then you can make a quick and accurate decision for the team members. But what if you don't have experience with the technology and your team members have limited exposure to this portion of the work? How can you make a wise decision based on the information in front of you? You can't!

You will need to invent some experience. As with any project, you should have a testing lab to test and retest your design and implementation. Encourage your team members to use the testing lab to try both sides of the equation to see which solution will be the best.

If a testing lab is not available, or the problem won't fit into the scope of the testing lab, rely on someone else's experience. Assign the team members the duty of researching the problem and preparing a solution. They can consult books, the Internet, or other professionals who may have encountered a similar problem.

Disciplining Team Members

No project manager likes the process of disciplining a team member—at least they shouldn't. Unfortunately, despite your attempts at befriending, explaining the importance of the project, or keeping team members on track, some people just don't, or won't, care. In these instances, you'll have little choice other than to resort to a method of discipline. The discipline should be based on the defined consequences of non-performance.

Within your organization, you should already have a process for recording and dealing with disciplinary matter; this is an example of an enterprise environmental factor. The organizational procedures set by human resources, the PMO, or management should be followed before interjecting your own project team discipline approach. If there is no clear policy on team discipline, you need to discuss the matter with your project sponsor before the project begins. In the matter of disciplinary actions, take great caution—you are dealing with someone's career. At the same time, discipline is required or your own career may be in jeopardy.

As you begin to nudge team members onto the project track, document it. Keep records of instances where they have fallen off schedule, failed to complete tasks, or have done tasks half-heartedly. This document of activity should have dates and details on each of the incidents, and it doesn't have to be known to anyone but you. Hopefully, your problematic team members will turn from their wicked ways and accept your motivation to do their jobs properly. If not, when a threshold is finally crossed, then you must take action.

Following an Internal Process

Within your organization, there should be a set process for how an unruly employee is dealt with. For some organizations, there's an escalation of a write-up, a second write-up, a suspension of work, and then ultimately a firing. In other organizations, the disciplinary process is less formal. Whatever the method, you should talk with your project sponsor about the process and involve them in any disciplinary action.

In all instances of disciplinary action, it would be best for you and the employee to have the project sponsor or the employee's immediate manager in the meeting to verify what has occurred. Not only does this protect you from any accusations from the disgruntled team member, it also ensures that you, the project team member, and the manager all hear the same conversation about the performance at the same time.

TIP Treat your team and people in the project with respect. And even when people don't treat you with respect, you still show kindness, grace, and professionalism. Your goal is to manage the project and get things done.

Removal from a Project

Depending on the situation, you may discover that the team member cannot complete the tasks required on the project, and removal from the project may be the best solution. In other instances, the team member may refuse, for their own reasons, to complete the work assigned and be a detriment to the success of the project. Again, removal from the team may be the most appropriate action.

Removing someone from the project requires tact, care, and planning. A decision should be made between you and the project sponsor. If you feel strongly that this person is not able to complete the tasks assigned, rely on your documentation as your guide. Removal of a team member from a project may be harsh, but it's often required if the project is to succeed.

Of course, when you remove someone from the project, you need to address the matter with the team. Again, use tact. A disruption in the team can cause internal rumblings that you may never hear about—especially if the project team member who was removed was everyone's best friend. You will have created an instant us-against-them mentality. In other instances, the removal of a troublemaker may bring cheers and applause. Whatever the reaction, use tact and explain your reasons without embarrassing or slandering the team member who was removed.

Using External Resources

There comes a time for every organization when a project is presented that is so huge, so complex, or so undesirable to complete that it makes perfect sense to outsource the project to someone else. In these instances, no matter the reason the project is being outsourced, it is of utmost importance to find the right team to do the job correctly.

Outsourcing has been the buzz of all industries over the last several years—and certainly many companies dealing with complicated IT implementations have chosen to "get someone else to do it." There are plenty of qualified companies in the marketplace that have completed major transitions and implementations of technology—but there are also many incompetents that profess to know what they're doing, only to botch an implementation. Don't let that happen to you.

Finding an Excellent IT Vendor

Finding a good IT vendor isn't a problem. Finding an excellent IT vendor is a problem. The tricky thing about finally finding excellent vendors is that, because they keep so busy (because of their talented crew), they are difficult to schedule time with. So what makes an excellent vendor? Here are some attributes:

- Ability to complete the project scope on schedule
- Vast experience with the technology to be implemented
- References that demonstrate customer care and satisfaction
- Proof of knowledge on the project team (experience and certifications)
- Adequate time to focus on your project
- A genuine interest in the success of your organization
- A genuine interest in the success of your project
- A fair price for completing the work

Finding an excellent vendor to serve as your project team, or to integrate into your project team, is no easy task. Remember, the success of a project is only as good as the people on the project team. It's not just the name of the integrator, but the quality of the individuals on the integrator's implementation team that make the integrator great (or not so great). Never forget that fact. Figure 7-3 demonstrates how a vendor can be integrated as your project team. Often, the success of the project is dependent on the vendor's implementation. When the vendor has completed their work, the buyer needs to complete the contract and pay the vendor according to the terms of the contract. The project manager should oversee the process as a contract that is binding for both parties—vendor and buyer.

Size doesn't always matter. Those monstrous integrators and technical firms that have popped up in every city over the past few years don't always have the best people. Some of the best integrators you can find are small, independent firms that have a tightly knit

Figure 7-3 Vendors need to adhere to the terms of the contract and support the project vision.

group of technical wizards. Do some research and consider these smaller, above-average tech shops. You may find a diamond in the rough.

To begin finding your integrator, you can use several different methods:

- **References** Word of mouth from other project teams within your organization, contacts within your industry, or even family and friends are often the best way to find a superb contractor. A reference does something most brochures and sales letters cannot: it comes from a personal contact and lends credibility.
- **Internet** If you know the technology you are to be implementing, hop on the Internet and see who the manufacturer of the technology recommends. Once you've found integrators within your community, peruse their websites. Use advanced searches to look for revealing information about them on other websites, in social media, or in newspapers and trade magazines. Know who you are considering working with before they know you. Google is your friend; look for reviews, articles, and testimonials.
- **Phone and web interviews** When all else fails, call and interview the prospective vendor over the phone or through web conferencing software. Prepare a list of specific questions that you'll need answered. Pay attention to how the phone is answered, what noise is in the background, and how professional and organized the individual on the phone is. Was the person rude? Were they happy to help? Take notes and let the other person do much of the talking.
- **Trade shows** If you know your project is going to take place in a few months, attend some trade shows and get acquainted with some potential vendors. Watch how their salespeople act. Ask them brief questions on what their team has been doing. Collect their materials and file them away for future review.
- **Previous experience** Never ignore a proven track record with a vendor. Past performance is always a sure sign of how the vendor will perform with your project.

EXAM TIP Always follow your organization's procurement process. Usually, this mean starting with a statement of work (SOW) and a request for quote (RFQ), request for proposal (RFP), or an invitation to bid. Remember, a quote and a bid provide a price for the work. A proposal is more elaborate, suggests ideas, and includes a price.

Interviewing the Vendor

Once you've narrowed your search to two or three vendors, it's time to interview each one to determine to whom the project will be awarded. In the interview process, which the vendor will probably consider a sales call, remind yourself that this is a chance to find out more information about the vendor.

Document all parts of the meeting: How difficult was it to arrange a meeting time? How polite was the salesperson? Did the salesperson bring a technical consultant to the meeting? All of these little details will help you make an informed decision. Again, as mentioned in Chapter 6, in such meetings, pay attention to the following regarding vendors' representatives:

- Do they pay attention to details? Are they on time? Dressed professionally and appropriately for your business? Are their shoes shined and professional? How vendors pay attention to the details in their appearance and presentation to win your business will be an indicator of how they will treat you once they've won your business.
- How organized are their materials? When a salesperson opens their briefcase, can they quickly locate sales materials? Are the brochures and materials well prepared and neat, and not wrinkled or dog-eared? Again, this shows attention to detail, something every project requires from the start.
- What is their body language saying? Pay attention to how they are seated, where their hands are, and how animated their answers become. A salesperson should show genuine interest in your project and be excited to chat with you. If they seem bored now, they will likely be bored when you call them to discuss concerns down the road.
- What does your gut say? Gut instinct is not used enough. The meeting with the salesperson should leave you with a confident, informed feeling. If your gut tells you something is wrong, then chances are something is. If you're not 100 percent certain (and you probably shouldn't be certain after only one meeting), do more research or ask for another meeting with the project integrators.

Looking for a STAR

When you are interviewing the potential vendors, you need to ask direct, hard-hitting questions to slice through their sales spiels and get to the heart of the project. One of the best interview techniques, especially when dealing with potential integrators, is the STAR methodology. Figure 7-4 shows STAR, an acronym for situation, task, action, result.

When you use the STAR method, you ask a situational question, such as, "Can you tell me about a situation where you were implementing a technology for a customer and you went above and beyond the call of duty?"

Figure 7-4 STAR is an interview methodology to gauge experience.

The vendor should answer with a specific *situation,* followed by the *task* of the situation, the *action* the vendor took to complete the task, and then the *result.* If the potential vendor doesn't complete the STAR, add follow-up questions, such as, "How did the situation end?" to allow the vendor to finish the STAR question.

This interview process is excellent, as it allows the project manager to discern fact from fiction based on the vendor's response. Try it!

After Hiring the Consultant

Consultants know what they know—and what they do not know can hurt them and your project. In other words, consultants need to learn about your environment, how your standard operating procedures work, whom they should talk to, and so on. Consultants need to know how to get things done within your organization. You cannot throw a consultant into your organization and expect them to have the same level of detail, same level of expertise, and same organizational knowledge that you have. It takes some time and some guidance.

For this reason alone, you should demand and require that the consultant attend project meetings, be involved with the project team, and take an active role in meeting the project team members and stakeholders. With the availability of web collaboration software, it's easier than ever to remove the locale constraint to hiring great consultants. Consultants need to get involved in order to be successful and productive. Most consultants and experts, if they are worth their salt, will be eager to follow these rules and requirements. Often it's the project manager who wants the consultant to feel comfortable and not get into the mix of things so quickly. But this limits the consultant's ability to contribute.

CompTIA Project+ Exam Highlight: Project Team Management

You'll be faced with several questions on managing and developing a project team for your CompTIA Project+ exam. Each organization is different—the approach you use in your work environment may be entirely different from one that works for that neat company down the street. The enterprise environmental factors, policies, and human resource procedures all affect how the project manager is allowed to manage the project team. On that note, you can expect the CompTIA Project+ exam to consider how every organization may have a different approach to human resources and how they'll stick to the core, generally accepted principles of project management and human resource management.

1.2 Compare and contrast Agile vs. Waterfall concepts Project teams in a waterfall, or predictive, project work differently than teams in an agile project. Predictive project teams report to the project manager and the team follows orders and the project plan. An agile project is decentralized and the project manager role is really split across three roles: the product owner, the scrum master or coach, and the development team members.

The team members in an agile project are self-led and self-organizing, meaning they don't need a project manager to tell them what to work on next. The coach or scrum master isn't there to delegate work, but to ensure that everyone is following the rules of the agile project management approach.

1.10 Given a scenario, perform basic activities related to team and resource management This was the primary objective covered in this chapter. Forming and leading a project team affects the entire project, as the people on your team are doing the work that can directly affect the quality, acceptance, and long-term support of the project deliverables. The scope baseline, and in particular the WBS, can help you identify the human resources you'll need on your project team. By examining the deliverables of the WBS and the related activity list, you can assess the types of skills the project team members will need to complete the project work. A resource breakdown structure allows you to map needed resources to the WBS components for planning, cost, and schedule duration estimating. You can also use the WBS to plan for resource availability; if a resource is already assigned to one project activity, that person can't complete another activity at the same time.

Resource management includes understanding the allocation of resources in the project. Recall that the project scheduler can help you identify overallocation, benched resources, shortages, and inter-project dependencies that you'll need to address in the project. You'll also need to work with management and other project managers for resources that are shared on other projects and operational activities.

Managing the project team starts with developing the project team and allowing the team to develop itself through the natural phases of team development. Team development moves through forming, storming, norming, performing, and adjourning during the project life cycle. Team development activities can help facilitate the team development process. Sometimes project team members may be in disagreement with one another over assignments, approaches to the project work, and even their availability among multiple projects and operations. Based on the scenario, the project manager can use one of six conflict resolution methods: problem solving, forcing, compromising, avoiding, smoothing, and withdrawing.

Virtual teams are project teams that are not co-located but work together through subteams, web collaboration software, teleconferences, and videoconferences. When managing a virtual team, the project manager must address challenges such as time zones, language differences, and cultural differences that can affect the project execution. For your CompTIA Project+ exam, you can expect questions about overcoming technical barriers and assumptions that can affect your ability to communicate with the project team and the team's ability to communicate with one another.

1.11 Explain important project procurement and vendor selection concepts It's not unusual for a project to have contractors serve as part of the project team. When hiring a contractor, you'll follow the procurement processes and procedures within your organization, just like you were purchasing physical resources. You'll often begin with a statement of work describing what the person will do in the project, and a request for quote, invitation to bid, or a request for a proposal for the vendor. Quotes and bids are price-only documents, while the proposal details possible solutions and price.

2.4 Given a scenario, perform activities during the project execution phase
The project team executes the project plan and the tasks required to create the product. If the team doesn't have the skills necessary, you'll need to train the team. Training helps the team members gain the knowledge and competence needed to do the project work, but also enhances the organization's ability to do future work. This exam objective also touches base on managing conflicts among the project team members. Recall that you can use problem solving, forcing, compromising, avoiding, smoothing and withdrawing-depending on the significance of the conflict and the urgency of the project.

Chapter Review

Teamwork is the key to project management success. As a project manager, you must have a team that you can rely on, while at the same time, the team must be able to turn to you for guidance, leadership, and tenacity. When creating a team, evaluate the skills required to complete the project and then determine which individuals have those attributes to offer. Interviewing potential team members allows you to get a sense of their goals, their work ethics, and what skills they may have to offer.

Subteams are a fantastic way to assign particular areas of an IT project to a group of specialists or to a geographically based team for implementation. When creating subteams, communication between the project manager and the team leaders is essential. Subteams require responsible leaders on each team and a reliable, confident project manager.

You may be serving as a project coordinator rather than a project manager. This role often performs the same duties as the project manager, but with more oversight from management, the project sponsor, the PMO, or even a senior project manager. In some instances, the project coordinator role may actually be an assistant to the project manager. In this capacity, the project coordinator will help the project manager with many project duties to keep the project moving toward completion.

Project schedulers are a resource for the project manager. A project scheduler helps the project manager develop and manage the project schedule—a boon to any busy project manager. A project scheduler works with the project team to identify activity duration, sequencing, management, risk, and opportunities. The project scheduler will also meet with the team on a regular basis to record progress, hours worked, and hours remaining and to make predictions about the project completion date.

Disagreements can flair up among team members, and you must have a plan in place before disagreement happens. Document problems with troublesome team members in the event that a team member needs to be reprimanded or removed from the team. The project sponsor should be kept abreast of the situation as the project continues.

Should the scope of the project be beyond the abilities of the internal team, the project can be outsourced. When outsourcing the project, you need to use careful consideration in your selection of an integrator. Project managers should rely on references of vendors, their ability to work with the technology, gut feelings, and word of mouth to make a decision.

Building a project team is hard work, but it is also an investment in the success of the project. Once again, the success of any project is only as good as the members on the project team.

Exercises

These exercises allow you to apply the knowledge you have learned in this chapter and are followed by possible solutions.

Exercise 7-1: Create a Web Developer Job Description

In this exercise, you will create a job description for a web application developer from a project scenario. You will be given prompts to guide you in the creation of the job description.

Scenario: You are the project manager for Cardigan Adhesives Corporation. You have been assigned as the project manager for the development of the new corporate website. The website should be easy to navigate for all guests, have video streaming features, allow users to chat with salespeople, and provide customer support for technical issues. In addition, the website will have an application that will query a database to report on inventory, cost of goods available, and online ordering.

Answer the following questions to begin creating the job description for a web developer. If you are uncertain of the answers, use the Internet to research web developers and the types of activities they are required to do.

Prompt	Your Answer
What is the primary purpose of this role on the team?	
What type of software will the team member be using to create the application?	
What type of database is being queried?	
To whom will the team member report?	
What other activities will the team member be responsible for?	
What are some personal traits that this person should have?	

Based on your answers, create a job description that is appropriate for a web developer on this project. A solution appears at the end of this chapter.

Exercise 7-2: Prepare for an Interview

In this exercise, you will create a 12-question interview to assess the skills of a prospective team member using the project scenario provided. Prompts will assist you in creating key questions for your interview. On questions that you find relevant, create reactionary questions for the interviewees' expected answers.

Scenario: You are the project manager for Cardigan Adhesives Corporation. You have been assigned as the project manager for the development of the new corporate Internet site. The website should be easy to navigate for all guests. In addition, the website will have an application that will query a database to report on inventory, cost of goods available, and online ordering.

Prompt	Your Question for the Prospective Team Member Interview
How important is the level of experience for this job?	
How important is it to you that this person knows web development software?	
Can you use STAR?	
How important is an industry certification for this role?	
How important is the ability of the individual to work with web design software?	
How important is experience with databases for this project?	
What are some web technology requirements and how does this relate to the interviewee?	
How important are security concepts on a web application?	
Will the e-commerce portion of the project be done in-house or outsourced?	
What type of personal traits are you looking for in this role?	
Create the next two questions that are relevant to the position on your own:	

Exercise Solutions

The following offer possible solutions for the chapter exercises.

Exercise 7-1: Create a Web Developer Job Description

Prompt	Answer
What is the primary purpose of this role on the team?	Web developer
What type of software will the team member be using to create the application?	PHP, Java, or any other web development program
What type of database is being queried?	Oracle, SQL, AWS, or others
To whom will the team member report?	The project manager
What other activities will the team member be responsible for?	Working with the database designer, the database administrator, the web developer, and other team members
What are some personal traits that this person should have?	Hard working, fast learning, dedicated, focused

(*continued*)

Prompt	Answer
Your job description should be something like this:	Web developer: This highly skilled, focused individual will be responsible for creating a commerce-enabled application that communicates with a SQL database. Experience in SQL development, cloud computing, and application design is a must. Experience working with web development software is a plus.

Exercise 7-2: Prepare for an Interview

Prompt	Your Question for the Prospective Team Member
How important is the level of experience for this job?	How many years have you been working as an application developer?
How important is it to you that this person know web development software?	What type of web development applications do you work with?
Can you use STAR?	Can you give an example of a particular web development project where you found a faster or better solution?
How important is an industry certification for this role?	Have you ever taken any classes on the applications you work with? Reactionary question: Have you earned any certifications from the associated vendor?
How important is the ability of the individual to work with web design software?	Have you ever designed any of your own web pages? Reactionary question: What web software did you use to design the site?
How important is experience with databases for this project?	Have you ever worked with SQL (or the database you specified for this job description)?
What are some web technology requirements and how does this relate to the interviewee?	What type of web servers have you worked with? For example, AWS, Google Cloud, UNIX, Linux?
How important are security concepts on a web application?	What type of security mechanisms have you worked with? For example, how did you address security with other commerce-enabled designs?
Will the e-commerce portion of the project be done in-house or outsourced?	Have you implemented e-commerce applications from the ground up? Reactionary question: If so, what are some examples?

Prompt	Your Question for the Prospective Team Member
What type of personal traits are you looking for in this role?	What's your average day like? Reactionary question: How often have you worked overtime?
You create the next two questions that are relevant to the position:	What's the best thing about web development? If you could change one thing in web design, what would it be?

Questions

1. When the project manager is creating a project team, why must they be aware of the skills of each of the prospective team members?

   A. It helps the project manager determine the budget of the project.

   B. It helps the project manager determine how long the project will take.

   C. It helps the project manager determine whether they want to lead the project.

   D. It helps the project manager assign tasks.

2. You are the project manager for your organization and you're creating the project team. Of the following, which two are methods the project manager can use to assess internal skills?

   A. Prior projects

   B. Reports from other project teams

   C. Recommendation of management

   D. Projects the project manager has worked on

3. When requesting an internal résumé to recruit team members, why must the project manager use extreme caution?

   A. Résumés have the connotation of getting, or keeping, a job.

   B. Résumés have the connotation of pay raises.

   C. Résumés have the connotation of relocating users.

   D. It is illegal within the United States to ask for a résumé after the individual has been hired.

4. Esperanza is the project manager for her organization, and management has instructed her to create some team development exercises. Of the following, which one is not an example of team development?

   A. Training for the project work

   B. Industry certifications

   C. Team events such as rafting

   D. Forming, storming, norming, performing, and adjourning

5. Marty is the project manager for his company, and he believes that he'll need to train the project team. Management is concerned about the cost of the training. Marty tells management that this training isn't really an expense for the project. When is training considered an expense?
   A. When the cost of training is beyond the budget of an organization
   B. When the time it takes to complete the training increases the length of the project beyond a reasonable deadline
   C. When the training experience does not increase the ability of the team to implement the technology
   D. When the training experience does not increase the individual's salary
6. The project team is in disagreement over which OS to use on a new server. The project manager tables the issue and says the decision can wait until next week. This is an example of which project management negotiating technique?
   A. Confrontational
   B. Yielding
   C. Coercive
   D. Withdrawal
7. As a project manager, you need to manage your project team, but you also need to lead your project team to finish the project. You need reliable, dedicated workers on your project team. What is the best way to create reliable, hard-working teams?
   A. Fire the team members who do not perform.
   B. Set an example by being reliable and hard working.
   C. Promise raises to the hardest-working team members.
   D. Promise vacation days for all who are hard working.
8. A project scheduler can help the project manager identify resource overallocation. What is the best example of resource overallocation?
   A. Too many project team members on one task
   B. Too many project team members on the project
   C. An individual with too much work in the project timeline
   D. An individual assigned more than 40 hours in a week
9. Why is the WIIFM principle a good theory to implement with project management?
   A. It shows the team how the success of the project is good for the whole company.
   B. It shows the team how the success of the project is good for management.
   C. It shows the team how the success of the project will make the company more profitable.
   D. It shows the team how the success of the project will impact each team member personally.

10. You are the project manager for your organization and will manage 23 people as part of your project team. Management has encouraged you to create subteams for this implementation project. What is a subteam?
    - A. A specialized team that is assigned to one area of a large project or to a geographical area.
    - B. A specialized team that will be brought into the project as needed.
    - C. A collection of individuals that can serve as backup to the main project team.
    - D. A specialized team responsible for any of the manual labor within a project.
11. What is the key to working with multiple subteams?
    - A. A team leader on each subteam
    - B. Multiple project managers
    - C. Communication between team leaders and the project manager
    - D. Communication between team leaders, the project manager, and the project sponsor
12. Of the following, which is a good team-building exercise?
    - A. Introductions at the kick-off meeting
    - B. A weekly lunch meeting
    - C. A team event outside of the office
    - D. Team implementation of a new technology over a weekend
13. Why should a project manager conduct interviews for prospective team members?
    - A. To determine if a person should be on the team or not
    - B. To learn what skills each team member has
    - C. To determine if the project should be initiated
    - D. To determine the skills required to complete the project
14. What is the purpose of conducting interviews of existing team members?
    - A. To determine if the project should be outsourced
    - B. To determine the length of the project
    - C. To determine the tasks each team member should be responsible for
    - D. To determine if additional project roles are needed
15. When a project manager sends a vendor an RFP, what are they asking for?
    - A. A proposal so that a decision can be made based on price
    - B. A proposal so that a decision can be made based on the proposed solution to the WBS
    - C. A proposal so that a decision can be made based on the proposed solution for the statement of work
    - D. A proposal so that a decision can be made based on the proposed schedule

Answers

1. **D.** The project manager should know in advance of the WBS creation the skills of the team members so that they can assign tasks fairly. A skills assessment also helps the project manager determine what skills are lacking to complete the project.
2. **A, C.** Experience is always one of the best barometers for skills assessment. If the prospective team member has worked on similar projects, that person should be vital for the current implementation. Recommendations from management on team members can aid a project manager in assigning tasks and recruiting new team members.
3. **A.** Résumés can show the skill sets of prospective team members, but asking for a résumé can imply that the employee is getting, or keeping, a job. In lieu of résumés, project managers can ask employees for a list of accomplishments and skills to determine the talents of recruits.
4. **B.** Industry certifications are a valuable source for proving that individuals are skilled and able to implement the technology. Certifications, on their own, however, do not provide team development.
5. **C.** Training is an expense rather than an investment when the result of the training does not increase the team's ability to complete the project. A factor in determining whether training should be implemented or not is the amount of time required for the training and its impact on a project's deadline.
6. **D.** This is an example of withdrawal. While this method may result in failure to come to an effective decision, it can be effective when a decision is not needed immediately. This approach can allow the team to "cool off" regarding the issue and move on to other, more pressing, matters.
7. **B.** A leader leads best by doing. By setting the example of being hard working, reliable, and available to your team members, you will show them the type of workers you hope they are as well. Leading by fear or through an iron-fist mentality should not be an option in today's workplace.
8. **D.** Resource overallocation means that a project team member has more hours assigned to project work than what is allowed in the organization for a given time period. Of all the choices presented, an individual with more than 40 hours of project work in a week is overallocated in the project.
9. **D.** WIIFM, the "What's In It For Me" theory, personalizes the benefits of the project for each team member. By demonstrating what the individual will gain from the project, you help increase that team member's sense of ownership and responsibility to the project.
10. **A.** Subteams are not less important than the overall project team. Subteams are collections of specialists who will be responsible for a single unit of the project plan or a geographic team structure.

11. **C.** When working with multiple subteams, the project manager and the team leader of each subteam must have open communications. The subteams should follow the chain of command through each team leader, to the project manager, and the project manager to the project sponsor.
12. **C.** A good team-building exercise is something out of the ordinary, such as participating in an event outside the office, that gels the ability of the team to work together. Luncheons and introductions at meetings are standard fare that don't always bring a team closer together.
13. **A.** Interviews of prospective team members allow the project manager to determine if those people have assets to offer to a project. Because the team members are prospective, the project manager can conduct a typical interview using formal or informal approaches to ascertain the level of skills from each prospect.
14. **C.** By interviewing the project team members, the project manager can determine which project team should be responsible for specific tasks in the project. This interview process can help the project manager determine skills that are present and which skills may need to be acquired to complete the project work.
15. **C.** An RFP is a request for a proposal to the statement of work (SOW). The RFP typically means the buyer is open to recommendations and solutions from the seller. The decision to buy is not made based on cost alone.

CHAPTER 8

Managing Teams

This chapter covers the following topics:

- Leading the team
- Mechanics of project management
- Working with agile project teams
- Maintaining team leadership
- Making decisions as a team
- Team commitment to the project

Vince Lombardi, arguably one of the greatest football coaches of all time, said, "The achievements of an organization are the results of the combined effort of each individual."

What a powerful idea!

Imagine any technology project you've been a part of. What made the project a success? Was it the technology? Was it all because of your efforts? Was the success all because of the project manager? Probably not. Most likely what made the project a success was the "combined effort" of the project team, as in Lombardi's quote. The team collectively worked toward the project vision and got the job done.

IT projects often require a wide range of skills and talent. As a project manager, you'll have to lead and inspire your team to work together toward the common goal of the project. Your team members will have to learn to rely on and trust each other. The individuals on your team will need you to help them complete their tasks, challenge their abilities, and provide opportunity for growth and achievement.

The collection of the individual skills working toward a common goal is a powerful force.

VIDEO For a more detailed explanation, watch the *Leading an Agile Team* video now.

Leading the Team

To lead the team, a project manager must first act like a leader. Think of your favorite historical leaders from politics, sports, or business. What are some of the attributes they possessed to lead and inspire? Chances are all the leaders had one common trait: an ability to motivate people to achieve and aspire. Now this is not to say you must be the next motivational circuit speaker, but it does mean you'll need to develop a method to connect with your team members to inspire them to work toward the vision of your project.

Agile projects embrace the concept of a servant leader. Servant leadership is characterized by carrying food and water for the team. This doesn't mean, of course, that you are literally carrying pizza and beer for your team, but rather that you are getting the team what they need to be successful. You are keeping people from disrupting the team, so the team can focus on their work. Servant leaders work to remove impediments or blockers so the team can move forward. While this is embraced by agile, servant leadership can be used in any project management approach.

The easiest and most direct path to making others develop a passion for your project is for you, yourself, to develop a passion for the project. Passion for the deliverable, excitement for the project work, and zeal for the success of the project is contagious. Consistently communicate the project vision with excitement, passion, and a sense of group ownership. A project manager who wants to lead the project team has to care not only for the success of the project, but also for the success of the individuals on the team. Take time to get to know the team members, learn what their passions are, and develop a relationship with them to work with them, not over or around them.

Establishing the Project Authority

With any project, regardless of the size, the project manager must establish their authority over that project. Usually, the project manager's authority is first defined in the project charter, but authority over the project is not the same as authority over the project team members. You, the project manager, will be responsible for the success of the project, so you need to take charge of the activities to finish the project. In other words, responsibility for the success of the project must be accompanied by an equal level of authority over the actions to create the deliverables.

Authority and responsibility are bound together in project management. The success of the project rests on the shoulders of the project manager. The project manager's career, opportunity for advancement, and reputation all rest on the ability of the project team to finish the project and create the deliverables. As Figure 8-1 demonstrates, if the project manager does not have the authority to assign tasks to the project team members, how can they ever reach the objectives of the project? The level of authority is relative to the autonomy assigned within your organizational structure. It's not impossible for a project manager to successfully lead a project in a functional or weak matrix environment. The authority of a project manager, in any organizational structure, must be leveraged with the respect of the project team.

Team members, of course, also have a level of responsibility for completing the work, and they have risks involved in the project as well. For example, a team member may be dedicated to the project because they can perceive the personal benefits of working on a successful project. A successful relationship between the team members and the project manager should be symbiotic, as Figure 8-2 depicts.

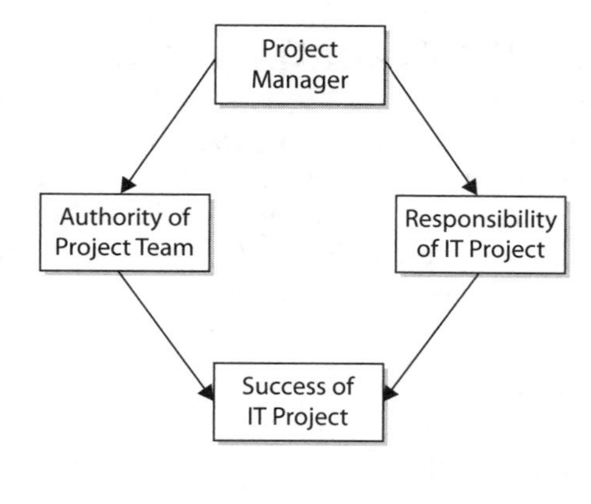

Figure 8-1
A project manager must have authority in proportion to responsibility.

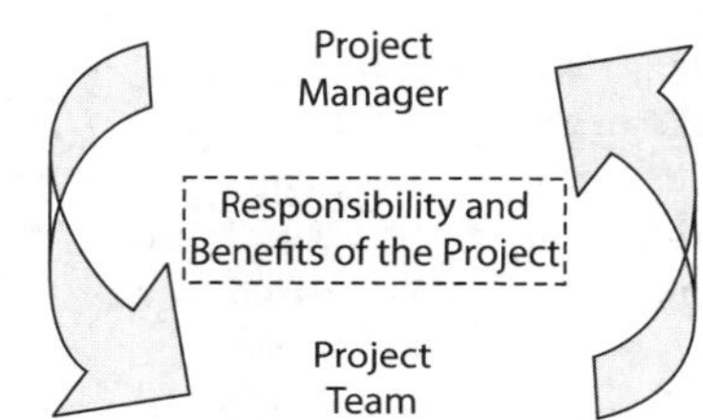

Figure 8-2
The relationship between the project team and the project manager must be mutually beneficial.

Team members must agree that you are the leader of the project and that they will support your decisions, your management of the resources, and your approach to managing the project. The project manager, though not the manager of the individual team members, still must exude a level of confidence and authority over the project team to gain their respect and desire to work on the project.

Many IT project managers stem from IT backgrounds. For these individuals to be successful, they must possess the following attributes:

- Organizational skills
- Passion for the team's success
- Passion for the project's success
- Ability to work with people
- Good listening skills
- Ability to be decent and civil
- Ability to act professionally
- Commitment to quality
- Dedication to finish the project

Mechanics of Leading a Team

There is no magic formula to leading a team. It is one of the unique qualities that some people have naturally, and others must learn. One of the best methods you can use to lead a team is to emulate the leaders you admire. By mimicking the actions of successful leaders, you will be on your way to being successful, too. Much of your ability to lead will come from experience and maturity. There are, however, certain procedures and protocols of project management that you must know to be successful.

Decision Making

Many new project managers are afraid to make decisions. They do not want to offend team members, make a mistake, or look bad in front of management. The fact is, your job as a project manager will require you to make decisions that may not always be popular with the project team. Figure 8-3 demonstrates the balance between acceptable risk and the safeguards of using experienced staff. The decisions you make will need to be in the best interest of fulfilling the project requirements, in alignment with the project budget, and in consideration of the project timeline.

Some decisions you will not have to make entirely on your own. The project team can make many decisions. For example, a company that is upgrading all of their workstations from one operating system to another will have many obstacles to pass. One of the primary questions that will need to be answered in the planning stage is how the new operating system will be deployed to the workstations.

Some of the project team members may be in favor of using disk imaging software. Others may want to use scripts and policies to deploy the image. Still other members may want to visit each machine and install from an optical drive or USB drive, assuming that workstations have an optical or USB drive. Obviously, many different approaches exist for installing this operating system, but the team needs to make a clear decision on what the best method is for the project—and why that method is preferred.

A project manager can lead the team through these decisions using the talents, experience, and education of each team member to come to a conclusion. To facilitate

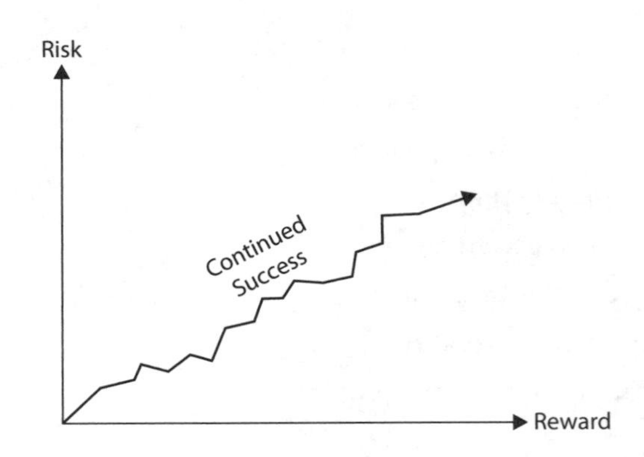

Figure 8-3 Project managers must balance risk and reward to be successful.

the discussion, the project manager may use three types of decision-making processes to arrive at a solution:

- **Directive** The project manager makes the decision with little or no input from the project team. Can you see the danger here? The project manager may be aware of the technology to be implemented, but they may not be the most qualified to make the entire decision. Directive decision making is acceptable, and needed, in some instances, but it isolates the project manager from the project team.
- **Participative** In this model, all team members contribute to the discussion and decision process. This method is ideal for major decisions such as the process to roll out an operating system, design a new application, or develop a web solution for an organization. Through compromise, experience, and brainstorming, the project team and the project manager can create a buzz of energy, excitement, and synergy to arrive at the best possible solution for a decision.
- **Consultative** This approach combines the best of both preceding decision-making processes. The project team meets with the project manager, and together they may arrive at several viable solutions. The project manager can then take the proposed solutions and make a decision based on what they think is best for the project. This approach is ideal when dealing with projects under tight deadlines, restrictive budgets, or complex technology. When there are many variables that can cause the project to stall, the project manager must assume more of the responsibility to safeguard the project.

Working with Team Members

During the process of arriving at a solution or after a solution has been made, some team members may simply disagree with you. Disagreements are fine and are encouraged (if respectful), as they show that team members are thinking and looking for the best solution to a project. In some instances, though, team members may create conflicts among themselves over differences of opinion. These internal conflicts can cause a team to break into cliques or uncooperative partners, and ultimately be a nemesis to the success of the project.

You will have to learn how to be diplomatic among the team members to keep the project moving toward its completion. You will encounter four types of team members in your role as a project manager:

- **Evaders** These team members don't like confrontation on any level. They would just as soon nod their heads, smile, and scream internally, "No, no, no!" These team members may be new to the company, shy, or intimidated by outspoken team members—including the project manager. When using the participative method to arrive at a decision, everyone's input is needed—including input from these people. You will learn very quickly who these people are on your project team, as they'll never or rarely offer a differing opinion or disagree with any suggestions made. To get these individuals involved, try these techniques:
  - Have each team member offer an opinion on the topic, and then write the suggestions on a whiteboard.

  - If possible, allow team members to think about the problem and then e-mail their proposed solution to you.
  - Call directly on the evader team members first when asking for suggestions.
- **Aggressives** These team members love to argue. Their opinions are usually in opposition of the popular opinion, they are brash in their comments, and they are typically smarter than anyone else on the project team—at least they think they are. These folks may be very intelligent and educated on the technology, but they play devil's advocate out of habit rather than trying to help the team arrive at the best solution. You'll know who these individuals are rather quickly—as will everyone on the project team. To deal with these folks, try the following methods:
  - Allow these team members to make their recommendations first before taking suggestions from other team members.
  - Ask them to explain their position in clear, precise reasoning.
  - If necessary, speak with them in private and ask for their cooperation when searching for a solution.
- **Thinkers** These team members are sages. They are usually quiet through much of the decision-making process, and then they offer their opinion based on what's been discussed. These team members are excellent to have on the project team, though sometimes their suggestions stem from other team members' input. Try to work these thinkers into the discussion by asking them questions or calling for their opinions early on if you think they should contribute early on in the process.
- **Idealists** These team members, while their intentions are good, may see the project as a simple, straight path to completion. They may ignore, or not be aware of, the process to arrive at the proper conclusion. Often idealists are well trained in the technology but have little practical experience in the implementation. These team members are usually open to learning and eager to offer solutions to the project.

Dealing with each of these personas takes patience, insight into their personalities, and knowledge of their motivations. You have to spend time with your team members, develop a relationship with them, and lead by example. You won't be effective leading your project team if your only time invested with them is talking about the project, their assigned work, and your review of how they're doing on the project.

EXAM TIP Don't worry about memorizing the different types of people in a project team. Chances are a team member will shift personality types throughout the project. Instead focus on how best to interact with different personality types to keep the project moving toward completion.

Working with Agile Project Teams

Agile project teams don't utilize a command-and-control concept that's prevalent in traditional, waterfall project management. Agile project teams are self-led and self-organizing. This means that the team decides who'll do what work in each iteration of the project.

The team's leader isn't necessarily the project manager (or scrum master or coach), but can be anyone on the project team. The leader isn't a formal position, but it's someone on the team who emerges as a leader that can propel the project forward and lead people on the team follow to the project vision. And the leader can change as the project evolves—the leader isn't an elected or designated person, but a natural leader based on performance and affinity within the project team.

In a scrum project, the team selects the amount of work they believe they can finish in the iteration. Recall that the selected work is called the *sprint backlog*. The sprint backlog is then broken down into activities and tasks that the team must complete in order to create the items in the sprint backlog. The team then divvies up the work—the activities—to get the thing done. This is the idea of the team being self-led, self-directed, and self-organizing. The project manager doesn't delegate to each person what they'll do in the iteration, the team members do this among themselves.

That's a big concept many project teams in agile projects struggle with, especially new agile teams. In the early phases of an agile project, the scrum master may have to coach the team more on the principles and processes of agile. Like anything new, it's helpful to have an experienced person guide the team on how the approach is supposed to work. With time, the team will learn the process and approach and look to the project manager less and less for guidance in the project.

Team Meetings

A project manager who wants to lead an effective team must be organized, prepared, and committed to a strict timetable. When you meet with your team members, they will be looking to you to lead the meeting in an organized, efficient manner. It is not necessary, or advised, to ramble on about the project and discuss issues that are not pertinent. Simply put, call the meeting to order, address the objectives of the meeting, and then finish the meeting. Time in meetings is time not spent completing the project.

Meeting Frequency

Decide at the onset of the project how often the team should meet to discuss the project. Depending on your project, a weekly meeting may be required; in other circumstances, a biweekly meeting is acceptable. The point is to decide how often the team needs to meet as a group to discuss the project as a whole and then stick to that schedule. The project meeting schedule should be documented in the communications management plan.

It is acceptable (and wise) to meet with some members of the project team if the agenda of a meeting is geared toward just those individuals. Project managers often feel the need to involve the entire team in every discussion related to the project—this is a waste of time. While the project manager should make an effort to keep the team informed and moving forward as a whole, there will often be instances when the objectives of a meeting are geared to just a few individuals. These meetings should be separate and in addition to the regularly scheduled team meetings.

TIP Avoid WOT meetings. WOT means Waste of Time. Don't go to meetings if you're not needed. Don't invite people to your meetings if they don't need to be there. The more time spent in a meeting the less time there is to do the work that needs to be done. A good way to spot WOT meetings is to look for those without a clear agenda or history of following through on action items. In case you can't tell, I despise meetings that accomplish nothing.

Meeting Purpose

Once you have decided to meet on a regular basis for the duration of the project, you must also decide why you are meeting at all. In other words, what is the purpose of the meeting? Typically, you will want to meet regularly with your project team to discuss the status of the work and concerns that may have evolved. Other ongoing topics include

- Review of tasks completed
- Review of action items assigned to the project team members
- Risks
- Recognition of team members' achievements
- Review of outstanding issues on the project
- News about the project

A project manager should create an agenda of topics that need to be discussed and then stick to the schedule. These regular meetings with the staff should usually consist of the same order of business, the same length of time, and the same participants. In a geographically dispersed project with subteams, teleconferences or videoconferences are ideal.

Using a Meeting Coordinator

A *meeting coordinator* runs the business of a meeting to keep the topics on schedule and according to the agenda. The project manager does not have to be the meeting coordinator. If you have a very eager team member who is excited about the technology and is ambitious, they may be an excellent meeting coordinator. This individual, like the project manager, must be organized, timely, and able to lead a team meeting. The meeting coordinator will work with the project manager to ensure that key points are covered in the meeting and that the agenda is followed.

When sensitive issues are discussed, the project manager may intervene for the meeting coordinator. If you decide to use a meeting coordinator, you must be certain they have certain attributes:

- Agreement to maintain the position throughout the project
- Willingness to learn
- Willingness to speak before the project team
- Organizational skills

- Time management abilities
- Commitment to gathering resources needed for the meeting

A meeting coordinator can be a great help to the project manager, and the associated responsibilities allow the designated meeting coordinator to gain some experience hosting meetings. You should, however, respect the position and not interrupt as they lead the meeting or take over the meeting. If the meeting coordinator needs help, then step into the role or meet with them outside the meeting to offer advice.

Meeting Minutes

IT projects require documentation on all activities, including meetings. Prior to the meeting, determine who will keep the minutes of the meeting. This does not have to be the same person each time, but it would be helpful if it's someone who can type up and distribute the minutes to the team members. The person who keeps the minutes of the meeting is called a *scribe.*

You need meeting minutes because they provide a record of the meeting, the problems and situations that were discussed, and documentation of the project's progress. Meeting minutes are an excellent method for keeping the team aware of what has already been discussed and settled, resolutions of problems, and proof of the attendees.

Agile Daily Coordination Meetings

Agile projects use a daily coordination meeting, a timeboxed meeting of 15 minutes, in which participants stand. It's a quick meeting specifically for the project development team, not all the stakeholders. Each person in the meeting defines what they have accomplished since the last coordination meeting, what they are working on today, and if there are impediments blocking progress. It's quick, simple, and direct.

The meeting model requires that only people with assignments report. The meeting also asks that those reporting communicate with the whole group, not just with the scrum master or project manager. Side conversations aren't allowed, and big issues are addressed after the meeting. It's quick, daily, and keeps everyone involved. It's also an opportunity to introduce any risks or issues that could prevent progress.

EXAM TIP Some organizations use a similar model to the daily coordination meeting called a *huddle*. Huddles let the team get together for a quick daily meeting about the project work, issues, and other pertinent information.

Virtual Team Communications

As more and more of us are working from home, we're using web conferencing software to communicate. While video and phone conferences can be great, we often lose the value of the face-to-face communication, immediate, clear feedback, and nonverbal clues. With technical tools to host and attend meetings, project managers need to ensure that their message is clear and understood. E-mail and SMS messages can often be misinterpreted. Have you ever sent a joke via text and the recipient didn't get the humor?

I have. The tone and inflection of our voices carry nuances that affect the message's meaning—something we lose in texts, e-mails, and chat applications.

Collaboration tools can be leveraged to help virtual teams work together even though they're far apart from each other. Collaboration tools for idea sharing include

- **Enterprise social media** This is a forum like LinkedIn or Facebook but just for your organization
- **Multi-authoring software** Multiple people can work on the same file at once.
- **Workflow and e-signature platforms** These tools are ideal for moving work through a defined process, like purchasing equipment or scheduling a vacation for approval.
- **Whiteboard apps** Draw, save, and print a whiteboard with multiple people contributing.
- **Cloud-based project management applications** Instead of a local version of the app, team members access the project management application via the Internet. Popular versions include Monday.com, Trello, Basecamp, and even Microsoft Project.

Like any tool, to be effective the team needs to learn how to use collaboration tools. I've seen many organizations make the mistake of buying a new app and expecting instant results with the application. Uh, no. That's like buying an airplane and expecting to be able to fly immediately; simply not how it works. It takes time, effort, education, and patience to become proficient in anything. Some apps are easier than others, but it will still take time and effort to become efficient at the new software.

Maintaining Team Leadership

Once the project has been launched, the meeting schedule has been established, and the project team has developed a routine for completing, reporting, and finishing assignments, it's tempting to relax and let the project take care of itself. Unfortunately, as Figure 8-4 shows, predictive projects won't lead themselves to the finish line—the project manager has to do this.

A constant flux of problems, scenarios, lagging tasks, and technology challenges will be lurking just beneath the surface no matter how calm things may appear. The project manager must lead the team around the pitfalls, past the traps, and over the hurdles to finish the project. This requires a project manager with many talents, abilities, and experience. So what is the ideal project manager as a leader?

Background and Experience

When it comes to information technology, experience means practically everything. An IT project manager may see project management as a logical segue into a management position with a company—and certainly it can be. But to be a great IT project manager, you need experience within the technology sector. By relating personally to

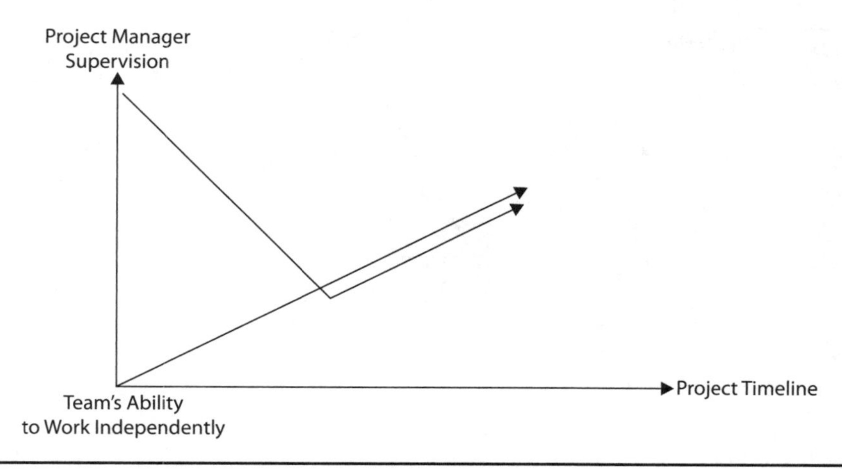

Figure 8-4 Projects require the project manager's constant attention.

the technology, you gain a level of respect not only from the project team but also from management. Your experience in the IT field will allow your guiding hand to nudge the project back into alignment with the project's goals. Some IT professionals, however, may be lacking in the interpersonal skills, tact, and charm that is sometimes required to motivate and lead a group of individuals.

For other IT professionals, it may be difficult to surrender the workload to the project team. They may discover that it's difficult to assert their authority and rely on someone else to complete the implementation. It may behoove these professionals to take a management course such as Dale Carnegie Training to enhance their ability as a manager.

But what of project managers who come from other industries? Perhaps they've joined a company to become a project manager within the IT field, but their experience is based on a traditional managerial background. How can these individuals relate to the technology and the type of people who excel with technology?

These individuals will have to rely on their skills as managers to relate to and lead a project team. They will need to rely heavily on the subject matter experts on their team to give them accurate information. They are also going to have to work with the project team to learn about the technology they are implementing to gain technology experience. It may be helpful for these professionals to enroll in a course on the technology being implemented to understand the process of the implementation from a technician's point of view.

Of course, the best project managers, who are always in demand, are the rare individuals who have years of technology experience but also have the keen ability to work with and lead the project team. Figure 8-5 demonstrates the ideal balance of managerial experience and technical background for any IT project manager. You can become one of these professionals regardless of where you are in your career right now. The secret is to identify the areas of your career where you may be lacking and then seek out projects and opportunities to gain experience with the technology or the management position.

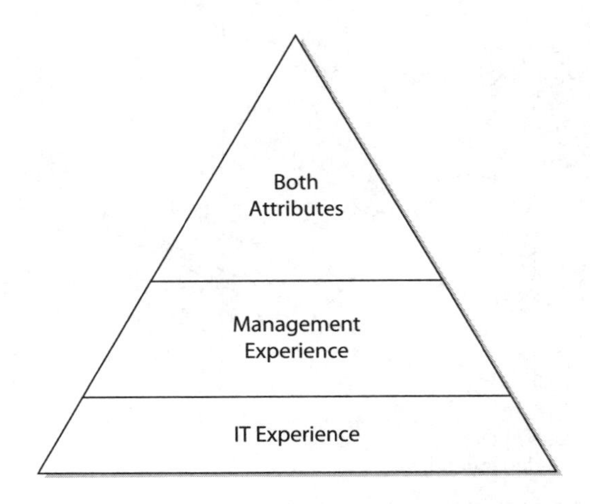

Figure 8-5 Managerial and technical experience is necessary for IT project managers.

Working Toward the Finish

A project requires many things: finances, hardware, time, and other resources. Chief among the required resources is a commitment from all parties involved in the project. This includes the project manager, management, the project sponsor, and the project team. You will need to create and maintain a relationship with each of these parties to ensure their continued support of the project and their commitment to seeing the project through. Project managers who isolate parties that are not actively involved with the implementation are doing their project and career a disservice. Management, project sponsors, and departments that are impacted by the technology implementation want to hear from the project manager on a regular basis, as Figure 8-6 shows. They want, and need, to be kept informed.

Here's a nifty formula to demonstrate the number of opportunities for communication to fail in a project: $N(N - 1)/2$, where N represents the number of stakeholders. For example, imagine a small project with just 20 stakeholders. In this example the formula would read $20(19)/2 = 190$ communication channels. That's 190 opportunities for communication to fail—and it's a sure sign that the larger the project is, the more communication the project manager must undertake.

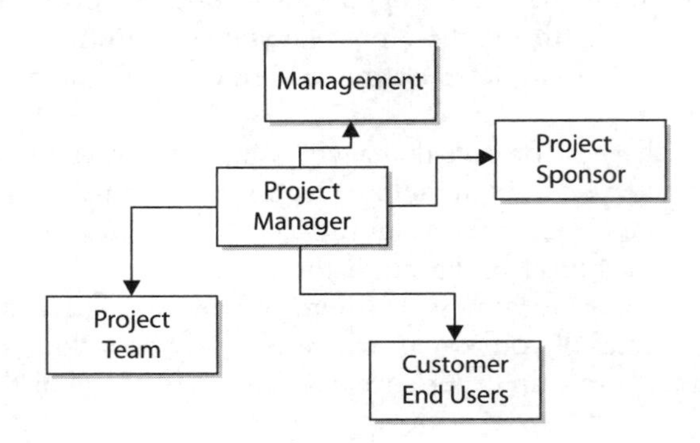

Figure 8-6 Project managers must keep many people informed on the project status.

EXAM TIP The larger the project the more you'll need to communicate with purpose and intent. Larger projects require more communication, more planning, and more stakeholder engagement than smaller projects.

Commitment from the Project Team's Managers

If you are working in a functional or matrix environment, managing a project team is a complex process that requires a commitment from the team members' managers. These managers may represent several different departments within the organization, or they could all work directly within the IT department. The structure of your organization will have a huge impact on the attitude and outlook of the project team on the technology project.

For example, if all the team members have the same manager, as is the case in a functional organization, it will be easier to coordinate activities and participation from all of the team members and the one manager. This scenario is typical in smaller companies or organizations with a very tightly structured IT department. In these instances, a relationship between you and the manager is easier to create than in a project that has team members from several departments with different managers.

Typically, your project team will consist of people from various departments who have an interest in the development and implementation of the technology. In these instances, you'll need to develop a relationship with each of their managers to relay to them what their employees are contributing to the project. A relationship is also needed so that the managers can see the importance of the project and the team members' dedication to it.

Project Completion and Team Members' Growth

As a project manager, your obvious goal is to complete the project as planned, on time, and on budget. As you begin to assign your team members to tasks, you'll have a serious challenge to conquer. Team members will look to you to assign tasks that allow them to grow and learn new skills. You, on the other hand, will be looking toward the project deliverables and will want to use the resources available to get there the best and fastest way you can. The paradox is your desire to assign the strongest resources to the critical path and the desire of the team members to learn new skills and improve their abilities. This is the concept of acceptable risk with regard to team development, as Figure 8-7 demonstrates.

The managers of the team members will want you to assign tasks to their employees fairly and according to their skills, but also to allow them to stretch their abilities. The team members, according to the WIIFM ("What's In It For Me") principle, will have a desire to complete the exciting parts of the project to gain valuable experience for their own career growth. You, of course, have a desire to complete the project smoothly and accurately from the start.

A project manager who never allows team members to attempt tasks that may be slightly beyond their grasp will not win the support of the project team. A project manager must give team members a chance to learn from the work and glean new skills and abilities.

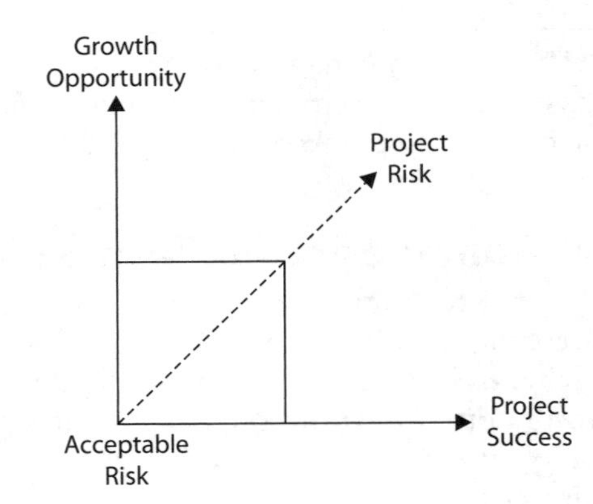

Figure 8-7 Team members' growth must be balanced with the project's health.

If you always assign the critical path tasks to the same technically advanced team members, they may become bored with always doing the same type of work, just as the less technically astute team members may be bored with their perceived menial duties.

A solution that you should try to incorporate is mentoring. Allow inexperienced (but willing to learn) team members to work with more advanced team members on the critical path assignments. By coupling team members on assignments through the critical path, you are accomplishing several things:

- Allowing the inexperienced team member to gain new experience
- Allowing the technical team member to share their knowledge
- Providing a degree of on-the-job training
- Ensuring the critical path will be completed accurately
- Satisfying the needs of management to allow team members to grow
- Allowing your resources to become more savvy for future projects

Motivating the Team

Your team looks to you for more than just directions on what tasks should be completed next, settlements of issues, and updates on the project. Your team also looks to you for motivation. Motivation is more than a pep speech and a positive quote in your outgoing e-mails. Motivation, in project management, is the ability to transfer your excitement to your team members and have them act on that excitement. This section is a reminder of the Management Theories discussed in Chapter 3.

No matter how wonderful your smile, your ability to talk with your project team, and your passion for the project, not everyone will be motivated. Much of the motivation of the project doesn't even stem from the project manager! The motivation and level of excitement will come from the company itself, the working atmosphere, and the overall commitment to the organization of each project team member.

Understanding Motivation

Fred Herzberg, a management consultant and business theorist, conducted a study in 1959 that resulted in his *Motivation-Hygiene Theory*. This study's results, as demonstrated in Figure 8-8, showed that workers are impacted by nontangible factors called *motivating agents* and *hygiene agents*. Motivating agents are elements such as opportunities to learn new skills, promotions, and rewards for our hard work. Hygiene agents are elements we expect in employment: a paycheck, insurance, a safe working environment, vacation time, and a sense of community. The presence of hygiene agents does nothing to motivate employees—only motivating agents motivate them. However, the absence of hygiene agents will demotivate workers.

Herzberg's theory also asserts that people are either motivation seekers or hygiene seekers. Hygiene seekers take comfort in

- Company policy and administration
- Supervision
- Salary
- Interpersonal relationships
- Working conditions

These employees like to feel safe, guarded, and secure in their job and their organization. They are not overly excited by opportunity, growth, or the challenge of the work.

Inversely, motivation seekers take comfort in the following five factors:

- Achievement
- Recognition
- The work itself
- Responsibility
- Advancement

The contrast between the two types of workers is startling. The hygiene seekers take comfort in, for example, the health insurance policies, the sick day allowance, and the

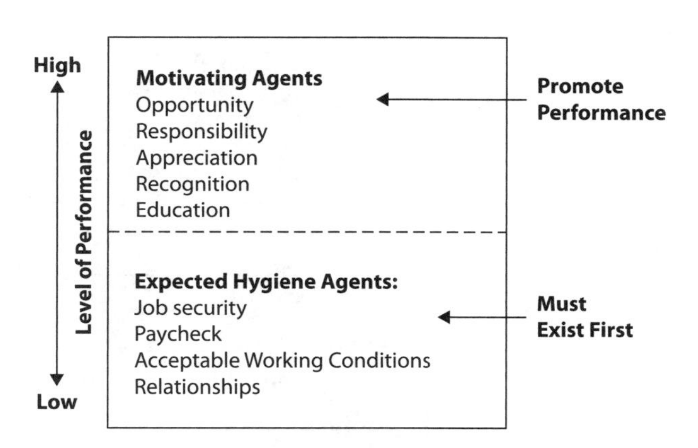

Figure 8-8 Hygiene agents must exist before motivating agents can be offered.

number of vacation days allowed per year. While motivation seekers appreciate the company policies, they find more comfort in the challenge of achievement, growth, and opportunity for advancement.

Which would you rather have on your project team? Chances are you'll encounter both types of workers, so the actual motivation for each type of employee will vary. Perhaps for the hygiene seekers, time off for work, a bonus, or the opportunity to travel on the project will be their reward. Motivation seekers will look for more long-term rewards than a free day from work and will be motivated by their achievements, their opportunity for advancement, and public recognition of the work they've completed.

In all of us, there is likely a mixture of both the hygiene seeker and the motivation seeker. The trick for you is to determine which personality type is predominant in your project team members and then act accordingly.

What Team Members Need

Another viewpoint on motivation is that people are motivated by needs. According to Abraham Maslow, people work in order to satisfy a hierarchy of their needs. The pinnacle of needs is self-actualization. People want to contribute, prove their worth, and use their skills and abilities. Maslow's hierarchy of needs, shown in Figure 8-9, states that people reach their full potential by fulfilling one layer of needs at a time.

1. **Physiological** People require these necessities to live: air, water, food, clothing, and shelter.
2. **Safety** People need safety and security; this can include stability in life, work, and culture.
3. **Social** People are social creatures and need love, approval, and friends.
4. **Esteem** People strive for the respect, appreciation, and approval of others.
5. **Self-actualization** At the pinnacle of needs, people seek personal growth, knowledge, and fulfillment.

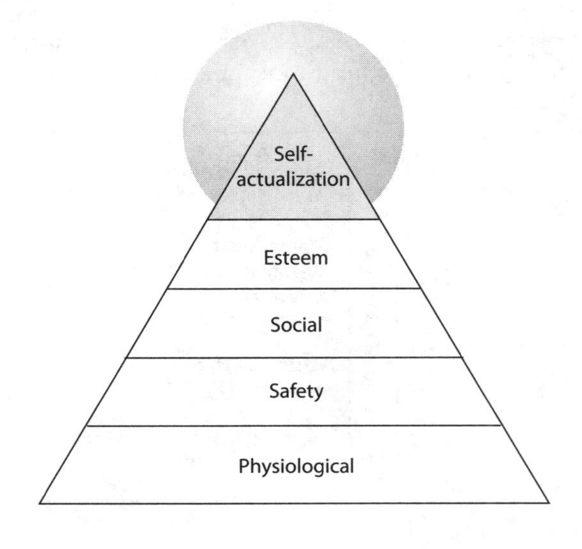

Figure 8-9 Maslow's theory states that people work for self-actualization.

In this theory, you need to satisfy the lower-level needs before you can ascend and satisfy the higher needs. While this theory may not help you become an instantaneously better project manager, it will help you understand what may be motivating your project team members. Another theory, also based on needs, is McClelland's Theory of Needs. David McClelland developed his acquired-needs theory based on his belief that a person's needs are acquired and develop over time. People's needs are shaped by life experiences and circumstances. This theory is also known as the Three Needs Theory, because there are just three needs for each individual, and one need is considered the driving motivation behind the actions people take. Depending on the person's experiences, the order and magnitude of each need shifts:

- **Need for achievement** People who have this dominant trait need to achieve, so they avoid both low-risk and high-risk situations. Achievers like to work alone or with other high achievers, and they need regular feedback to gauge their achievement and progress.
- **Need for affiliation** People who have a driving need for affiliation look for harmonious relationships, want to feel accepted by people, and conform to the norms of the project team.
- **Need for power** People who have a need for power are usually seeking either personal or institutional power. Personal power generally means wanting to control and direct other people. Institutional power means wanting to direct the efforts of others for the betterment of the organization.

McClelland developed the *Thematic Apperception Test* to determine what needs drive individuals. The test comprises a series of pictures, and the test-taker has to create a story about what's happening in the pictures. Through the storytelling, the test-taker will reveal which need is driving their life at that time.

Theorizing on Management and Leadership

Management is about getting things done and creating key results. Leadership is about aligning, motivating, and directing people. As a project manager, your focus will be on the management side of the house much of the time, and you'll lead part of the time. When it comes to management, you should consider Douglas McGregor's Theory X and Theory Y. McGregor states that management believes there are two generic categories of workers—basically, good and bad. The theory states that management actually shifts from position to position, depending on the worker and the conditions that warrant the behavior of the manager:

- *X is bad.* These people need to be watched all the time and micromanaged; they cannot be trusted. X people avoid work, shun responsibility, and lack the aptitude to achieve.
- *Y is good.* These people are self-led and motivated, and they can accomplish new tasks proactively.

In reality, it's not that management truly believes that all employees are either good or bad, but that how you manage the employees will shift according to circumstances, conditions, and your history with the person. With some employees, you can get out of the way and let them do their work, while other employees want you to micromanage them, direct their every action, and provide your constant input. According to McGregor's theories, you'll identify the type of employee and then adjust your management style accordingly.

Another theory is William Ouchi's Theory Z. This isn't directly related to McGregor's theories, but it is based on the participative management style of the Japanese. This theory states that workers are motivated by a sense of commitment, opportunity, and advancement. Workers in an organization subscribing to Theory Z learn the business by moving up through the ranks of the company. Ouchi's theory also credits the idea of "lifetime employment." Workers will stay with one company until they retire because they are dedicated to the company that is, in turn, dedicated to them.

Finally, Vroom's expectancy theory states that people will match their behaviors to what they expect as a result of their behaviors. In other words, people will work in relation to the expected reward of the work. If the attractiveness of the reward is desirable to the worker, they will work to receive it—and people expect to be rewarded for their effort.

CompTIA Project+ Exam Highlight: Project Team Management

The CompTIA Project+ Exam will test your knowledge of managing and leading project teams. Your primary focus as a project manager is to manage the project team to complete the project work in order to satisfy the project objectives. By understanding the type of personalities on your project team, the interests of the project team members, and things that motivate project team members, you can adjust your management style.

1.8 Compare and contrast communication management concepts When it comes to leading and managing the project team, you'll be using communication. Communication, it's been said, is 90 percent of a project manager's job. To get the team to follow the rules, work together, and report on their project accomplishments, they'll need to communicate with you, the project manager. You'll also be communicating with project stakeholders, as they expect updates on the project. Communication must be precise, appropriate, and honest; skewed or sloppy communication can have ramifications on the project work and the attitude of the people within the project.

1.9 Given a scenario, apply effective meeting management techniques Meetings are an opportunity to quickly and effective bring together the required project stakeholders and share information about the project. Meetings should have an agenda that's distributed to the needed meeting participants before the meeting begins. All meetings should be timeboxed, have a meeting leader or coordinator to keep things on track, and have a scribe to document the meeting minutes. Efficient meetings are appreciated, while rambling, sloppy meetings are frustrating, as participants have been invited away from their work for information that's not relevant to their goals or the vision of the project.

1.10 Given a scenario, perform basic activities related to team and resource management The project team needs to know who's in charge of the project—the project manager. The organizational structure will affect the amount of power the project manager has over the project. The project team needs to know what the project ground rules are, what your expectations of the project team are, and how they can operate within the project. By involving the project team in the decision-making process, you'll garner their trust and ownership of the project. The participative and consultative decision-making approaches get the team involved in the project direction. By getting the team involved in the project decisions, you'll create respect, synergy, and a sense of ownership in team members of the project work and the project creation.

3.2 Compare and contrast various project management productivity tools Teams need to be able to communicate with one another, document their project work, update task information, and find information through the project management information systems. While some may argue that a whiteboard, a notebook, and a stack of markers are all that you really need to be effective, that's not realistic for most projects (or your CompTIA Project+ exam). Communications tools, like web conferencing software, internal wikis, calendaring tools, and cloud-based solutions, help the team work collaboratively even if they are not physically located together. Communication with non-co-located teams, or virtual teams, requires more diligence to be effective.

Chapter Review

As the project manager, you will find yourself managing things, such as quality, scope, and risk, but leading people to achieve the project goals. Through your actions, your example, and your excitement for the project, you will find solutions to motivate, inspire, and urge your team to complete the project according to plan. Adaptive project teams are self-led and self-organizing, as the team members will determine who'll do what work based on the backlog.

You can nearly always succeed by emulating the activities, processes, and solutions of other successful project managers and leaders you admire. By mimicking the abilities of successful leaders, you can begin to absorb and implement the same talents with your team.

IT project managers who come from a technical background may need to learn interpersonal skills, organizational abilities, and management techniques to best lead the team and benefit the organization. On the other hand, project managers who come from a more traditional management background will need to gain experience and education in the technology being implemented to be most effective.

You must allow the team to make decisions as a group. Discussion from each team member's perspective is required and will contribute to the overall good of the project. You'll also need a method for dealing with team members who don't want to contribute, or who become confrontational when others disagree with their suggestions. There must be a balance of input, compromise, and some disagreement in discussions by the project team.

Managers of the project team must be as committed to the project as the project manager and the project team. To be a successful project manager, you have to make an effort to create relationships with management and share the news of the project status.

The team members must also be given an opportunity to increase their value and their skills by working on important assignments. This can be accomplished by partnering less experienced team members with more advanced professionals.

Finally, you need to recognize the two primary personality types—hygiene seekers and motivation seekers—and learn to use what excites them to move them forward on the project. Motivation seekers are likely to be the more exciting personalities. These folks are achievers, entrepreneurial in nature, and excited by learning.

Leading a project team is one of the project manager's toughest assignments. Developing a personal relationship with each team member, establishing mutual respect, and motivating each person to succeed not only will help you complete the project but will also make you a better person.

Exercises

These exercises allow you to apply the knowledge you have learned in this chapter and are followed by possible solutions.

Exercise 8-1: Making Project Decisions with the Project Team

In this exercise, you will be presented with a project management scenario. You will then be asked a series of questions based on the scenario.

You are the project manager for the upgrade of 1,200 workstations. Of the workstations, 730 will be new computers with no operating system installed, while the remaining 470 computers will be operating system upgrades. You have decided that all of the workstations, regardless of the new equipment or the upgraded operating system, will need a uniform installation. Domain names, installed software, and security policies should be the same across all of these workstations.

Your team has discussed the various methods that could be used to install the operating system image. Here is a quick record of their recommendations:

Team Member	Recommendation
Javier	I think that we should deliver and place all of the workstations first and then configure the workstations according to guidelines. Applications can be installed via a Windows system policies capability.
Chen	I think the machines should all be configured prior to delivery to the clients. We could use imaging software to deploy the images to the workstations and then deliver them once all of the workstations have been configured.
Henry	Imaging software is definitely the way to go. But I think we should map out the physical departments where the machines are to be delivered. Once we've identified all of the machines in each area, we can put the image on the PC and then move it to each department.

Team Member	Recommendation
Shelly	What we need is a combination of solutions. Let's take Chen's imaging idea and combine it with Henry's mapping of the physical placements of the machines. We should take the machines and physically place them on the network. Then, over the network, push the image to the machines through the imaging software. This way we can be delivering the PCs, installing the PCs, and installing multiple PCs over the network all at once.
Faaris	Okay, that's a good idea. The only problem is it's going to be hard to deliver 730 workstations during the workday—or even over a weekend. We're going to have to do this after hours and in sections. Let's take Shelly's idea and combine it with Henry's idea to deploy the workstations over several weekends.

Complete the following questions to finish the exercise:

1. What type of decision-making process was your team participating in?

 ____

2. What are advantages of using this process?

 ____

3. Are there any disadvantages? If so, what are they?

 ____

4. Which team member may feel that their contribution was not valuable?

 ____

5. How could you respond to this team member to encourage that individual to continue to participate?

 ____

6. Which team member could be considered the thinkers personality type?

 ____

7. Now that you've heard what the team members have to say, what method would you recommend for this installation process?

 ____

8. Why would you deploy the workstations in the method you just described?

 ____

Exercise 8-2: Defining Project Team Member Personality Types

In this scenario, you will be presented with three descriptions of team members on your project. To complete the exercise, answer the questions after each team member description.

Scenario: You are the project manager for a Microsoft Office installation. The software will need to be customized to fit your environment and will be installed on 1,700 workstations throughout your company's network. Here are your team members:

John Umphreys John is a Microsoft Office Specialist. He has worked with Microsoft Office for several years and is generally considered to be the lead technician on Microsoft Office development. John is, however, becoming bored with Microsoft Office and would like to move into management, networking, or advanced application development after this project.

1. What methods can you use to motivate John on this project?

 ____

2. What personality type do you think John is—a hygiene seeker or a motivation seeker?

 ____

3. Why do you think that John is this type of personality?

 ____

4. How does this information help you in motivating John on the project?

 ____

Sarah Williams Sarah is the new member of the IT department. She graduated last year from the University of Illinois with a degree in computer science. She does not have much experience in Microsoft Office but is very eager to learn. She considers herself a quick learner but likes to experiment with the technology before she is willing to try anything in production. She feels she must master the entire product before her contributions will be of value.

1. What methods can you use to motivate Sarah on this project?

 ____

2. What personality type do you think Sarah is—a hygiene seeker or a motivation seeker?

 ____

3. Why do you think that Sarah is this type of personality?

___

4. How does this information help you in motivating her on the project?

___

Nasir Ahmadi Nasir is an experienced network administrator. He has earned a Microsoft Certified Solutions Expert status and has worked with Microsoft Office. He has limited experience with Microsoft Office development work. Nasir is not very interested in learning the development side of Microsoft Office, as he'd rather stay in his comfort zone with network administration. His biggest concern is that the project will not go smoothly and his job as a network admin will be compromised.

1. What methods can you use to motivate Nasir on this project?

___

2. What personality type do you think Nasir is—a hygiene seeker or a motivation seeker?

___

3. Why do you think that Nasir is this type of personality?

___

4. How does this information help you in motivating Nasir on the project?

___

Exercise Solutions

The following offer possible solutions for the chapter exercises.

Exercise 8-1: Making Project Decisions with the Project Team

1. Your team was using the participative decision-making process. This is evident in how each team member contributed to the discussion and offered solutions.
2. This process allows team members to brainstorm for a solution and interact with different solutions to create an ideal resolution to the implementation challenge.
3. The disadvantages are that team members may not be comfortable offering their opinions on the technology; some may be confrontational, while others may evade confrontation and side with the majority.

4. Javier's opinion on the installation was immediately dismissed and never referred to again in the discussion, so he may feel his opinion isn't valuable.
5. You could invite Javier back into the discussion by asking for his opinion of the other solutions. Javier may need some encouragement to participate in the discussion after his suggestion had been eliminated.
6. Faaris did a good job listening to the other solutions and then offering a combination of all the proposals. Team members who fall into the thinker category are often great at analyzing problems and then creating a solution from what others have proposed.
7. Faaris's solution is probably the wisest, as it addresses the bulk of the machines to be delivered, the process of installing and configuring the operating systems, and the desire to use imaging software.
8. A tiered deliverable is ideal in this situation due to the number of PCs involved. It would be practically impossible to deliver all of the workstations in one step. By delivering the PCs over several weekends or after hours, production is not interrupted and the network traffic will be minimal for the network push of the disk image.

Exercise 8-2: Defining Project Team Member Personality Types

Answers for John Umphreys

1. John is likely a motivation seeker. His desire to move into management after this project may be an excellent motivational tool. By assigning John more responsibility on this project, he can learn new skills to aid his career.
2. John is a motivation seeker.
3. John is a motivation seeker because of his desire to achieve and his apparent boredom with his talents with Microsoft Office.
4. Knowing what excites John allows the project manager to assign him to tasks that John will find interesting and challenging.

Answers for Sarah Williams

1. First allow Sarah to get comfortable in the project by assigning her to tasks that are not in the early stages of the critical path. By assigning her to tasks that she will be able to complete successfully, her confidence will grow. Once she is comfortable, she can be grafted into tasks on the critical path to underscore and test her abilities.
2. Sarah may be either a hygiene seeker or a motivation seeker.
3. From the information provided, she has a desire to learn and to achieve but also seeks comfort in having a complete understanding of the product before completing any tasks.

4. One method a project manager could use with Sarah is to assign her to tasks not in the critical path. Upon successful completion of the tasks, she could enroll in a technical course on the product. A reward of learning may excite Sarah and enable her to accomplish more later in the project.

Answers for Nasir Ahmadi

1. Assign Nasir to projects that you both feel he is comfortable completing. His forte is likely in the server side of the project, so tasks within that path would be best for Nasir and the good of the project.
2. Nasir is likely a hygiene seeker.
3. Nasir is likely a hygiene seeker because of his desire to stay put in his current role and his fear of risk in the new project.
4. A project manager can use Nasir's fear of failure to assign him to tasks that are closely related to his own area of expertise. By assigning these tasks to Nasir, he will likely do his best to complete the assignments correctly and efficiently not only because he is qualified and talented to do so, but also because of his apparent fear of failure on the project.

Questions

1. You are the project manager of the NYQ Project. You suspect that some project team members aren't interested in the project. What is the best path to make others develop a passion for the project?

   A. Offer bonuses if the project is completed on time.

   B. Remind the team members that the project is essential to their careers.

   C. Become passionate about the project yourself.

   D. Remind the team members that mistakes they make on the project will be documented.

2. Why must the project manager have authority over a predictive project?

   A. To ensure that the team will complete the work as dictated

   B. To ensure that the actions required to complete the project will be enforced

   C. To ensure that the budget required to complete the project is available

   D. To ensure that the organization realizes your potential as an effective manager

3. What management theory suggests that people will perform better when they are offered a reward that motivates them?

   A. Herzberg's Motivation-Hygiene Theory

   B. McClelland's Theory of Needs

   C. Maslow's hierarchy of needs

   D. McGregor's Theory X and Theory Y

4. Complete this sentence: A successful relationship between the team members and the project manager should be _______________.
   - **A.** Orderly
   - **B.** Symbiotic
   - **C.** Symbolic
   - **D.** Relaxed
5. Of the following, which are two skills that a project manager must have to be successful?
   - **A.** Public speaking abilities
   - **B.** Organizational skills
   - **C.** Ambition for a successful career
   - **D.** Passion for the project
6. McClelland's Theory of Needs suggests that people are driven by one of three needs. Which one of the following is not one of the three needs in McClelland's theory?
   - **A.** Power
   - **B.** Affiliation
   - **C.** Reward
   - **D.** Achievement
7. Of the following, which decision-making process is reflective of the directive decision?
   - **A.** A project manager who makes the decision with no team input
   - **B.** A project manager who makes a decision based on team members' counsel and advice
   - **C.** A project manager who allows the team members to arrive at their own decision
   - **D.** A project manager who allows the team members to arrive at their own decision with the project manager's approval
8. Of the following, which decision-making process is reflective of the consultative decision?
   - **A.** A project manager who makes the decision with no team input
   - **B.** A project manager who makes a decision based on team members' counsel and advice
   - **C.** A project manager who allows the team members to arrive at their own decision
   - **D.** A project manager who allows the team members to arrive at their own decision with the project manager's approval
9. Why is disagreeing considered an effective part of team discussions?
   - **A.** It keeps the team members competitive against one another.
   - **B.** It allows the project manager to pit team members against each other to keep the project moving.

C. It shows that the project team is thinking and considering alternative solutions.

D. It allows team members to become passionate about their decisions.

10. Of the following, which is *not* a method you should employ when working with evaders during a team discussion?

A. Have each team member offer their opinion on the topic, and then write the suggestion on the whiteboard.

B. If possible, allow team members to think about the problem and then e-mail their proposed solution to you.

C. Have the evader listen to all of the comments and then make their decision.

D. Call directly on the evader team members first when asking for suggestions.

11. You are a project manager for an application development project. You have eight team members who are all adding opinions about the web connectivity feature of the application. Of the following statements, which one most likely came from a team member with the aggressive personality type?

A. I've done this before so I know it works.

B. We could model our application after a project we've made before.

C. I agree with Susan.

D. I think we should take two different approaches and let the users decide which is most effective.

12. How often should a project manager meet with the entire project team?

A. On a regular basis, as warranted by the size and duration of the project

B. As directed by management

C. At least weekly

D. At least monthly

13. Of the following, which is the best choice for conducting meetings when the project team is dispersed geographically?

A. Have the entire project team travel to a central location to discuss the project.

B. Have the team leaders from each subteam travel to a central location to conduct the project. Team leaders would then report the results to the team when they return from the trip.

C. Have the project manager travel to each location and meet with the project team there.

D. Use videoconference software to link to all of the different teams to discuss the project.

14. Why should project team meetings have minutes recorded?
    - **A.** It allows management to have a record of each meeting.
    - **B.** It allows the project sponsor to have a record of the meeting without having to attend.
    - **C.** It allows the project manager to maintain a record of what was discussed.
    - **D.** It allows subteams to have details on what is happening in the other subteams' meetings.
15. Of the following, which two are attributes of a successful project manager?
    - **A.** Experience working with technology
    - **B.** Experience working with the project team
    - **C.** Experience working with the project sponsor
    - **D.** Experience as a traditional manager

Answers

1. **C.** The best method you can use to make others become passionate about a project is to become passionate about it yourself. Your excitement and desire for success are contagious.
2. **B.** A project manager needs authority over a project to ensure that the team will complete the tasks to finish the project. Because a project manager is responsible for the success or failure of a project, they should also be given a level of authority to ensure their success.
3. **A.** Herzberg's Motivation-Hygiene Theory states that performance can be promoted through motivating agents, which are often rewards. Hygiene agents, such as a paycheck and benefits, are needed before motivating agents are effective.
4. **B.** The relationship between the project team and the project manager should be mutually beneficial, or symbiotic.
5. **B, D.** Organizational skills and a real passion for the project are two attributes every project manager requires.
6. **C.** Reward is not one of the three needs of McClelland's Theory of Needs. His theory states that people are driven by power, affiliation, or achievement.
7. **A.** When a project manager makes a decision with no team input, but based on their own experience, research, or intuition, they are making a directive decision.
8. **B.** A consultative decision allows the project manager and the project team to work together to make a decision. This is ideal for situations where the project manager may not be well versed in the technology but must make a decision to safeguard the project.
9. **C.** Some disagreements in a team meeting are healthy because it shows the team is working toward the best solutions. Compromise and openness to different ideas are fantastic components to any project.

10. **C.** Evaders are team members who avoid confrontations and do not want to contradict any other team member. By allowing the team member to wait until they've heard all of the other team members' suggestions, their true opinion may be masked by the desire to agree with the majority of opinions.
11. **A.** Aggressive team members are typically experienced but have no qualms about stressing their ability to always be right.
12. **A.** Project managers should determine a schedule that is appropriate for the project. A team that meets too often will waste valuable work time. A team that meets infrequently will miss an opportunity to discuss and resolve issues on the project.
13. **D.** Videoconference software is an excellent solution for all of the team members to participate in a discussion of the project without losing days to and incurring expenses for travel.
14. **C.** Minutes need to be kept in team meetings so that the project manager has a record of the discussion of the meeting. It allows the project manager another avenue to document the details of the meetings.
15. **A, D.** A project manager who has experience either in working with technology or in the role as a traditional manager has an advantage over unproven project managers. Technical experience plus a traditional management background is a great combination for any IT project manager.

CHAPTER 9

Implementing the Project Plan

This chapter covers the following topics:

- Reviewing assignments with the project team
- Leading team meetings
- Tracking progress
- Tracking financial obligations
- Determining actual project costs
- Calculating value of work performed
- Determining the cost performance index

By this point of your project, things should be getting very exciting. You've researched the project, planned for success, and created (and survived) the budget process—now you're ready to set your plan into action. Your team is eager to get moving on the project, and there is electricity in the air, as all your hard work is about to come to fruition. Don't break out the champagne yet. You can congratulate yourself for successfully completing a solid foundation for your project, but you're going to have to coach your team to follow the project plan through the execution to have a real reason to celebrate.

In this portion of your project, your team will create the components defined in the work breakdown structure (WBS) and follow the sequence of activities in the project network diagram (PND). You will interact with your team members to ensure their successes as they complete the tasks. You'll create a work authorization system to allow work to continue given past results. You'll add quality control mechanisms, and you'll continue to increase your communication among the project team and stakeholders. You'll be deeply involved with tracking the actual costs of the project and comparing them to the budgeted costs. Finally, you'll track the implementation against your PND, cost, and time estimates and then implement corrective actions as needed to keep the project on track and on budget. Of course, you'll document what's working on the project, what's not working, and other things you've learned in your lessons learned documentation.

In an adaptive project, there is not a project management plan like there is in a predictive project. Adaptive projects, as you know, plan each iteration in a type of rolling wave planning. The project focuses on the product backlog, the sprint backlog, and the finished business value. This doesn't mean that the team doesn't plan. Planning happens at the launch of each iteration and throughout the iteration as needed. Because the work is completed in short chunks, planning is also completed in smaller instances. Like all projects, however, there is ample planning at the launch of the project to determine the direction of the project, the architecture of the systems needed, and the framework the project team will follow.

Recall that the items in the product backlog are called *user stories*. User stories are considered for their level of effort and are assigned user story points. The amount of work the team can complete in an iteration is expressed as *X* amount of story points, such as 35 user story points. Over time the number of story points the team can create normalizes, and this value is called the *velocity*. Velocity can help predict when the project will be done based on the number of story points in the product backlog and the duration of each timeboxed iteration in the project.

This part of project management will test and challenge all your planning, research, and ability to lead and react to situations that may impact your project completion. Keep your cool, analyze problems when they arise, and always remember that this project is yours to control, to complete, to savor.

VIDEO For a more detailed explanation, watch the *Earned Value Management* video now.

Reviewing Assignments with the Project Team

An ideal project team would consist of people who are confident in their work, their abilities, and their commitment to the project. They would always complete their tasks on time and without flaw, and they'd happily report to you that everything is perfect and on schedule. "In fact," they'd say, "we're a little ahead of schedule." Now let's step into the real world. Team members will be nervous about their duties. They won't always complete their tasks on time or without flaw. Team members may report to you that everything is fine, when, in fact, it's far from that.

One of your responsibilities as the project manager will be to mold this team into a reliable, interdependent collection of professionals who can rely on each other, themselves, and you. Through regular team meetings, outings, and one-on-one conferences, you'll develop a working relationship with each member and will learn how to motivate, inspire, and lead each individual. You'll track the work, understand what's remaining, and even graph assignments in burnup and burndown charts for stakeholders.

In an adaptive project, the team is self-led and self-organizing, which means the team determines who'll do what assignments in the iteration. During the daily standup meeting, the team members communicate with each other what they've accomplished since the last meeting, their work for the day, and if there are any impediments to their progress.

So while there is accountability and clear communication regarding who's doing what, there's not a command-and-control approach that you'll see in a predictive project.

Focusing on the Work

One message all project managers should convey to the project team members, but often don't, is simply to focus on their work, their tasks, and their responsibilities. If team members would ignore the superfluous activities of the project and home in on what their responsibilities are, the project would scream along with few interruptions. Of course, this depends on the level of detail you have completed in the planning processes and the people you have to manage on your project team.

This is not to say that team members shouldn't be involved in project planning—they should! However, once the plan has been created, the team should just get to work, ignore the gossip and the details that don't involve them, and focus on their duties to complete their tasks. Planning is not, however, a group process that's completed and never revisited. Planning is a set of iterative, integrated processes that needs to happen throughout the project. As issues and needs arise, which they will, the project shifts back into planning mode to discover the best responses to issues and concerns that have happened within the project.

Part of planning is to find the most appropriate resource for each project task. The assignment of resources to tasks allows the team members to know what they have to do and when they need to do it. If team members could just ignore activities that are not related to them, the attraction of other technology, and the world of office politics, what a fantastic team they'd make! As a project manager, you should directly encourage your team to focus on their individual tasks. Encourage them to focus on their duties and their commitment to completing their assignments and ignore what anyone else may or may not be doing.

Like a machine, your team is collectively working toward deliverable results, but within the machine are many moving parts to make the deliverable happen. Each component of the machine is responsible for only certain tasks; one component cannot do everything, and all components are required to make the machine work. The same is true within your project. The team is a collection of individuals who need to work together, but who also can work independently as their tasks require it.

Adaptive project teams are sometimes called *generalizing specialists*. This means the people on the team can do more than one type of work no matter what their role in the team is. For example, a developer could help do testing as needed in the project. This helps get rid of the silo mentality of people only doing one type of work regardless of what the project needs.

Task ownership doesn't mean that team members should not help each other with tasks because the work may not be assigned to their realm of responsibility. The goal is that team members know their responsibilities, focus on them, take pride in them, and complete them successfully. If other team members need help with tasks, the team should, by all means, be fluid enough to help a colleague and keep the project moving toward its completion.

Hosting a Project Status Meeting

Within your communications management plan, you'll define which people need what information, when they need it, and how they'll get the information. Communication can be both formal, like a report, and informal, like a hallway conversation. In project management, both methods of communication are needed. The type of communication you'll use should be appropriate for the message you send. In other words, communicate major decisions using formal means, such as reports, project plan updates, or memos.

When it comes to informal communications, project managers should be near their project team. Managing a project where you and your project team are in separate buildings is not conducive to effective communication. If you can be seated physically close to the project team, you should. This allows for informal meetings and conversations to pop up, and it lets you get involved with the team and really lead the work. Virtual teams create new challenges for status meetings and communication. You'll need an approach that allows you to communicate effectively with remote teams on a regular basis. Teleconferences, videoconferences, web-collaboration software, and good old travel are a few options. While nothing beats face-to-face communication, the reality in today's world is that it's just not always an option.

Virtual teams create challenges for effective project management that need to be addressed in the communications management plan. Consider things like time zones and technological factors, such as web collaboration software and calendaring. For international projects, you'll also need to address language barriers and cultural differences that may affect the project work and overall progress. You'll also need to address the preferences of the project team and how they'll communicate, though the criticality of the project may supersede preferences, something else you'll need to communicate with the project team.

Regardless of where your office is, close to the team or far away from your virtual team, you will need to create a regular schedule to meet with your project team. Regular meetings, whether daily, weekly, biweekly, or on your own custom schedule, will serve several purposes:

- They enable a team member to report on their activities.
- They underscore the project vision.
- They allow the project team to report what's been accomplished since the last meeting and what they're working on now.
- They enable the team to identify problems.
- They enable the project manager to lead the team without hovering.
- They create a sense of ownership of the project.
- They create a sense of responsibility to the project for team members.

To host a team meeting, you need one thing: preparation. Create an agenda, even if it's a quick list of what needs to be discussed, and then follow it. Also, set a time limit for these meetings—and make it snappy. Perhaps the number one complaint among project teams is WOT (waste of time) meetings. For every meeting, your agenda should include at least these points:

- The objectives accomplished since the last team meeting
- Discussion of any situations impacting the entire team
- Acknowledgment of major team member accomplishments
- Overall project status—good or bad
- Pending risks, issues, and upcoming activities
- The objectives in queue before the next meeting

By starting with a review of the objectives of the past week, you are allowing team members to report on their activities and update the team on the state of the project. You are also allowing the team members to shine in front of their peers by reporting what tasks they have successfully completed. Don't be surprised if your team applauds when key events are finished.

Another aspect of having your team report what objectives were accomplished is that it creates, again, a sense of responsibility. A team member who knows they'll have to report that an assignment was not completed on time in front of their peers may be inspired to complete the task. The goal is not to embarrass team members, but to keep team members on schedule and committed to the project.

Should there be any outstanding unresolved issues from the last meeting—for example, the delivery of hardware, software, or other resources—the project manager, or the responsible party, should quickly update the team on the status. The issues discussed should be only those that affect the entire team. Anything impeding the project progress is a team issue.

EXAM TIP Issues and risks are not the same thing. *Risks* are uncertain events that could happen: the vendor could be late, a task may not meet a deadline, the software could conflict with the operating system, and so on. *Issues* are risk events that have happened: the vendor didn't deliver on time, the software won't work with the operating system, a new firmware upgrade is causing printer problems, and so on. Risk goes in the risk register; issues are documented in the issue log.

If team members have completed a major challenge or tasks within the project plan, call attention to that in front of the entire team. Offering a public acknowledgment to the team members is simply giving credit where credit is due. Always acknowledge major completions and, when necessary, acknowledge team members who may not have major tasks but are doing a great job. This heartfelt thank-you is an excellent way to boost morale and show your team that you do care for them and their success—not just the project's. Team morale has a huge impact on the project's success. Remember that acknowledgments, thank-yous, and kudos are not something you can fake—develop a true sense of care and compassion for your team.

Next on your agenda should be a quick capsule of the overall status of the project. If the project is on target and moving along swimmingly, this is easy. However, if the project is lagging and the team members aren't completing their assignments, you need to let them know. Being a tyrant and reprimanding a group is not the goal. The goal is to make adjustments to get the team back on track and focused on their duties and to avoid a pile-up of work and a completion date that has to change. Don't be shy in expressing your discontent; just use tact and express your passion for the success of the team.

You'll also want to review risks that have passed, been mitigated, or are pending within the project. Risk management calls for a review of risks and a reminder to the risk owners of triggers that may signal a risk is coming into play. You're not micromanaging the project team over pending risks, but rather bringing everyone's attention to the identified risks, making them aware of the risk response plans you've created, and assigning ownership to the project team. If there has been any change to a risk, that update should be documented in the risk register.

Finally, hold a round-table discussion against your PND to review assignments in the queue for the upcoming week or weeks. Ideally, you should have each team member verify what tasks they are working on before the next scheduled meeting. This does not need to be a long, drawn-out discussion—just a confirmation of duties. It is also a way to ensure this team meeting ends with your team members knowing what tasks are required of them before they meet with you again.

The foundation for successfully implementing an IT project is not the project plan; the technology being implemented; or the speed of the network, processor, or disk. The foundation is the ability of the team to communicate honestly to the project manager about how the actual work is going, the challenges the members are facing, and how they've reacted to those challenges. That type of communication is going to come from their ability to trust you. You have to earn that trust, and, hopefully, if you've involved your team members in the planning phases and kept them informed of how the project will develop, you will have it.

Tracking Progress

For several reasons, you must have a formal process for tracking progress. At the top of the list of reasons: Tracking project progress will help you take corrective actions and preventive actions to the project. *Corrective actions* allow you to direct the project team to fix an error to get the project back in alignment with the project objectives. You might also know this as *defect repair*. *Preventive actions* are steps you and the team take to prevent a possible error from ever entering the project—or learning a lesson and not repeating an error again.

You will need to develop an internal process for your team to report completed tasks so that you can reflect the project progress in an electronic form and analyze the team's work, the budget, and the days until completion. This also allows you to report accurately to management how the project is moving along. When errors, delays, and defects creep into the project, they have a direct effect on the project's ability to reach its target objectives. No surprises there, I hope.

Creating a Reporting Process

You should create a mechanism that allows your team to report the status of assigned tasks on a regular schedule. In some organizations this is a formal *work authorization system.* This system requires project team members to report activities completed with given metrics so that downstream activities can begin. A *project management information system (PMIS)* can streamline this process—so long as the quality of data is there and you can verify that the work is actually being completed.

Some project managers like team members to report as each milestone is reached, while others prefer weekly status reports on the tasks completed over the last seven days. Whichever method you choose or develop, it is important that your collection of data be on a consistent schedule. Although it's not well advised, you can start a project and collect status reports weekly one month and then biweekly the next. Develop a schedule that works best for you and fits the project and stick with it. Of course, you'll document this schedule of status reporting in your communications management plan.

EXAM TIP Changes to the project scope, risks that happen, and shifting project resources can all affect the project schedule. While it's ideal to say the schedule won't change, in reality it probably will change. It's good to create a schedule baseline, but as events alter the schedule, update and version the baseline for future reference. You don't want to get to the end of the project and have no documentation or history of why the project took longer than you predicted.

Determine the format for how work should be reported. Ideally, you should base this on the number of hours or days assigned to the task. For example, during the activity duration estimating process, say you allot 56 hours for testing a new application and assign Rick to the task. When Rick reports his progress, he should indicate the number of hours into the testing phase, in addition to a percentage of the total completed work Rick believes is done. As Rick moves closer to 56 hours, he should be moving closer to 100 percent completion of the task. Figure 9-1 shows the impact of exceeded hours on the budget and the overall time of the project duration. The actual collection of work

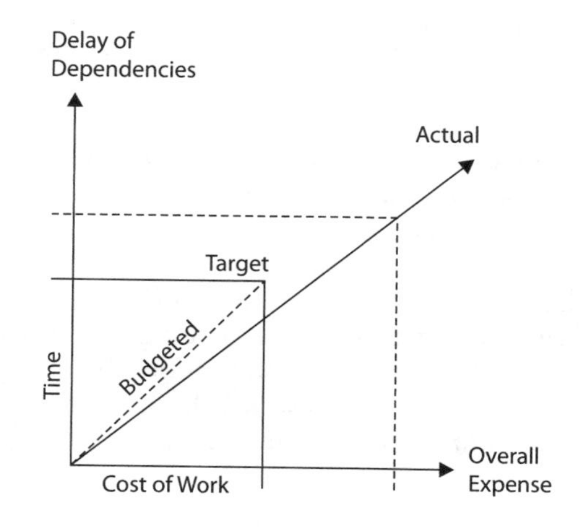

Figure 9-1 Tasks that exceed their durations impact both the budget and the project.

completed will allow you to see how the progress is going and to make adjustments to the project schedule to keep things on track.

When problems arise in the implementation phase, the number of hours assigned to a task will no doubt increase. For example, Rick's testing of the new application is taking longer than the 56 hours assigned to the task because of the discovery of a hardware conflict. Drawing on your communication with Rick, what you have learned through your regular meetings, and the hours reported by Rick through your reporting process, you should be able to ascertain quickly if more hours will be required for the testing activity. In other words, it shouldn't be a surprise when Rick reports he'll need some more time to complete his task.

To react to this problem, you need to analyze how additional hours will impact the following:

- Dependent tasks in the PND
- Other, nondependent tasks in the PND that Rick has been assigned to
- The critical path
- The budget
- The project completion date
- The management reserve
- Additional resources
- Risks
- Communication needed regarding the additional hours

To resolve a problem, analyze each of the facets of the project plan impacted by the new requirement for additional time. If dependent tasks are being held up by the problem, you need to find a solution to resolve the problem as quickly as possible. Generally this means you'll have to do one or more of the following things, discussed in the following sections:

- Implement schedule duration compression
- Invoke management reserve
- Reassign the work unit

Implement Schedule Duration Compression

Schedule duration compression is a fancy way of saying, "Hurry up!" You often need to compress the schedule when things are running late in your project. One of the easiest methods to implement is to examine your activities in the project network diagram for activities that have a finish-to-start relationship. You might enable some of these activities to overlap with their predecessors by adding *lead time* to the activities. For example, an activity to create a new software application might precede the activity to write the user manual for the software. It's feasible that the technical writer for the user manual could write portions of the manual before the entire software application is compiled, packaged, and ready to ship. Lead time would allow one activity to overlap slightly with its predecessor to finish sooner than was originally planned.

A similar approach to lead time is *fast tracking*. While lead time focuses on individual activities in the project, fast tracking focuses on entire phases or individual activities of the project. Fast tracking allows phases to overlap slightly so that all of the activities in the subsequent phase can finish sooner than what was planned if the phase didn't start until after the previous phase had finished. Fast tracking is considered risky, however, because if there are errors in the previous phase, they can have huge ramifications on the work already completed in the second, overlapping phase.

Another approach to schedule compression is *crashing*. Crashing allows you to add more effort to an activity or phase to get the work done faster. Crashing a project increases the labor to complete effort-driven activities, but this approach can also increase the cost of the project because someone is paying for the labor you've added. There's a special law of economics you have to consider, too: the law of diminishing returns. This law states that you can't add labor to reduce the duration of the work. You can't add 100 electricians to pull 100 network cables through an office plenum and be done with the job in just a few minutes. Furthermore, the law states the cost of the added labor may exceed the yield the labor creates. For example, you can add 100 electricians to reduce the amount of time for the network cable installation, but it probably won't be very cost effective, and there's only so much work each electrician can do in the task. Finally, when you add too much labor, the laborers are going to get in each other's way and things can become counterproductive.

In theory, assigning additional resources to a task should reduce the amount of time required for the task to complete. In reality, this is not always the case. For example, when installing an operating system that takes one hour to install, assigning two team members to the task doesn't mean that the installation procedure will take only 30 minutes. These activities are fixed-duration as opposed to activities that are effort-driven. In some instances, however, assigning an individual who is more experienced in the technology may cost more per hour (as in the case of a consultant), but that person can finish the task in less time, saving overall costs and preventing the delay of dependent tasks.

EXAM TIP *Effort-driven activities* can be completed faster with more effort applied, such as adding more people to do the work. *Fixed-duration activities* won't get done faster by adding labor, such as waiting for the software to install.

Invoke Management Reserve

Recall that *management reserve* is a final task in the critical path of the PND. It is an artificial task that is generally 10 to 15 percent of the total amount of time allotted for all tasks. When tasks exceed their allotted time, you assign the overrun to the management reserve task. For example, Rick is testing the software and will overrun the allotted time by 24 hours. A project manager could assign the 24 hours toward the completion of the management reserve and allow the critical path to continue as planned—assuming no other constraints are affected by the delay of the activity.

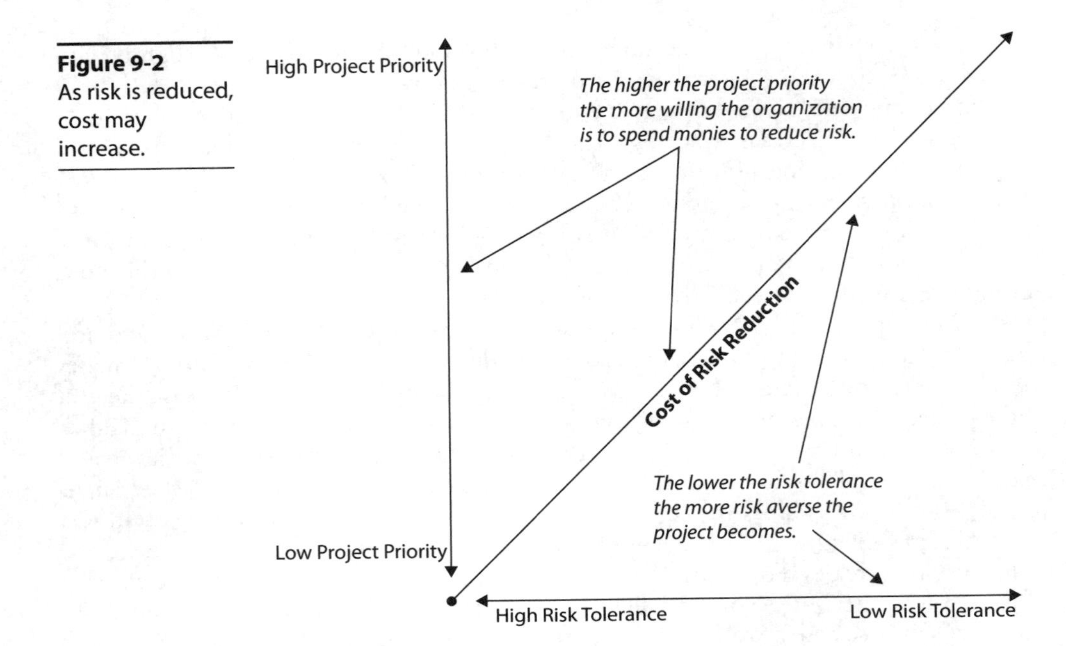

Figure 9-2 As risk is reduced, cost may increase.

You can also use a combination of additional resources and management reserve. For example, assigning an additional team member to assist Rick may reduce the time overrun from 24 hours to 16 hours. The 16 hours would then be applied to management reserve.

Reassign the Work Unit

Finally, you can choose to reassign the task to someone more qualified in the procedure. If Rick has exasperated himself and cannot resolve the issue with the software, then you may elect to hire a consultant or assign another team member to the task. The result of this solution is generally less risk of additional hours spent by Rick, but additional financial costs applied to the budget, as Figure 9-2 shows.

Status Collecting Tools

As a project manager, you may not always have the time to chat with each team member every week to get a verbal confirmation on the progress of each task. You will need a process to streamline the collection of hard numbers on the hours and percentages of the work completed. There are several methods you can use to collect this information from your team:

- **E-mail** A simple solution is to have your team members e-mail the hours of the work they've completed on their assigned task. This simple solution is not very automated, but at a minimum it allows for easy collection and accessible reporting for all team members. Of course, you'd then have to transfer the information into Microsoft Project, Excel, or another project management software program.

- **Information radiators** Adaptive projects utilize an information radiator, which is a collection of charts, project information, and status updates in a public place where everyone can access the information. An information radiator typically includes a burndown chart of the product, schedule forecasts, a product roadmap, issue tracking, a Kanban board, and any other significant information that's needed for the project. It's just a public way to be transparent, but also a centralized location for project information as stakeholders need.
- **Spreadsheets** A slightly more advanced method to collect and compute project status is through a weekly spreadsheet each team member would complete to report the tasks that person has been working on. You could create a template that lists the tasks, allotted hours, and hours actually worked on the tasks and include an area for any comments from the team member. Once you receive the spreadsheet, electronically of course, you can have macros and formulas retrieve the team members' information and dovetail it into a master spreadsheet.
- **Web forms** An aggressive approach is to create a web-based reporting system that would allow team members to report their activities and the hours committed to each via a web page form. The submitted form could be configured to automatically calculate percentages of work, overruns, and impact on the critical path. This would be ideal for a geographically dispersed team.
- **Project management information system (PMIS)** Any of the preceding methods can be used in conjunction with a PMIS, such as Microsoft Project Server, Jira, Basecamp, MS Teams, or Monday.com. A PMIS can enable collaboration via e-mail, web interfaces, and other applications such as Excel. If you are using a PMIS from the onset of your project, you will have created the WBS, the Gantt chart, and the PND within the application. Calculating task completions, overruns, and assignments of additional resources is easier to do with a collaborative PMIS. Server-based products allow for true collaboration between the project manager and the project team via web technologies.

Whichever method you choose or develop for your project, it is imperative that you create and document a detailed schedule for collecting project status. You must have periodic project updates, or the project will grow stale, you'll miss opportunities to make adjustments to prevent overruns, and team members may lag behind on tasks. A consistent, persistent project manager is needed to keep a project team dedicated to tasks and collect information about the completion of each task.

Communicating Project Status

How will you communicate project status to the project stakeholders and to the project team? Many project managers use a weekly status report that provides an update of what's been accomplished in the project over the past week, any issues that need to be addressed, status of risks, overall project completeness, and project performance. While status reports are handy, they aren't always the best method to provide information about the project.

Status reports are *push communication*—you are pushing the report to the stakeholders via e-mail. A *pull communication* is a reporting feature that's growing in popularity. A pull communication for project status lets stakeholders visit a project website or information radiator to communicate what's happening right now in the project. It's pulled from a central repository that is updated frequently based on project information.

As a project manager you'll use many different tools to manage communications within a project. Common communication tools include the following:

- **Dashboards** A dashboard is a one-stop window into the project, a console containing all the project starts and performance results. For example, your dashboard might show the project tasks completed, tasks remaining, number of items currently in progress, schedule and cost performance, total hours of labor by the team, and labor hours for each worker. The dashboard might have filters and toggle switches so users can compare and contrast information. For stakeholders, the dashboard could be read-only and allow them to see only certain parts of the project information based on security and configured rules for the project.
- **Knowledge management tools** Knowledge management tools are a way to find and filter information in the project. These are usually accessed by a stakeholder through a web form to get information such as project costs, progress, labor hours, pending risks, and risks resolved. These tools could also be configured to allow people to see the project scope, WBS, budget, project charter, and other documents in the project—again, based on the user's role in the project, access rights to the information, and what the project manager wants to share. These tools also help with reporting and archiving project information for future usage.
- **Performance measurement tools** Later in this chapter, we'll explore earned value management, a suite of formulas used to show project performance. Your dashboard, project website, or information radiator could show these values and chart out the project performance for your project. You can choose to show cost, schedule, and likely future performance based on the current conditions of the project. You can also track and share expenditures that the project has taken on or will take on so management can plan accordingly.

Tracking Financial Obligations

In addition to collecting information on the status of tasks from your team members, you will have a responsibility to collect information on the financial aspect of your predictive project. You will need, and generally be required, to meet with your project sponsor to report the status of the cost of the project. When variances exist, you'll likely have to document them in a *cost variance report.* This report explains the variance, the cause of the variance, and its impact on the project's success. You will need to report on a regular schedule the following items:

- The amount of budget spent on the project to date (cumulative costs)
- Any cost variances
- Actual costs versus the budgeted costs

- Value of work performed
- Cost variances and offset
- Suggestions, when necessary, to reduce the cost of resources

Tracking Actual Costs

To track the actual cost of the project, collect the amount charged on invoices from vendors and consultants and the dollar amount assigned to the team members' hours or the tasks they are completing. The ongoing sum of this collection is the actual cost of the project. This includes rework due to lack of quality; waste from materials; and purchased time from consultants, subject matter experts, or vendors.

The invoices you receive from vendors will be, obviously, a result of the goods or services rendered. The deliverables (service or goods) stem from a *commitment document*—which is a generic way of referring to a contract, a purchase order, or a letter of intent. This leads to the committed cost. A *committed cost* is the amount of money approved and assigned to a portion or the entirety of a project. On a regular schedule, you apply the committed cost to the actual cost. As Figure 9-3 demonstrates, the process of applying the committed cost to the actual cost should result in a balance based on the original budget creation.

The comparison of the actual cost and the committed cost should reflect the cumulative budget cost for the entire project. If there are inconsistencies, a line item comparison of the goods and services delivered against the cumulative budget may be required. Discrepancies between the actual cost and the committed cost may arise from flux in hardware prices, additional services or features added, or an error on behalf of the vendor (at least you can call it an error). Refer to your contract for details on price overruns.

An *expenditure* is the actual spending of project money. When you purchase goods or services, management will want to see when the expenditure happened and reconcile it with the invoice or purchase order attached to the expense. Your project's budget will need to track the expenditure as the item or service purchased is logged against your project. In other words, as you spend money on the project, the available balance of funds for your project is lowered. You might also need to report an expenditure that has happened or that is about to happen. You don't want to surprise the key stakeholders with a pending expense and stall your project.

The key to controlling the finances within your project is to safeguard your budget and react to cost variances as soon as they appear. This requires a routine to confirm the cost of goods or services delivered and the cumulative budgeted costs. You can automate this procedure within Microsoft Excel and Microsoft Project.

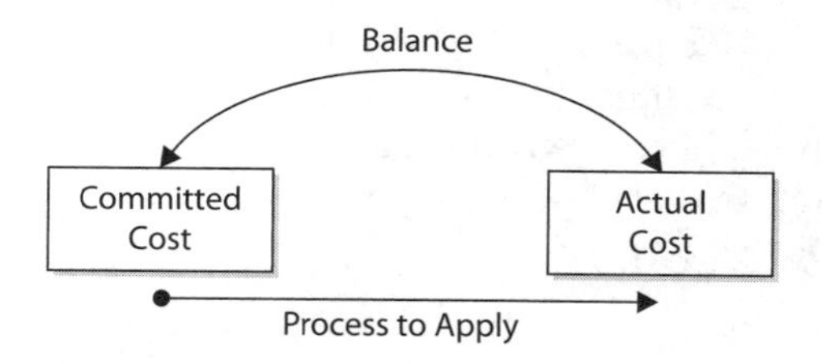

Figure 9-3 The committed cost and actual cost should be in balance.

When cost variances happen—and sooner or later they will—you will need a plan to analyze the costs and see what offsets may be made to control the total actual costs for the project. In other words, spending $5,000 for a consultant's time that was not budgeted will leave your budget $5,000 in the red. (*In the red* means a negative balance. *In the black* means a positive balance.) You will have to either find a solution to reduce other costs or approach management for additional funding.

Your first approach is to examine the budget to see how the extra expense can be reduced. You can reduce expenses by

- Using less-expensive resources
- Assigning additional resources to a task to complete it sooner and reduce its overall labor costs
- Arranging the PND so that tasks are start to start (SS) or finish to finish (FF) rather than finish to start (FS)
- Reducing the cost value allotted to the management reserve
- Reducing the size of the project scope

Determining Earned Value

Earned value is an excellent system to test, in an ongoing process, whether the work completed on a project is in alignment with the budgeted costs for a project. Earned value is a measure for project performance. This approach to financial management is ideal for hourly workers such as consultants, application developers, and resources that have a fixed hourly rate.

Earned value project management evolved from the early 1900s, from the factory floors. Industrial engineers created a formula to predict the value of a factory. Their formula has three variables: earned standards (what the factory actually produced), the accumulative costs incurred, and the original budgeted costs. To use the formula, the engineers took the earned standards number and compared it to the actual costs amount. Then they compared the earned standards with their planned goals for the factory output.

The comparison of values allowed the engineers to predict the profitability for a company from the output and costs of the workers and machines within the factories. Based on these figures, changes could be made to streamline production, address actual costs, or determine a plan of action for a less-than-profitable output.

To apply this formula to today's world, a project manager first needs to have completed all of the planning stages. Specifically, the project manager must have the WBS completed with accurate predictions of the amount of time and cost estimates required for each of the work packages. While some of the time required may be little more than estimates, there must be a serious attempt to calculate time for each work unit, as addressed in Chapter 5. Without an accurate account of time for each task within a project, earned value is not reliable because it compares the current output with the predicted output.

Controlling Finances

Your organization probably already has set processes for handling requests for payments, purchase orders, and payment on invoices. If you are not familiar with the internal flow of paperwork, the approval of funds dispersed, or the procedure to supply purchase orders versus payment on invoices, speak with your project sponsor or company comptroller, who'll be happy to educate you on the process.

In addition to just knowing where to route papers and whom to call when invoices are due, you'll need a formal approach to tracking and analyzing the actual costs. You can use a number of ways to create a system for collecting this information, though Microsoft Excel and Microsoft Project are two of the best tools available to project managers. In addition, you'll need organization and a regular schedule to update the expenses on your project.

Earned value management (EVM) has a few fundamental values:

- **Earned value (EV)** Earned value is representative of the work completed to date, regardless of how long it took to accomplish it. For example, if a project has a budget of $100,000 and the work completed to date represents 25 percent of the entire project work, its EV is $25,000.
- **Planned value (PV)** Planned value is how much the project should cost to get to a specific point in the schedule. For example, if a project has a budget of $100,000 and month six represents 50 percent of the project work, the PV for month six is $50,000.
- **Actual costs (AC)** Actual costs are the actual amount of monies the project has required to date. For example, if a project has a budget of $100,000 and $35,000 has been spent on the project to date, the AC of the project would be $35,000.
- **Cost variance (CV)** A cost variance occurs when the actual cost of the project work is more or less than the EV. For example, your EV is calculated to be $25,000 but you had to spend $35,000 to get there. This would result in a negative $10,000 cost variance.
- **Schedule variance (SV)** A schedule variance occurs when the EV is more or less than the PV. For example, the project is supposed to be worth $75,000 in month six; however, at month six your EV is only $45,000. You've got a whopping SV of negative $30,000.

To implement EVM, the project manager collects the status of the computed percentage of tasks completed. For example, as team members report their status of hours applied to their assigned duties, your project management software can report a percentage of the task completed. Each work unit that has a predicted number of hours can be assigned a dollar amount as well. For example, if Marcella the programmer needs 36 hours to reach a particular milestone, and her hourly rate is $130, the dollar amount assigned to the work is $4,680.

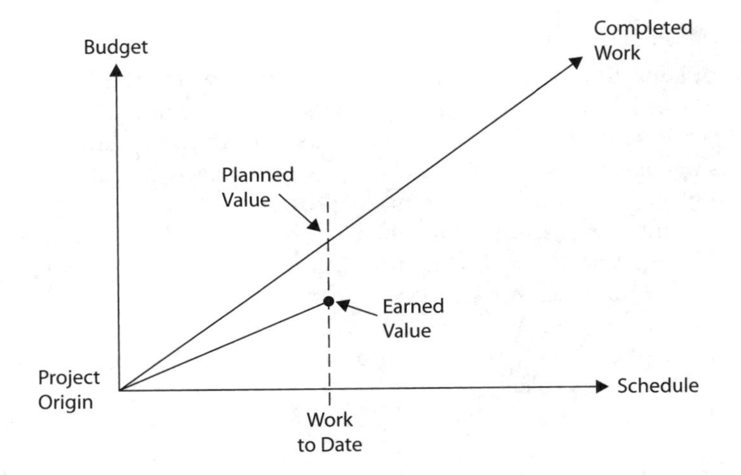

Figure 9-4 Earned value can predict if a project will be financially strapped.

If Marcella has completed 12 hours of the work and is on schedule now at a third of completion, the earned value is 33.33 percent of $4,680 (the total cost of the work unit), which is $1,560. However, if Marcella has completed 12 hours of the work but reports that only 20 percent of the project is completed, the earned value is now out of sync with budgeted costs. The cost of the work unit has just risen to $7,800 if Marcella stays on this schedule of production.

TIP Earned value management is a fancy and mathematical way of showing project performance, but it's not great for smaller projects. Larger projects, with fat budgets, are more inclined to use EVM. In this discussion, I'm sharing the most common EVM formulas—they are more for conditions like front-loaded projects where you have to spend a large chunk of the budget on equipment, for example.

The primary benefit of predicting earned value is that a project manager can predict if the project is going to be in financial trouble early on in the implementation phase, as Figure 9-4 demonstrates. Unfortunately, many IT project managers simply do not take the time necessary to predict the earned value of their project as they implement it. It is not a difficult process and should be, quite frankly, mandatory to keep expenses in alignment.

Let's take a look at the EV formula in action. This example is for a project that has a budget of $250,000. The project is 15 percent complete, but it should be 20 percent complete by this point in the calendar. In addition, the project manager has had to spend $43,000 of the budget just to get to this point in the project.

Term	Value	Definition
Budget at completion (BAC)	$250,000	This is the expected cost of the project.
Percent complete	15 percent	This is the actual percentage of the work completed as reported by the project team.

Term	Value	Definition
Earned value (EV)	$37,500	This is simply 15 percent of the BAC.
Planned value (PV)	$50,000	This is what the project work should be worth by this point in the project.
Actual costs (AC)	$43,000	The actual costs reflect the amount of funds the project manager had to spend to get to this point in the project.
Cost variance (CV)	–$5,500	This is found by subtracting the AC from the EV. This project is off budget.
Schedule variance (SV)	–$12,500	This is found by subtracting the PV from the EV. This project is off schedule.

In this book's online resources, you will find an Excel file named EV Worksheet that you can use to calculate your own EVM figures for your projects. You will be working more with computing earned values in an upcoming exercise. (Refer to Appendix C for accessing the online resources.)

Calculating the Cost Performance Index

The *cost performance index (CPI)* is a reflection of the amount of actual cumulative dollars spent on a project's work and how close that value is to the predicted budgeted amount.

For example, as Figure 9-5 depicts, a total network upgrade project has a budget at completion of $209,300, and to date the project has spent $34,500 on actual costs. Based on the percentage of the completed project, which is 15 percent, the EV is $31,395. The planned value, however, is $36,000. The project also has a CV of –$3,105.

To compute the CPI for this project, the earned value, $31,395, is divided by the actual costs, $34,500. This results in 0.91, which means the project is 9 percent off the target rate of spending for this stage in the project. The project manager can use this information to reschedule resources; adjust schedules; reassign tasks; and, if worse comes to worst, ask for additional funding.

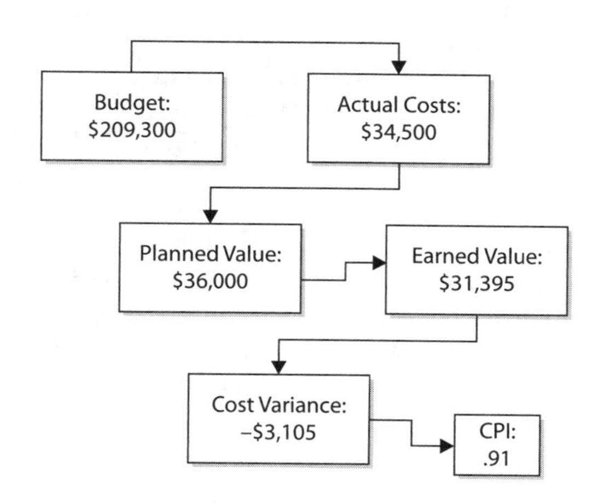

Figure 9-5 CPI reflects how closely the project is following the budget.

$$SPI = \frac{\text{Earned Value}}{\text{Planned Value}}$$

Figure 9-6 SPI is the ratio of the work performed (earned value) to the work planned (planned value).

Calculating the Scheduled Performance Index

The *scheduled performance index (SPI)* is a formula to calculate the ratio of the actual work performed versus the work planned. The SPI is an efficiency rating of the work completed over a given amount of time. It is not a dollar amount, but rather a percentage of how closely the completed work is to the predicted work. The formula to calculate the SPI is fairly simple, as Figure 9-6 shows.

If the result of your formula is 1, you are on schedule. If the result is less than 1, you are behind schedule. Of course, if the result is greater than 1, you are ahead of schedule. For example, if the EV is $18,887 and the PV is $20,875, the SPI is 0.90, which is less than 1, so this project is behind schedule.

EXAM TIP The Project+ exam objectives specifically name the schedule and cost variances as two key objectives. Be familiar with these other earned value management techniques, but really understand schedule and cost variances for your exam.

Predicting Project Performance

The *estimate at completion (EAC)* is a prediction of what the total cost of the project will be, given current performance factors. As your project progresses, there may be some variances between the cost estimate and the actual cost to reach milestones within your project. The difference between these estimates is the cost variance. These cost variances affect the overall performance of the project—and it's what the CPI is really based on.

There are several different formulas for calculating the EAC, but the most common formula is EAC = BAC / CPI. Consider a project where the BAC is $200,000 and the CPI is 0.80. In this instance, the EAC is $250,000. In other words, this project isn't doing so hot, so if things keep trekking along as they are, the project will actually cost $250,000.

Using this information, you can use another fun formula to predict how much more money the project will need in order to reach the end of the project—also known as the *estimate to complete (ETC)*. The ETC is a pretty straightforward formula: ETC = EAC – AC. Let's say our EAC was calculated to be $250,000, and our AC is currently $25,000; our ETC would then be $225,000. That's how much more the project will need to reach its closing.

Calculating the To-Complete Performance Index

Once you've calculated the CPI and the SPI and you realize your project may be late, your first thought is, "How can we get back on schedule?" Okay, that's not your first thought, but pretend it is.

Imagine a formula that would tell you if the project can meet the budget at completion based on current conditions. Or imagine a formula that can predict if the project can even achieve your new estimate at completion. Well, forget your imagination and just use the *to-complete performance index (TCPI)*. This formula can forecast the likelihood that a project will achieve its goals given what's happening in the project right now. There are two different flavors for the TCPI, depending on what you want to accomplish:

- If you want to see if your project can meet the budget at completion, you'll use this formula: TCPI = (BAC – EV) / (BAC – AC).
- If you want to see if your project can meet the newly created estimate at completion, you'll use this version of the formula: TCPI = (BAC – EV) / (EAC – AC).

Anything greater than 1 in either formula means that you'll have to be more efficient than you planned to achieve the BAC or the EAC, depending on the formula you've used. Basically, the greater the number is over 1, the less likely it is that you'll be able to meet your BAC or EAC, depending on which formula you've used. The lower the number is from 1, the more likely you are to reach your BAC or EAC (again, depending on which formula you've used).

To predict how much harder you and your project team will need to work to finish the project within budget, you'll need to calculate the TCPI. At the end of the formula, which is shown in Figure 9-7, if the number is greater than 1, you'll have to buckle down and work harder. If the result is 1 or less, breathe a sigh of relief; you can make it on your current schedule.

Let's consider a project that has a BAC of $750,000 and is 30 percent complete, although it was planned to be 50 percent complete by this point. In addition, the project has spent $240,000 so far in the execution. The formula we'll use is TCPI = (BAC – EV) / (EAC – AC). In this example, the formula would read TCPI = (750,000 – 225000) / (800,000 – 240,000), which equates to 0.97. Given this project's EAC, the project manager is likely to hit the new EAC.

Figure 9-7
TCPI is a formula to predict the ability of a project to stay within budget.

Numerator is how much is left in the budget.

$$TCPI = \frac{\text{Budget} - \text{Earned Value}}{\text{EAC} - \text{AC}}$$

EAC is the Estimated Cost at Completion.

Communicating the Project Burn Rate

Just like it sounds, a *burn rate* describes how quickly you're burning through the project budget. The burn rate shows the consumption of funds by the project and the anticipated costs of the remainder of the project. Imagine a project that has a budget of $500,000. As the project team works on the project, there is a cost associated with their work—their hourly wage. (Your projects may or may not include the costs of the employees in your budget, but let's say that it does for this example.) As the project team works and accumulates costs, it's reflected in the burn rate for the project. So if the project team has five people and their total hourly rate has cost the project $120,000 to date, the burn rate is $120,000 against the $500,000 budget. Add to that any other expenditures to date, and you'll quickly see how you're burning through the project budget.

EXAM TIP The burn rate has traditionally been associated with project costs. Adaptive projects use burn rate to discuss how quickly the team is burning down the number of requirements and user stories they have completed. On the CompTIA Project+ exam, be sure to check out the context of the question in regard to whether it's a financial burn rate or an adaptive project burn rate.

With this information, you can forecast costs based on the predicted work remaining in the project and the future purchases your project will make. Mistakes and rework can drive your burn rate through the roof—and blow your project budget. By tracking expenses with a burn rate, you can see whether you'll have enough funds left to finish the project work, assuming that everything in the project goes according to plan. Tracking the project burn rate will also help you uncover the risk of depleting the project budget before the project is actually finished.

CompTIA Project+ Exam Highlight: Initiating Processes

In order to execute, predictive projects need a plan and adaptive projects need a product backlog of requirements. Predictive projects require a project management plan and are resistant to change. Most of your planning questions on the CompTIA Project+ exam will reflect predictive projects, but adaptive projects use planning too, obviously, but not with the same formality of predictive projects. While predictive projects create a project management plan for all of the project management knowledge areas, adaptive projects focus on getting the work done in short intervals of planning and doing.

Executing the project is all about getting the work done. It's the project team's opportunity to fulfill the project plans you and the team have created together. On the CompTIA Project+ exam, you'll have a few questions on project execution, but not as many as you might expect. Consider how project execution focuses on the actual work of the project and how each type of project work is different.

1.6 Given a scenario, apply schedule development and management activities and techniques A project is a temporary endeavor, so it requires a schedule to reach the business value. When the schedule is longer than warranted, the project manager and the team can work together to implement schedule duration compression. Fast tracking allows phases of the project work to overlap, but increases risk in the project. Crashing adds resources to effort-driven activities to finish the work faster. In some instances, contingency reserves can be utilized to allow for tasks that are taking longer than anticipated. Finally, we touched based on the idea of user story points the team can complete in an iteration, which is called the team's velocity. Velocity helps predict when the adaptive project will be finished.

1.7 Compare and contrast quality management concepts and performance management concepts Earned value management is a technique to measure overall project performance through a suite of formulas. Reference the file called EV Worksheet in this book's online resources for an approach to memorizing these formulas for the CompTIA Project+ exam. Here's a quick recap of each formula and how it is calculated:

Formula Name	How Calculated	Hint for Remembering
Planned value	Percent of where the project should be	Someone always has to tell you where you are in the project.
Earned value	Percent of where the project is	This is the percent complete times the BAC.
Cost variance	EV – AC	EV always comes first.
Schedule variance	EV – PV	Variance is something minus something.
Cost performance index	EV / AC	Index is something divided by something.
Schedule performance index	EV / PV	Schedule always uses planned value. Costs are always actual costs.
Estimate at completion	BAC / CPI	Account for pennies lost on the dollar.
Estimate to complete	EAC – AC	How much more do you need?
To-complete performance index (BAC)	(BAC – EV) / (BAC – AC)	What's left to do divided by what cash is left.
To-complete performance index (EAC)	(BAC – EV) / (EAC – AC)	What's left to do divided by the predicted amount of cash left.
Variance at completion	BAC – EAC	How far is the project likely to be upside down?

1.8 Compare and contrast communication management concepts As the project team executes the project plan, they'll need to report to you on their status of completing their assignment. You should host regularly scheduled status meetings for the project team. When you're working with virtual teams, as may often be the case, you'll need to adjust your communication efforts to ensure that the remote project team members are involved in the project execution and communication. Vendors that play a role on your project team will also need to be included in regular communication to be active parts of the project team, responsible for completing their activities.

1.10 Given a scenario, perform basic activities related to team and resource management Projects are done by people, not computers, equipment, and software. It's the management of the people, the most important resource in the project, that requires finesse, emotional intelligence, and leadership to get the project done. Human resource management is about communicating project status and letting people experiment and develop keen methods to get the work done, while still controlling the boundaries of the project work. Meetings are one of the most effective ways to discuss how the team is progressing, what's been accomplished and left to do, and any impediments, risks, or issues that should be addressed.

Chapter Review

As the project manager, you are responsible for all facets of the project planning and implementation. If you've built a solid foundation, and surely you have, the implementation will follow your work breakdown structure and the project network diagram. Predictive projects plan upfront and try to predict what will happen in the project. Adaptive projects plan just a little upfront, with the majority of planning done throughout the project, and expect requirements and direction to change.

On a regular schedule, you'll meet with your team members to review their work, congratulate them on successful milestones, and prep them for the next project activities. A regular, efficient meeting keeps the team focused on the project vision and accountable for the tasks to be completed. Part of keeping the project on track will be to create a process to collect work information from your team. You'll need a regularly scheduled method to request and retrieve information on the team members' progress for assigned tasks.

Just as you hold your team members accountable for their actions, management will hold you accountable for yours. Specifically, management is interested in your ability to control the finances and deliver expected results on time. You will need an education on how your organization processes payments, creates purchase orders, and reviews budget adjustments. You will also need a process to keep tabs on the finances and hours committed to a project and a method to track the value of work.

Project management is a long process that requires dedication from you and the project team. Your job is to ensure that the team stays dedicated to the project, that finances are always in order, and that you are always ready to react appropriately to any situation. Sounds easy, right?

Exercise

This exercise allows you to apply the knowledge you have learned in this chapter and is followed by the possible solution.

Exercise 9-1: Calculating Earned Value Management

In this exercise, you will compute the earned value management for a project and review the EVM formulas. If you have Microsoft Excel installed, you will be able to use a template to complete the exercise. If you do not have Microsoft Excel, you will be able to complete the exercise but will need a calculator to finish.

Exercise 9-1a: Follow These Directions if You Have Microsoft Excel Installed

1. In this book's online resources, open the Microsoft Excel document named EV Worksheet. (If you need assistance accessing this file, refer to Appendix C.) The document is a spreadsheet that will automatically calculate the earned value management values based on the information you supply.
2. To begin, confirm that you're working on the worksheet named EVM Formulas. This worksheet lists all of the EVM formulas and how they operate. You can hop back to this worksheet as you move through the exercise.
3. Move to the worksheet named EVM Actions by clicking the worksheet name at the bottom of the workspace.
4. You'll enter values into the green highlighted area at the top of the worksheet. You won't need to edit any of the values that are highlighted in yellow.
5. Here's your scenario: You are the project manager of the APPDEV Project. Your project has a BAC of $550,000 and is expected to last one year. As of now, your project is 25 percent complete, but it should actually be 40 percent complete. Due to some incidents early on, you've already spent $225,000 of your project budget. Enter the appropriate values from this scenario into the appropriate cells in the green highlighted area in the worksheet.
6. What is your earned value?
7. What is your SV?
8. What is your CPI? What does this value mean?
9. What is your SPI? What does this value mean?
10. Good news! You've just learned that your project is actually 40 percent complete, not 25 percent. What does this do to your project's EVM values?

Exercise 9-1b: Follow These Directions if You Do Not Have Microsoft Excel Installed

1. Here's your scenario: You are the project manager of the APPDEV Project. Your project has a BAC of $550,000 and is expected to last one year. As of now, your project is 25 percent complete, but it should actually be 40 percent complete. Due to some incidents early on, you've already spent $225,000 of your project budget. Enter the appropriate values from this scenario into the appropriate cells in the following table. Grab your calculator and complete the table.

Term	Formula	Result
Budget at completion	The project's given budget	
Percent that should be complete	The scheduled project completion	
Planned value	Planned %Complete × BAC	
Actual costs	What the project has actually spent	
Earned value	%Complete × BAC	
Cost variance	EV – AC	
Schedule variance	EV – PV	
Cost performance index	EV / PV	
Schedule performance index	EV / PV	
Estimate at completion	BAC / CPI	
Estimate to complete	EAC – AC	
To-complete performance index	(BAC – EV) / (EAC – AC)	
Variance at completion	BAC – EAC	

Exercise Solution

The following offer possible solutions for the chapter exercises.

Exercise 9-1: Calculating Earned Value Management

The solution to Exercise 9-1a can be found in the file Completed Exercise 1a, which is in this book's online resources.

The solution to Exercise 9-1b is as follows:

Term	Value
Budget at completion	$550,000
Percent that should be complete	40%
Planned value	$220,000

Term	Value
Actual costs	$225,000
Earned value	$137,500
Cost variance	–$87,500
Schedule variance	–$82,500
Cost performance index	0.61
Schedule performance index	0.63
Estimate at completion	$900,000
Estimate to complete	$675,000
To-complete performance index	1.27
Variance at completion	–$350,000

Questions

1. You are the project manager of the NYQ Project for your organization. Your project is slipping on some tasks and you would like to implement duration compression for this project. Which one of the following is a duration compression technique that generally does not increase risks within the project?
   A. Crashing
   B. Fast tracking
   C. Resource leveling
   D. Lead time for all fixed-duration activities
2. What should be the goal of the individuals on the project team?
   A. To make the company more profitable
   B. To help each team member finish their tasks
   C. To focus on completing their own tasks
   D. To finish their work as quickly as possible
3. Why should a project manager host a regularly scheduled team meeting?
   A. A meeting allows team members to report on their activities.
   B. A meeting allows the project manager to make changes to the project scope.
   C. A meeting allows team members to air grievances about other team members.
   D. A meeting allows the stakeholder to identify requirements.

4. What concept is defined in the law of diminishing returns?
   A. Work expands to fill the amount of time allotted to it.
   B. Padding the time surrounding activities causes project risks.
   C. Adding labor increases project costs.
   D. You cannot continually add labor to reduce project duration.
5. What is a work authorization system?
   A. A method to approve work that's been completed
   B. A method for the project stakeholders to approve the project work
   C. A method to allow successor work to begin based on the completion of predecessor work
   D. A method to allow the project manager to track uncompleted milestones
6. Why should a project manager allow team members to report on completed milestones in a team meeting?
   A. To create team development through project work
   B. To create a sense of pride in the work accomplished
   C. To create internal competition on the project
   D. To create peer pressure to outdo the other team members
7. Why should you review the upcoming assignments in a project team meeting?
   A. To remind the team members of the work they must complete
   B. To remind the team that you are in charge of the project
   C. To confirm that team members know their duties for last week
   D. To confirm that the project is moving and on track
8. Your project has a budget of $280,000 and is 30 percent complete. You have spent $90,000 on your project, however, the project is overbudget and late due to some rework and additional time from a vendor. Your project is supposed to be 50 percent complete by this time. What is your earned value for this project?
   A. $56,000
   B. $84,000
   C. $140,000
   D. 0.93

9. Your project has a budget of $280,000 and is 30 percent complete. You have spent $90,000 on your project, however, due to some rework and additional time from a vendor. Your project is supposed to be 50 percent complete by this time. What is your CPI for this project?

   A. 0.93

   B. 93

   C. 0.60

   D. $6,000

10. Your project has a budget of $280,000 and is 30 percent complete. You have spent $90,000 on your project, however, due to some rework and additional time from a vendor. Your project is supposed to be 50 percent complete by this time. What is your SPI for this project?

    A. 0.93

    B. 0.60

    C. 0.77

    D. 1.01

11. Your project has a budget of $280,000 and is 30 percent complete. You have spent $90,000 on your project, however, due to some rework and additional time from a vendor. Your project is supposed to be 50 percent complete by this time. What is your EAC for this project?

    A. $300,000

    B. $210,000

    C. 1.03

    D. –56,000

12. Heather is working on an operating system rollout to 1,256 workstations. The rollout is completed through imaging software, but there are scripts that have to be run at each workstation to complete the installation. The task has been assigned 400 hours to complete. Heather reports that she has committed 200 hours to date but is only 30 percent complete on the assignment because the process is taking longer than originally planned. What should a project manager do in this instance?

    A. Remove Heather from the assignment and reassign another team member.

    B. Add resources to the assignment to decrease the length of time to completion.

    C. Add time to the critical path.

    D. Remove all of the time from the management reserve and apply it to this assignment.

13. When a project is taking more hours to complete than originally planned, which of the following is a viable solution to reduce the hours required while maintaining costs?
    - **A.** Use less-expensive materials.
    - **B.** Fast-track the project.
    - **C.** Assign additional resources.
    - **D.** Apply management reserve.
14. True or False: Assigning additional resources to a task will always reduce the amount of time required to complete the task.
    - **A.** True, added labor reduces task duration.
    - **B.** False, added labor doesn't affect fixed-duration activities.
    - **C.** True, added labor is called project crashing.
    - **D.** False, added labor, or project crashing, won't reduce project duration.
15. What is the risk in reassigning a lagging task to a consultant you've hired to complete the task?
    - **A.** Additional time
    - **B.** Additional costs
    - **C.** Demoralization of the project team
    - **D.** Decrease in management reserve

Answers

1. **A.** Crashing generally does not add risks to the project, but it does add costs because of the added labor.
2. **C.** The goal of project team members should be to focus on completing their own tasks. If each team member would focus on completing the assignments as planned, the project would flow smoothly.
3. **A.** Regular team meetings accomplish many different tasks—including allowing team members to report on their activities and allowing the team to solve project problems as a group.
4. **D.** The law of diminishing returns addresses the concept of adding labor to reduce the duration of activity. While you may be able to add labor to decrease the duration of a task, you cannot continually add labor to reduce the duration of the task. In addition, added labor may become counterproductive for the project. Finally, the cost of the labor added may not be worth the yield (the value) of the work the added labor is able to perform.
5. **C.** A work authorization system is a formal process where project team members report the completion of their tasks so that downstream activities may begin.

6. **B.** By requiring team members to report verbally on the status of their assignments, they are held accountable for their activities. In addition, this practice creates a sense of pride for the team members who have accomplished major milestones in the project.
7. **D.** By reviewing the assignments with the project team, you are ensuring that the project is moving and on track. At the same time, you are confirming that the team members know their duties for the week and reminding them what their assignments are—but overall, the purpose is to keep the project moving.
8. **B.** The project's EV is found by multiplying the budget at completion by the percent of the project that is complete. In this instance, the EV is $84,000.
9. **A.** Your cost performance index (CPI) can be found by dividing the earned value (EV) by the actual costs (AC). In this instance, the CPI is 0.93.
10. **B.** You can find your scheduled performance index (SPI) by dividing the earned value (EV) by the planned value (PV). In this instance, the SPI is 0.60.
11. **A.** To find the estimate at completion, you use the formula budget at completion (BAC) / cost performance index (CPI). The CPI for this project is 0.93, and the budget is $280,000. The EAC for this project, based on current performance, is $300,000. Note that, without rounding, $280,000 / 0.93 is $301,075.
12. **B.** The project manager should assign additional resources to the project if at all possible. By adding resources to the project, the PM enables Heather and another individual to launch the installation process on more workstations simultaneously throughout the network. The length of the actual installation process will still take the same amount of time, but the number of workstations involved can increase.
13. **B.** This question is asking for a schedule duration compression technique, which will not increase project costs. The only viable option is to fast-track the project; this approach allows entire phases of the project to overlap.
14. **B.** False. Simply adding resources to all tasks will not always decrease the amount of time a task requires. For example, when installing an application, the time of the application installation can be streamlined through policies or scripts, but the installation time is limited by the speed of the workstation, not the number of team members installing the application.
15. **B.** When reducing risk, project managers usually increase cost. By hiring an expert consultant to complete a lagging task, the project manager will most likely have to pay a higher hourly rate for the consultant to be involved in the project.

CHAPTER 10

Revising the Project Plan

This chapter covers the following topics:

- Defining the need for revision
- Addressing operational change control process
- Implementing change control
- Implementing changes to the project plan
- Welcoming changes in adaptive projects
- Delaying a project
- Rebuilding management support

Have you ever taken a wrong turn? One minute you're cruising along listening to your favorite radio station, windows open, and you're enjoying a perfect summer drive. The next thing you know you're on I-90 when you should really be on I-94. You've cruised along for 20 minutes and now you're frantic. Not only are you now 20 minutes out of your way, but you've got to drive those 20 minutes again in the opposite direction just to get back to where you should have been to start with. So what do you do? Click off the music, put the windows up, and grit your teeth as you pretend you meant to screw up all along.

Project management can be like that unfortunate summer drive. No matter how much research you do, how many times you test a process, or how detailed your plan, no one can predict the future. Project managers can, and often do, start in one direction and, by chance or design, come to realize they've been going in the wrong direction. In some instances, they discover a better method or product in the early stages of the implementation. In others, a request from management or the customers to change the deliverables of the product can alter a project's direction. And in some cases, the cause for change rests on the shoulders of the project manager.

In this chapter, you'll examine the process a project manager can use to decide if a change in a project is feasible—and which system to put in place to review, approve, or decline change requests. Of course, adaptive projects welcome and expect change, so we'll examine that mindset and the process of welcoming changes in an adaptive project. Every project, and every change request, is different, so the project manager must react with caution to difficult situations, and try to keep their project, and wits, together. We'll also look at the operational change control for project management—your project may have large, sweeping changes based on how the organization operates. Pull the car over and silence the music; it's time to get to work.

VIDEO For a more detailed explanation, watch the *Managing Project Changes* video now.

Defining the Need for Revision

English writer Arnold Bennett said, "Any change, even a change for the better, is always accompanied by drawbacks and discomforts." How true that is!

In the world of IT project management, change is not, and generally should not be, an easy process to incorporate. Every project, as you know, needs a scope statement. The *scope statement* defines what will and will not be delivered as part of the project. The scope baseline is a point of reference for all future project decisions. Recall that the project scope is all of the required work—and only the required work—to complete the project. Once the scope has been created and agreed upon by the stakeholders, it must be protected from superfluous changes.

EXAM TIP Know how changes are managed in your organization. The details on change management in this book are based on the generally accepted practices, but your organization may have different rules and procedures for introducing changes to the project.

IT projects, however, are particularly subject to change due to the nature of the industry. Patches, service packs, new releases of software, bugs, threats, security issues, and new wishes from stakeholders can all task an IT project on a daily basis. Each change request must be evaluated for cost, time, risk, and other repercussions. In addition, each change request must be documented, tracked, and implemented in the plan or denied. When a change request is declined, the project manager must communicate the status of the change request and record the change request as declined in the change log. This helps during the final scope verification so that customers aren't puzzled as to why their change wasn't implemented.

But what happens in many projects? Change is forced into the project scope, even if it's a complete redesign on the deliverables, and then a project manager tries to shoehorn the project plan into the new and improved requirements. This rarely works. Instead, team morale declines, frustration ensures the deliverables aren't met, and the project manager loses control. To prevent this, you must have a process to control change and implement change when it is needed.

Addressing Operational Change Control

There is a big difference between a project and operations. Projects are temporary while operations go on and on. Creating new software can be a project. Supporting the software, help desk support, and backing up data from the software are operational duties. When we discuss change control, it's important to acknowledge if the change is an operational change or a project change. While your exam is on project management, you'll be tasked with addressing operational change control too, because these changes can affect your project.

Consider IT infrastructure changes: planned and unplanned downtime, scheduled maintenance windows, rollback plans, and validation checks that happen as part of operations. When an IT project is chugging along, the IT project manager must be aware of these interruptions in the infrastructure because they'll directly affect the project. Just imagine a project manager that has the team all ready for a big weekend rollout of some new software. The team charges in early on Saturday morning ready to rock-n-roll, only to discover the infrastructure team has planned for maintenance that weekend, so the project is stalled. The team is frustrated, their Saturday plans already interrupted, and now they'll have to schedule another weekend to do the work. Nobody is going to be happy with the project manager.

Planning for Software Change Control

Often when we think of software development for predictive and agile projects, we think of a brand new piece of software. However, it's often the case that the software development project is to update existing software. You've probably experienced this: You've mastered the current version of some software only to have a new version, a new update, blasted onto your computer. Now some of your keyboard shortcuts do different things, buttons on the toolbar are in different places (and I'm talking to you, Apple iPhone developers), and where you used to find settings, such as Do Not Disturb, are tucked inside of other settings that aren't logical.

In a project, changes to software may have a direct effect on your project. Consider how your project's deliverables may interact with the software, additional disruptions to the stakeholders, and support of the solution you create along with the changes to the software. You'll also need to schedule and budget for testing and staging the changes and how it affects your solution. As you know, everything in IT is related, and what appears to be a tiny change can have huge ramifications on the solution you create. All of IT comes down to four domains: software, hardware, network, and data. An external change as part of operations can cause internal changes to the project you're managing.

Establishing Change Control

Change control is an internal process an organization can use to block anyone, including management, from changing the deliverables of a project without proper justification. Change control requires the requestor to have an excellent reason to attempt a change, and then the proposed changes are evaluated in regard to their impact on all facets of the project. The Change Control System (CCS) will guide you and stakeholders through the change control process for the project.

Change control in project management is a documented, formal process for proposing, reviewing, and allowing changes within a project. The change control process presents how changes are reviewed for their value, costs, schedule impact, risks, and feasibility. The project must have a method to enter, track, and record the approval or denial of proposed changes.

Changes fall into two broad categories to consider: product change and project change. *Product change* means the features and functions of the end result, the requirements of the project, have changed. For example, the customer requests additional capabilities for the software the project aims to create. The client wants to change the number of servers from two to six. Your manager asks that you add 47 laptops, five different printers, segment the network differently, and three coffee makers to the project deliverables.

Product changes almost always introduce project changes, as product changes directly affect the project scope. *Project changes* are any changes to the project: scope, schedule, costs, quality, resources, communication, risk, procurement, and stakeholders. Project changes are integrated: a change to the project schedule can affect several areas of the project. The schedule delay may cause costs to increase because the current resources won't be available later in the project and you may have to procure different contractors to do the work. You'll need to communicate the delay and cost concern with stakeholders. You'll need to examine how the change may introduce or affect risks in the project. And you'll need to discuss quality. A simple change can have big, frustrating ripples in the project.

In many organizations, the change control system includes a *change control board (CCB)*. This board completes the review and analysis of the proposed changes to determine their worthiness and justification. Your organization may call the CCB an engineering review board, technical review board, or even technical assessment board. The CCB reviews and approves, declines, or defers proposed changes.

There are dozens of reasons why management or the project customer may want to change a project's deliverables. The worst scenario is when the recipient of the project deliverables drops by your office one day deep into the implementation plan and says, "Hey, I forgot to mention that this project thingie you're working on also needs…."

Another situation, equally painful, is a change that stems from your project team. In these instances, someone on your team discovers that the technology you are implementing really doesn't fit the bill. The technology won't actually deliver, the new technology will conflict with existing technology, or it becomes outdated during the course of installation. When this happens, you can almost hear your plan being sent through a paper shredder.

Or it could be that your team is pulled in so many directions that it's impossible for them to keep the project on schedule. In organizations that are short on IT staffers, 60- to 80-hour workweeks are not uncommon. They'll be working on so many different implementations, development projects, and daily duties to put out fires that it is physically impossible for them to keep pace with your project. In these situations, nothing short of additional resources or additional time will help.

As a general rule, it's easier to change the project deliverables at the beginning of the project than at the end. In other words, as the project moves closer to completion, the willingness to change the project deliverables wanes. The best method to avert serious change is prevention through serious planning. Again, as with most aspects of project management, a solid preproduction of research, planning, and interviews with the users impacted by the project is crucial. You can avoid the preceding situations using these methods:

- Interviewing the client (or end user) of the product, in detail, as part of the initiation and planning phases to ensure the requirements are well defined.
- Researching and testing the technology thoroughly before the implementation phase. A testing lab or project simulation that emulates the working environment is a must in many IT projects.

- Examining the required resources prior to the implementation. A reality check is needed to see if the existing staff has the time or knowledge to implement the proposed technology.

Change Control Process

Change control is a process that reviews and then approves or declines the proposed change. All changes must pass through the change control process, but in some instances, the change isn't proposed—it must happen or it already has happened. The change control process is still enacted to see the full effect of the change on the project. That's right! Even if the change has already occurred, you'll still need to evaluate the change to determine how you'll deal with it. Here are some examples of changes that happen without approval:

- **Organizational change** Mergers, acquisitions, and organizational splits—there's no change request for your project when these organizational changes happen. When companies merge or are busted apart, it's likely that there'll be an effect on your project.
- **Business processes** Your organization could change the processes you use in your project. Consider changes in purchasing, time-keeping, reporting structures, and other processes you use as part of the norms where you work. Changes to these items may affect how quickly you're able to manage the project, and that'll affect your project performance.
- **Reorganization** It's not unusual for companies to go through reorganizations. People are let go, are reassigned, or voluntarily leave during a reorg process. Organizational changes like this may mean your stakeholders, management, and project team can all change, which will require additional efforts from you in planning, risk management, and communications.
- **Relocation** It's possible that your organization may move its headquarters or parts of the organization. While the changes to human resources can be easy to spot, you may also need to consider the physical environment changes and the effect these changes can have on the project.
- **Outsourcing** Hiring contractors can create new challenges for project managers. Contractors have to learn your organization's processes, preferences, lingo, and approach to doing the project work. This change can create new challenges for the project manager.

The change management process, which I'll discuss coming up, needs to be established early in the project, communicated, and used for all changes in the project—whether they are proposed or have already happened. Changes can affect the entire project—not just the project scope, costs, or schedule—so it's vital to project success to allot time to analyze and consider the whole effect of the change.

Impacts of Change

The one thing that always stays the same in project management is change. Sooner or later, something will happen that will blindside your plan of attack and force you to change your plans. When a scope change request is introduced to the project, the request should be reviewed for its impact on the entire project—not just the scope. Integrated change control examines the change's effect on all areas of the project, including

- Scope
- Schedule
- Cost
- Quality
- Resources
- Communication
- Risk
- Procurement
- Stakeholders

All change requests must pass through integrated change control—it's part of project integration management. If a change bypasses the formal review of its effect on the project, chances are some part of the project will be wrecked. Integrated change control is mandatory when it comes to project changes.

EXAM TIP The examination of a change and how it affects all parts of the project is called integrated change control.

A formal change to the project plan, regardless of who's responsible for the change, is serious business—no matter how seemingly small or innocent. At this point of the project life, your project network diagram (PND) should be tight and solid. Recall that your PND is a visual representation of the flow of the work. It defines the paths to completion of the project and when tasks begin, and it identifies the critical path. The critical path is the longest path of tasks within the PND. There should be little room for additional deliverables without expanding the project finish date—not something that is always acceptable.

In addition, new deliverables cost money. A change in the project scope may mean additional resources—internal or external—and your budget may not be able to afford them. The changes can mean additional hardware and software expenses. Typically, additional funds will be required if the project scope is to change.

Team morale may plummet. Facing your team members and telling them that all of the planning, research, and work so far is about to get additional criteria for completion is not good news. You'll need to handle the news with grace and tact.

Change Request Form
Submitted by:
Phone:
Date Submitted:
Request ID:
Summary of Desired Change:
Purpose of Change:
Cost:
Time:
Risk Assessment:
Recommendation:

Figure 10-1 The Project Change Request form is a formal proposal that enters the scope change control system.

Project Change Request

As you can imagine, in a predictive project you want to control and restrict changes to the project scope. When changes are inevitable, you need a formal process to incorporate these changes into the project plan. This formal process begins with the *Project Change Request form.*

As you can see in Figure 10-1, the Project Change Request form formalizes requests from anyone to the project manager. The form can be electronic or paper based. The requestor, the project manager, and even the CCB will contribute to the form as they consider the change. The Project Change Request form requires that the requestor not only describe the change but also supply a reason why this change is appropriate and needed. Once the requestor has completed the form, the project manager, project sponsor, and other relevant stakeholders can determine if the change is indeed needed, should be rejected, or should be delayed until the completion of the current project.

Change requests, in a predictive project, must follow a certain process to determine if they should be incorporated into the current plan. Figure 10-2 follows the path of the change request from start to conclusion.

For example, a sample IT change goes through these steps (refer to Figure 10-2):

1. The project management information system collects and stores the documented change requests.
2. Integrated change control is a project-wide method of researching and documenting the effect of change on the entire project. It reviews the effects of change on scope, schedule, costs, quality, resources, communication, risks, procurement, and stakeholders. When a change request is being considered, there should be communication between the project manager and the appropriate stakeholders.

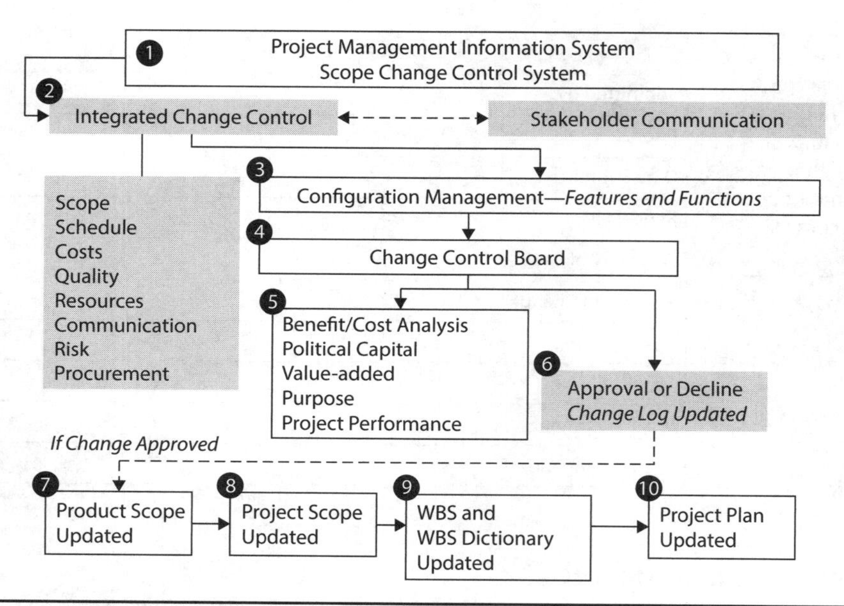

Figure 10-2 Change control follows a predetermined path.

3. Because the project scope is actually based on the product scope, it's necessary for configuration management to be invoked. This part of the scope change control system reviews and documents the scope change's effect on the features and functions of the project's deliverable. For example, the technical requirements of the software solution, if changed, will need to be reviewed, documented, versioned, and updated in the documentation of the product.
4. The change request may move on to the change control board (CCB) if one exists for the project; if a CCB isn't used, the project manager, project team, and key stakeholders will make a determination of the change. The CCB is composed of organization decision makers for the project and may include the key project stakeholders such as the project manager, the project sponsor, and the project customers.
5. The CCB completes or directs the benefit/cost analysis, the value-added benefits of the proposed changes, and other factors about the change and the project. The CCB may rely on expert judgment from those experienced with the technical nature of the change, such as subject matter experts, other project managers, and the project team. If needed, the change should be worked through in a lab or scenario environment to determine the actual impact on the project if it is approved.
6. Changes are either approved, declined, or deferred. The decision is documented in the change log and communicated to the appropriate stakeholders.
7. If the scope change has been approved, then it's likely that the product scope will need to be updated to reflect the new change.
8. When the product scope is updated, so, too, must the project scope statement be updated to include the change.

9. The WBS is updated to show the new deliverables the approved scope change request has generated. When new items are introduced to the project's WBS, they must also be included in the project's WBS dictionary. This will also cause the activity list and project network diagram to be updated as needed to reflect the new additions (or deletions) to the project.
10. The project management plan is updated to account for any additional considerations for the scope change request. The project plan updates could reflect additional costs, schedule changes, quality demands, resources, communication, risks, procurement, and stakeholders.

Consider a project to release a new OS to hundreds of users based on a common image. One department, however, needs additional software installed that other departments do not need. An OS image is an exact replica of an entire disk, generated using an imaging application such as Ghost, FOG, or Clonezilla. Such applications allow an administrator to capture the entire image of a disk and then disperse it to multiple machines quickly and easily. Because the one department needs an image that's slightly different from the others, it's outside of the predetermined project scope.

In this example, the department that needs additional software installed would basically require a different image of the disk to complete the job quickly and easily. While the request for the installation is valid, it does not fit within the current project scope. This request, as innocent as it sounds, may be better served as a separate project dependent on the completion of the current project. As Figure 10-3 demonstrates, project managers rejecting changes to a current project prevent runaway projects and advance change requests into the process to determine if a new project is warranted.

Figure 10-3 Change control can spur new projects.

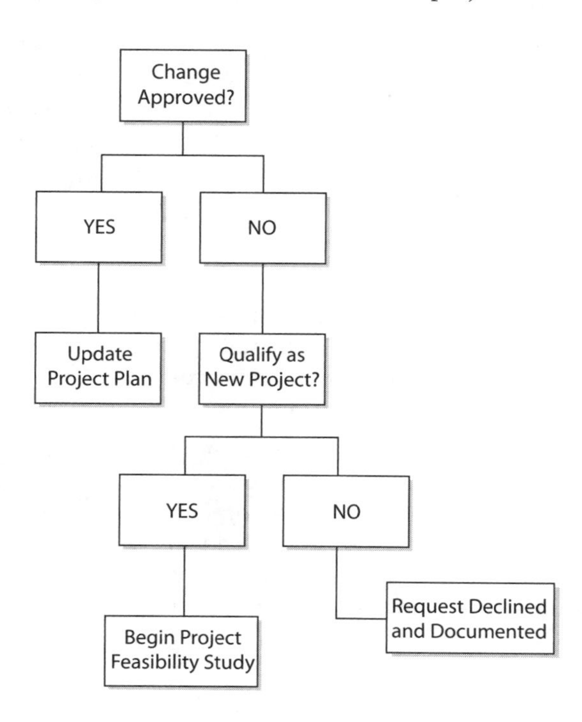

If the research of the change does not prove that the change can or should happen, the change should be rejected. Reasons for rejection could be lack of time, technology, funds, or resources, or the complexity of the request qualifies it for a separate project.

Finally, if the change is approved for project incorporation, the project manager must begin a plan to incorporate the change into the project schedule. The project manager, the project sponsor, and the project team must address the incorporated change. An examination of the PND and a review of the resources and budget will be most helpful when determining where, when, and how the change will fit into the current project.

Change Impact Statement

A *change impact statement* is a formal response to the Project Change Request form. It summarizes the proposed actions to incorporate the changes. Usually this is a listing of the paths and trade-offs the project manager is willing to implement. In some instances the change impact statement is given back to the customer with several responses the requestor can choose from. There are seven different responses a project manager can use on the change impact statement:

- *The proposed change is not approved.* Sorry! The change cannot be incorporated into the project scope. These add-on wishes cause runaway projects, scope creep, and a waste of funds, time, and resources.
- *The proposed change can happen within the current timeline, with the current resources.* Good news! The change is simple and won't require additional resources or time. This can be something as simple as changing the name of a domain or a server, or another variable.
- *The proposed change can happen with the current resources but will require an extended timeline.* The change request will take additional time to finish the project, but the current resources are able to complete the additional activities.
- *The proposed change can happen within the current timeline, but additional resources are required.* Depending on the change and the project, the deadline may not be movable. Therefore, in order to complete the change on time, additional resources will be needed to incorporate the additional work.
- *The proposed change can be completed, but the timeline will need to be extended and additional resources are required.* Phew! Given the change request, the timeline is no longer realistic, nor is it achievable for the current project team to complete the change. This stems from adding a component that requires skills beyond those of the current project team.
- *The proposed change can be completed, but the deliverables will be produced in a tiered strategy.* This reaction to the change accepts the proposal, but the deliverables will be released in priority sequence according to the customer. For example, if an OS rollout was to be for just a few departments, but the rest of the company was added to the plan, this solution could address the change. Management could choose the department order in which the OS rollout would occur.

- *The proposed change cannot occur without considerable changes to the project plan.* Bad news! The proposed change to the project is so significant it would render the current plan obsolete. The changes must have an excellent justification for scrapping all of the hard work, time, and funds committed to the project to date. An example could be a shift in business cycles, a company buyout, a new technology, or change in management.

Changes to your project might warrant the creation of a regression plan. A *regression plan* defines how the project team may undo the work a change brings into the project. Regression plans are usually created for larger, significant changes, where the outcomes are somewhat unknown. Regression plans are used as a conditional approval for significant changes to the project.

EXAM TIP Regression plans are also known as fallback plans. They are instructions for how to fall back to the last known good state of the environment. Staging areas are often used to test the solution before it enters production.

For example, suppose your project is installing a new network topology for your organization's 5,000 users on one campus. Management asks that you incorporate a change for the network topology to also include wireless access points for the network. This simple change request could be approved, but a regression plan could be created for how to "unapprove" the change if schedule, costs, or other problems are experienced during the implementation.

In agile projects we have a larger need for speed than we do in predictive projects. Change is expected and welcome, but that doesn't mean changes just fly into the product every day, all day. The role of the product owner is responsible for approving, declining, or deferring change and this role serves as a type of Change Control Board. When changes are introduced in an agile project, the product owner will insert the change into the product backlog and prioritize it accordingly. While the development team is in the working portion of the iteration, changes to the iteration backlog cannot be introduced. All changes go through the product owner and are entered into the product backlog first.

Internal Project Trouble

The most difficult changes to the project plan, unfortunately, happen from within. These changes are not always changes brought about by discovery of a new technology, a flaw in the project plan, or a conflict with the implementation. These changes are brought about by the one variable in any project that remains constant: the human element.

The human element is the predictable problem that arises from team members who fail to complete their assignments, fail to communicate troubles or flaws, or lose interest in their work. These blunders are the epitome of leadership failure. You, as project manager, must have such an active role in the implementation phase that you can sense trouble brewing like a firefighter can smell smoke. You need to spring into action, address the issue at its inception, and squelch it before it erupts into a full-fledged delay!

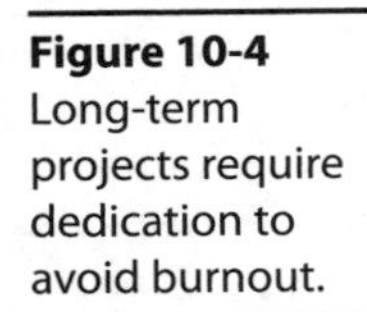

Figure 10-4 Long-term projects require dedication to avoid burnout.

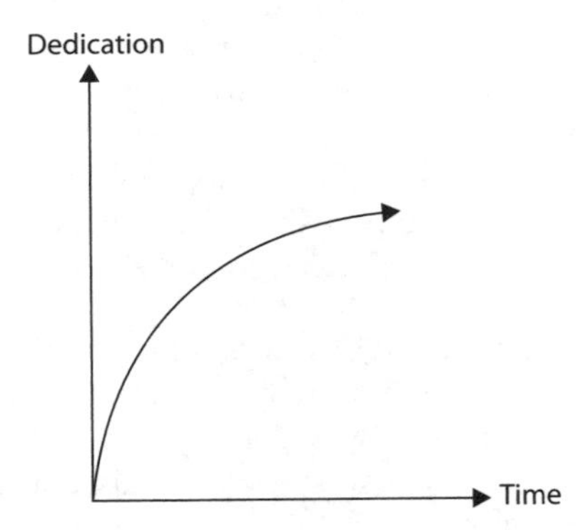

On long-term projects, it is easy for everyone, including the project manager, to get burned out on the implementation. As Figure 10-4 shows, the longer a project lasts, the easier it is for the project team and the project manager to lose interest and focus. Once a team member gets burned out on a project, they lose interest, care, and motivation. It's difficult to spark a team member's drive once they've reached this point. A project manager should sense a team member's dedication waning long before it actually happens.

Another problem IT project managers often are faced with is employee turnover. As you may know, within the IT industry, professionals are constantly climbing their own ladders of personal achievement. Workers come and go as they shift from organization to organization and move up within their own organizations. Economic conditions can cause layoffs, downsizing, and corporate takeovers that can wreck the project team.

When a team member leaves the team because they resign from the organization (or move within the organization), you must act immediately to find a replacement. This is no easy task. If you're lucky, someone within the organization can join the project team and begin where the original team member left off.

As Figure 10-5 depicts, the longer a project team position is vacant during the implementation phase, the longer the project's delay and the more it costs. In addition, costs may rise on the overall project as hours increase and IT consultants/contractors may be brought in to fulfill the duties of the missing team member.

Figure 10-5 Vacant team member positions cause delays and increase costs.

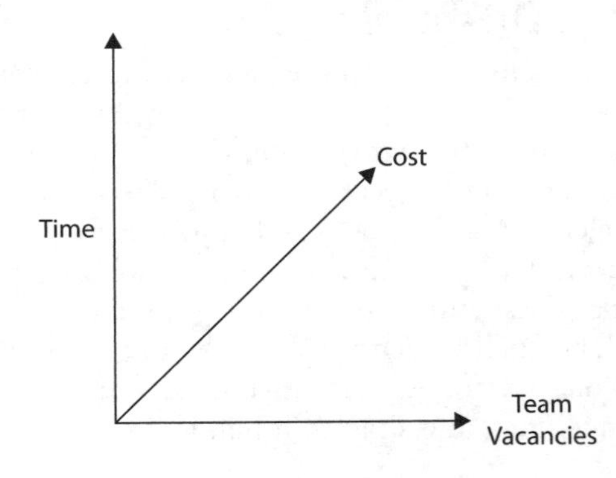

What's likely is that a new team member will join the team, and you'll be faced with assessing that person's skills and then rearranging the resources to complete the project on schedule. For example, if you're involved with an application development team and your SQL guru leaves the company, you may not have to hire a new SQL pro. Instead, you may have to move one team member who knows SQL into the role of the SQL pro and promote a new team member into the now-vacant app developer role—assuming that the person's skill sets match your requirements.

You may also have no other choice but to hire an independent contractor, a consultant, or an integrator to help the team finish the project. This independent contractor will most likely cost more than the hourly wage and benefit costs of the team member who has left the company. Additional funds may be required to complete the project.

Implementing Project Changes

Your project is always in a state of flux. Changes are called for from management, team members come and go, and new technology sprouts up along the way to completion. All around you are temptations to shift the focus of the project, to change your vision, and to broaden the scope just a bit at a time.

You must resist these temptations. Little, innocent changes pile up and result in scope creep. Changes to the project, no matter how small they may seem at first glance, are always major changes! Stay firm and require management, the customers, and the project team to stay focused on the original vision. Scope creep occurs when your project scope is defined and then it grows a little at a time. If a change to the project scope must happen, whether because of internal or external forces, you will need to enter the proposed changes into the CCS.

Changes from Internal Forces

When delays caused internally by the project team happen, due to the team's inventive changes, a lack of quality in the product, or a failure to complete assignments—and inevitably they will—the project manager can do several things to resolve the problem and stay on schedule. While these are not always changes to the project deliverables, they are changes that jeopardize the project from being completed on time, on budget, and with the expected level of quality. *Scope creep*, also known as *project poison*, refers to undocumented changes that can sneak into a project. Scope creep often happens when stakeholders request a change directly to the project team rather than following the prescribed change control system. The project manager must take corrective actions to get the project back on track:

- Using preventive actions such as training to avoid scope creep
- Applying corrective actions to remove non–value-added changes
- Hiring additional resources to complete the project on schedule
- Changing FS (finish to start) tasks within the PND to be SS (start to start) or FF (finish to finish) so that tasks can happen in tandem rather than in sequence, when possible

- Reassigning highly skilled resources to the critical path to speed the completion
- Reassigning tasks evenly among the remaining project team to stay on schedule
- Applying management reserve to lagging tasks
- Removing a portion of lag times to take up float within the project
- Requesting that the project scope be reduced in size

Changes from External Forces

When delays to the project are caused by external forces, such as the customer, management, new or pending regulations, or business cycles, the project manager can do all that they can to ensure the project will finish on schedule and on budget, but often delays or expenses are unavoidable. In these instances, the project manager must rely upon their negotiating skills to use leverage to secure additional finances, time, or both.

The Triple Constraints of project management (which CompTIA sometimes calls the Iron Triangle of Project Management) come to mind. The Iron Triangle is an equilateral triangle. The sides of the triangle (time, cost, and scope) must all be in balance with one another for the project to be successful. If a new deliverable is added to the project, there will likely be a need for more time, more money, or both. For a project to be successful, all three sides must remain in balance. You can't expect a $500,000 project scope to be met with a $300,000 budget.

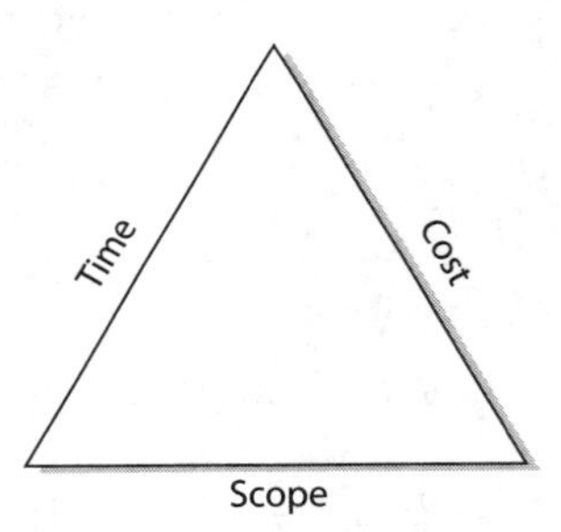

If the deadline of the project cannot be moved, but the newly incorporated deliverables have put an unforeseen strain on the project team, new resources may be required. There are only so many hours in a day, and it is not reasonable for a project manager to ask, or require, the project team to work all of them. In these instances, the most direct route to satisfying the demands of the project change is to add resources.

This usually means additional funding. The new resource may be a consultant, an independent contractor, or an internal resource that the project team absorbs. Whatever the solution, the project manager must work quickly to educate the resources on the project plan, their requirements for the plan, and when their assignments are due.

The project manager should make an effort to make the resource feel comfortable and welcome to the project. Poor project managers add team members, show them their assignments, and leave them to figure things out for themselves. You must welcome the new team member, introduce that person to the team, and explain why they were brought on board. A comfortable, happy team member will be more productive than one who is confused, misinformed, and uncertain of why they are on the team.

In some instances, changes to the project will not require additional resources, but just more money. For example, the new project deliverable may change the number of workstations, servers, or application licenses that are needed to complete the plan. These all will require additional funds. In such cases, if the deliverable must be met, then additional funds will have to be assigned to the project. There is no negotiation, as it's simple arithmetic: additional technology equals additional funds.

Adaptive projects use a similar approach as the Iron Triangle, but it's inverted. With the inverted Iron Triangle, the time and cost are fixed and the scope is flexible. So as new additions are added to the project scope, through the product backlog, the most important items are at the top of the list and the least important items, the items with the lesser business value, are at the bottom. With this approach, it's possible that some things won't be created in the project as time and money are consumed. At first glance that seems awful, but it's not really. The most valuable items are created first, so the items that are pushed out of delivery aren't as valuable as those features that have been created.

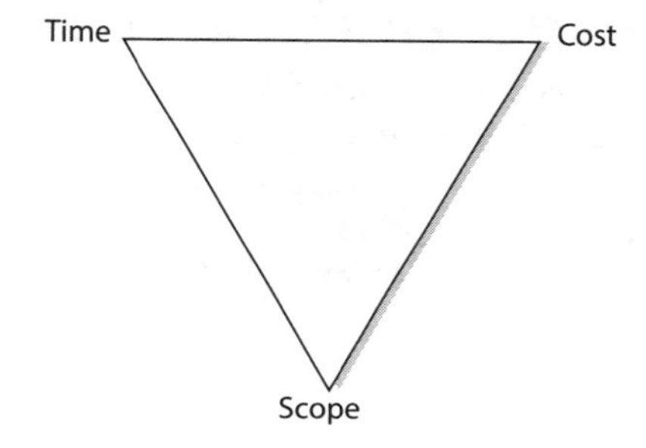

So what happens to all those requirements that didn't make it into the final project? These can be the start of a new project, like a new version or update to the software. Or the organization can shrug these off as not being worth the time and cost to include them in the product. The point is, these items aren't as valuable as the features that were created in the product.

Negotiate for Tiered Structures

A tiered structure allows the IT project manager to agree to meet the new requirements of the project scope, but the deliverables are released over multiple dates. For example, if the project deliverables have changed to be a database-driven website with e-commerce support, the IT project manager can bargain to meet the new scope but with multiple release dates. First the website can be released, then the tie-ins to the database, and finally the hooks into the e-commerce solution. Of course, the website would be updated on each release to reflect each new add-in.

With this type of hybrid approach, the project manager wins, as they've been given more time to ramp up their team to deliver the project. And the organization wins, as there are usable resources at each delivery. You may find it beneficial to create an *operational transfer plan,* which defines the timings, processes, and approach to transfer project deliverables and benefits from the domain of the project manager into organizational operations. The operational transfer plan includes project communications, project execution, and project scope management.

Extension of Time for Delivery

At the very least, when a project scope has undergone serious changes and no additional resources are available for the project, a delay is likely. What else can you do when significant deliverables have been added to your project? At first, you should examine the PND to determine if any float exists or whether some tasks can begin SS rather than FS.

After those adjustments, you must take a serious look at the project delivery date. If the date is not feasible, then additional time must be added. Neither you nor management should consider the project late—additional deliverables were added to the scope and additional time is required to meet those changes.

Holding Issue Management Meetings

An issue is a risk event that has occurred. Issues are also matters in the project where two or more people disagree over the direction the decision should take in the project. An *issue management meeting,* as its name suggests, is a meeting to resolve problems and issues as they arise on a project. Every project, no matter who planned it, will encounter troubles—they just always do. A delay in the critical path is a delay you cannot afford. For example, if a critical task is delayed because of a software compatibility problem, that's a delay for all dependent tasks in the project timeline. Pushing your project timeline beyond the targeted completion date, as shown in Figure 10-6, will cost additional funds because of the hours required to complete the project.

An issue management meeting allows you to address problems with your team, the vendor, or key personnel to find solutions. You can address an issue, brainstorm for different ideas, and drop in plans of variance in your PND. An issue management meeting is a serious piece of business that allows you to get your project back on track quickly and accurately.

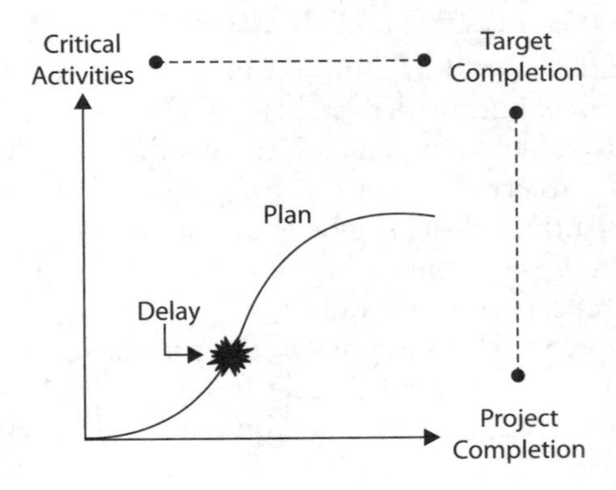

Figure 10-6 Delays will ultimately result in budget overruns.

At the end of your issue management meeting, which may comprise several meetings that run over days or weeks, you should have a resolution to the issues and a plan on how to attack. This plan allows you to address the problem with a solution and address how the solution will impact the remaining tasks in the project. Meet with the entire project team, discuss the solution you've arrived at, and explain the justification for the plan. And then assign the task to the appropriate team members to implement.

Delaying a Project

There will come a time in every organization when someone proposes that a project, specifically your project, be cut. Your hands will get sweaty, your gut will develop a sinking feeling, and you won't sleep well at night. Your poor, poor, project! The one you've invested months of your life in, the one you've inspired your project team to work so hard for, and the one you've sold management on how valuable it will be to the organization. Sooner or later one of your projects may face its demise.

There is, however, a tactic that you can try to save your project from being cut from the organization's plans altogether: convince the organization to delay, not cut. Delaying a project is different from cutting a project. A delay is a nod from management that your plan is still active, is still worthy, and will be resurrected at some point in the future.

TIP As a general rule, the longer a project lasts, the more it will cost.

The typical cause of healthy projects being cut is a lack of funds. Is your business experiencing a downturn? Are layoffs imminent? How's your company stock doing? These are all signals that a project cut may be visiting your team in the near future. Management has a responsibility to manage their project portfolio. If your project is low in priority, it's not difficult for management to give it the axe. Some project managers, when they sense these changes coming, will commit as many funds available to the project immediately to secure the implementation of the technology. For example, a project manager is leading a project to upgrade servers over the next eight months when they sense cuts may be looming. Rather than being stuck with a project without servers, they immediately order all the hardware necessary to finish the plan.

If this sounds sneaky to you, that's because it is. This is not a good project management technique because

- It does not follow the project plan or the project budget.
- The hardware may be present, but the team may not be available to implement it.
- It is a waste of the company's money, as the technology purchased today may drop in prices over the upcoming months.
- If management wants to rescind approval of the plan, they have the final say.
- It is a reaction to gossip, and not to facts.

You should continue on your project until you are told the project has been halted. There is nothing wrong with talking to your project sponsor about the possibility of your project being cut if you two have a strong working relationship. When the word comes that your project is to be halted, first find out the reason why the project must be stopped. If the reason is financial, and usually it is, make a request to change the project status to inactive rather than retired. Make a request to revisit the project in three to six months or after certain conditions are met to reactivate the project implementation.

Once the project has been officially delayed, edit the project to formalize the three-to-six-month delay. You will then have to break the news to the project team. Share with the project team the same reasons you were given, if you're allowed, as to why the project has been delayed and when the project will reconvene. Make backup copies of any current work done on the project and store them for safekeeping.

Coping with Vendor Delays

If you have outsourced all or a portion of your project, you may be faced with a vendor that cannot complete tasks according to the original schedule. In these instances, you will have to be stern, diplomatic, and reasonable all at once. Some methods you can use to work with vendors include the following:

- Review your contract with the vendor to determine what repercussions you may have. If you're using an incentive-fee contract, the vendor may face penalties for delivering their work late.
- Review their practices and recommend methods they can implement to improve the lagging schedule.
- Clearly identify to vendors the impact of the delay on the project scope and the completion date.
- Once a compromise has been reached, share this news with management and the project's customers.

Rebuilding Management Support

Here you are in front of management. Your predictive project is a month off schedule, and your budget is nearly gone. You need more time, and you need more money to finish. Gulp. If you've blown the opportunity to complete the project, don't be surprised if you are removed from the project. Sorry. But wait a minute, is this really your fault? Well, there is that "poor planning results in poor implementation" argument, but let's take the benefit of the doubt. Has the project scope changed? Have crucial members of your team left the project? Did you ever really have enough money to begin with?

If you are not to blame, things can be a little easier. Of course, management doesn't want to hear the bad news about your late project and the lack of cash to finish it, but someone has to tell them. The best method is to cut to the chase and tell management what the problem is. Prepare this ahead of a meeting and document your needs. You do not want to go before the company board, the partners, or an executive committee without some plan to present to them. Never go to management with a problem unless you also have a solution. Management is looking for value and solutions in their day—not your project problems.

Tell management what the problem is in clear, direct terms. Then, in equally clear terms, explain to them why the problem has occurred. If it was your fault, say so—honesty is always the best policy. If the problem has presented itself because of the project scope being changed or because of other circumstances clearly outside of your control, let management know—with your supporting evidence.

EXAM TIP Never go to management or the customers with a problem without also having a proposed solution to the problem. These stakeholders are looking to you to solve issues, not put the disruption in their lap.

Once you've broken the harsh news and presented the honest, direct reason why the problem has arisen, present a plan to resolve the issue. You are now selling management the sizzle of the project once again. Remind management of the value of this project, of the investment it means for the organization, and of the time and dollars already committed to progress the project to this point. While the monies that have been invested are sunk costs, it's hard to ignore the commitment invested in a project.

If these executives agree to continue to support your project, you will now have to rebuild their trust. To rebuild their trust, you will have to prove to them that you are capable of leading the project team to the conclusion of the project. You'll have to prove to them that you are capable of managing their investment. And, finally, you'll have to prove to yourself that you are capable of finishing this project after all you've been through thus far.

To begin once more, meet with your project team and discuss why and how things got off track. Look for reasons why you let down the team, or the team members let each other down. Address these issues and then promise they won't happen again. You'll need to return to your WBS and the PND to determine what tasks are lagging and how the team can regroup and attack the plan with gusto once again.

Focus and Refocus

Okay, your project has changed either through internal or external forces. You've made changes to the project plan, worked out details of the change, and are ready to move forward with the project. Put the frustration of the project change process behind you, rally your troops, and charge ahead into the project plan.

You now have to take on the role of an even more active project manager and ensure that the team is not discouraged with the change of the project plan. You will need to speak with the team members who are most impacted by the change to ensure their commitment and ongoing support.

You will also need to increase your level of communication with the team members and management. You want to keep all parties informed of the process of the project and your continued dedication to it. Some project managers are tempted to keep a low profile after the change process—out of either embarrassment or frustration. Hiding is not a good decision, as now you need to be seen and heard.

Renewal of Commitment

A change in the project plan, whether your fault or not, requires a change in you as well. After changing the plan, which is one of the most frustrating aspects of project management, you need to rekindle your excitement for the original project vision. Often, especially on long-term projects, it's easy to drift away from the excitement that surrounded the kickoff meeting, the first few team meetings, and reaching the first milestone.

The fact is, the business of project management is not always the most exciting business. But, the day-in, day-out tasks and review of work completed and work that needs completing are what get the project from an elaborate plan on a whiteboard to a living portion of a business. Rekindle the excitement, renew your commitment to the project, and lead your team to victory!

CompTIA Project+ Exam Highlight: Managing Project Changes

In project management, change is inevitable. There are value-added changes, organizational changes, technical changes, errors and omissions, and external changes such as laws and regulations. Changes can haunt a project unless the project manager and the stakeholders follow a systematic, documented change control process. All scope change requests must be documented and entered into the scope change control system so they can pass through integrated change control, be reviewed by the CCB, and then detailed in the project's change log.

1.3 Given a scenario, apply the change control process throughout the project life cycle In a predictive project, change requests should be documented and entered into the project's project management information system as part of the project's scope change control. The rules and procedures of the scope change control system must be followed by all stakeholders—including the project team and the project manager. The CCB reviews the value and purpose of the change and makes a determination to implement or deny the request; either outcome is communicated to the stakeholders and documented in the project's change log.

When change is introduced into the project, the integrated change control process is invoked. This process examines each change request and the potential impact on the project's scope, schedule, costs, quality, human resources, communications, risk, procurement, and stakeholders. A change in one area of the project can have ramifications on other areas of the project. When change is approved in the project plan, it's possible that all project subsidiary plans must also be updated as a result of the project scope change. Integrated change control is a mandatory process for considering any scope change requests in a project.

Some changes can happen without entering the change control system: scope creep or work that has been implemented incorrectly. The project manager must document the error and determine if corrective action should be implemented to fix the defect. Corrective action brings the project experiences back into alignment with the project scope and the project plan. Changes that do follow the project's scope change control procedures may be approved, but their approval can also mean updates to the project scope baseline and the project management plan.

The Triple Constraint of Project Management, also called the Iron Triangle, is the balance of time, cost, and scope in a project. When a change is introduced to the project scope, additional time and/or money likely will be needed to achieve the project scope due to the recently approved change. Due to poor project performance, the scope could be reduced to equate to the project budget and time available to complete the project by a given deadline. In the change control process, it's also possible for the budget of the project to be reduced or the schedule for the project to be changed. When the budget or the schedule changes, the scope is likely going to have to change, too, to achieve the new cost and time objectives.

Adaptive projects welcome and expect change to happen. Changes in an adaptive project are entered into the product backlog, where the product owner will prioritize the changes. The more business value and importance a change has, the higher the change will be in the product backlog. Adaptive projects use an inverted Iron Triangle, where time and costs are fixed and the scope is flexible. Some requirements may be shifted out of the project achievability because there isn't enough time or money to deliver everything, but the items that are shifted out of the project are of less business value than the items that the team creates.

1.5 Given a scenario, perform issue management activities Issues are risk events that have occurred in the project. If the issue is a conflict between two or more people, the goal is to peacefully resolve the disagreement and keep the project moving forward. The project manager will record the issue in the issue log, track its progress and resolution, and confirm that the agreed resolution is being followed.

Part of issue management is the consideration of escalation when the issue can't be resolved. In these instances, management or the project sponsor will have to decide on the best course to take in the matter. You'll consider the severity, impact, and urgency of the issue before escalating the issue to the project sponsor or to management for resolution. You don't want to go to management with every issue—it's your job as project manager to resolve issues and keep the project moving toward its goal.

4.5 Explain operational change-control processes during an IT project All projects, predictive or adaptive, are about change: the thing you create will add, move, change, or delete something in the environment. Other changes in the environment, part of the operations, may have a direct influence on your planning and execution of the project. You must keep abreast of what's happening in the IT sphere of your organization and how these changes can affect your project.

This means having insight to schedule maintenance and downtime of IT services, customer notifications of maintenance, and validation checks for the changes. When changes are performed as part of operations, you'll need to review, test, and determine if the changes have any effect on your project. You'll do this in a beta or staging environment before releasing your products to production. You don't want to release a solution that conflicts with changes implemented as part of ongoing operations.

Chapter Review

Whoever said, "The more things change, the more they stay the same," never worked in project management. No matter how much planning, preparing, and strategizing you invest in a project, change can happen. A change in deliverables can be requested from management or the customer receiving the deliverables. Business cycles, new requirements from the customers, or new management can all lead to changes in a project plan.

As the project manager, you must implement a change control system to formalize a change request, and then react to the change in a formal statement as well. A request to change the project deliverables, no matter how seemingly small, is a significant one that requires a business justification to begin implementation.

A change in the project deliverables may require additional resources, additional funding, additional time, or all three to complete the project. Changes to the organizational structure can also affect the project. Management or the customers must be willing to comply with your requests to complete the project. Solid research of the change will provide evidence of the new requirements to produce the deliverables.

Should change come about because of internal forces, such as lack of focus, change in resources, work units not being completed, or improper funding, you must take the lead on rectifying the problem. You have to face management and explain the problem and offer a solution to correct the problem and get the project back on track.

Once the project scope has been modified, you and your project team must renew your dedication to the vision of the project. A new sense of responsibility, dedication, and channels of communication must emerge to keep the project moving to completion.

Exercises

These exercises allow you to apply the knowledge you have learned in this chapter and are followed by possible solutions.

Exercise 10-1: Complete a Change Request Form

In this exercise you will complete a Project Change Request form as if you were the customer of a project.

Scenario: You are the sales manager for Carlington Enterprises. You are working with Carla, an IT project manager, on a new sales automation application. The highlights of the requirements for the software her team is developing are as follows:

- Contact management features
- Database searchable by any property of the client (sales, birth date, last contact, city, and so on)
- Ability to block salespeople from viewing records they did not enter
- Ability for the sales manager to reallocate leads to sales reps
- Ability for the contact information to be downloaded to a mobile phone
- Fax and e-mail ability
- Ability to log sales call activity to the contacts by the sales staff

The project is moving along when you suddenly realize that the software requirements are missing a major feature. You would like the software to have the ability to retrieve a contact's record through caller ID when a contact calls into the center.

Complete the following Project Change Request form based on the new request:

Project Change Request Form
Name of Project:
Your Name:
Date:
Summary of Desired Change:
Reason for Desired Change:

Exercise 10-2: Determine a Change Request Outcome

In this exercise you will approve or disapprove a Project Change Request form.

Scenario: You are the project manager for an operating system upgrade of all workstations throughout the company. The key points of your rollout plan are as follows:

- All workstations will be configured with a new operating system.
- The workstations will be deployed through imaging software.
- The workstations will be configured through scripts and system policies.
- The OS upgrade will be released to employees after they complete a four-hour training session.

Based on the following Project Change Request form, determine if the change is valid and should be approved. Complete the answers after the form to justify your decision.

Project Change Request Form	
Name of Project:	Workstation OS Upgrade
Your Name:	Chris Turner
Date:	December 30
Summary of Desired Change:	I would really appreciate it if you could upgrade the computers in my department (marketing) ASAP. If you could bypass any other departments en route to us in your rollout plan, that'd be great.
Reason for Desired Change:	We've added new software that we must use in the new year, and it is not compatible with our current operating system. If you want to chat about this give me a call at x232. Thanks!

Question	Your Response
What risks could this change have on your project?	
Should this change be incorporated into your project plan?	
Why or why not?	
Will any aspect of the change request require additional resources? (If so, what aspect?)	
Will any aspect of the change request require additional funding? (If so, what aspect?)	
Will any aspect of the change request require additional time? (If so, what aspect?)	
What areas of the change request have the most impact on your project?	

Exercise Solutions

The following offer possible solutions for the chapter exercises.

Exercise 10-1: Complete a Change Request Form

Project Change Request Form	
Name of Project:	Sales automation software
Your Name:	*Your name*
Date:	*Today's date*
Summary of Desired Change:	I would like the software to have the ability to use caller ID to retrieve the contact's information before a salesperson answers the phone.
Reason for Desired Change:	This feature would eliminate the delay it takes for the salesperson to look up the client's record. Not all of the salespeople can type very fast, and it's embarrassing for the sales rep and rude to clients to have them spell out their last names. An automated feature like this would increase sales and help the sales team develop better relationships with our clients.

Exercise 10-2: Determine a Change Request Outcome

Question	Your Response
What risks could this change have on your project?	The marketing department will need to be upgraded as part of the rollout. There's no real risk in moving the department to an earlier time in the project implementation. This is with the assumption that all of the marketing laptops are available on the day of the rollout.
Should this change be incorporated into your project plan?	Yes
Why or why not?	It is essential the marketing department use the new software in the new year. The OS upgrade will allow them to get to work on their software.
Will any aspect of the change request require additional resources? (If so, what aspect?)	The one drawback of the change is that the OS upgrade is dependent on the training sessions. The users in the marketing department will need to complete the training first before the OS upgrade will be released. No additional resources are required.
Will any aspect of the change request require additional funding? (If so, what aspect?)	No additional funding is needed, just a shift in the release.
Will any aspect of the change request require additional time? (If so, what aspect?)	No additional time will be required.
What areas of the change request have the most impact on your project?	The users in marketing will need to complete the training sessions before other users in the company. The class dates are still the same, except the users in this department will complete the training sessions first in order to receive the OS rollout first.

Questions

1. What is a scope change control system?
   - **A.** A formal process to review and then decline changes to a project
   - **B.** A formal process to decline a project without management interaction
   - **C.** A formal process to manage, review, approve, or decline changes to a project
   - **D.** A formal process to decline all changes to a project without additional funds committed to the project

2. Pierre is the project manager of a project to upgrade all of the print servers and printers within his company. The deliverables require that each floor have a pool of print devices, with the exception of the graphics department. They will have several different types of printers, including two high-end color printers. Martha, the sales manager, requests that her department receive a color printer as well. What is the first process Martha should follow?
   A. Submit funding for the color printer.
   B. Submit a Project Change Request form.
   C. Submit a change impact statement.
   D. Submit a proposal on how the new printer will help the sales department be more productive.
3. You are the project manager of the NHQ Project for your organization. A new change request has been introduced, and one of the stakeholders says that it's because of your lack of planning. Of the following reasons to change a project's deliverables, which is a result of lack of planning?
   A. A discovery that the technology is not compatible with the workstations' OS
   B. A request from the client for additional features within an application
   C. A request from management to finish the project earlier than the set date
   D. A request from the project team to delay the project by a week
4. You are the project manager for your organization. A change has been introduced that will include major revisions to the project scope statement. In the scope change control system, what component will control and document changes to the product's features and functions?
   A. Configuration management
   B. Product scope statement
   C. Project scope statement
   D. Function analysis
5. What term is assigned to small, undocumented changes to the project scope?
   A. Change management
   B. Corrective actions
   C. Defect
   D. Scope creep
6. What is one of the most dangerous things that can happen when the project scope is changed due to a request from the customer?
   A. The project cannot be completed on time.
   B. The project team's morale may plummet.
   C. The project manager loses interest.
   D. Management takes over the project.

7. What is the purpose of the Project Change Request form?
   - A. It allows changes to be easily melded into the project.
   - B. It allows the project team to request changes to the project based on discoveries in the field.
   - C. It allows the project sponsor to formalize and control changes from external sources.
   - D. It allows the project manager to determine if proposed changes are valid or not.
8. If a proposed change to a project does have merit, what must the project manager do in the change control process?
   - A. Implement the change.
   - B. Update the PND.
   - C. Research the proposed change.
   - D. Assign the change to a new resource.
9. You are the project manager for your organization. A new change has been approved for your project by the change control board. Where are all the changes recorded?
   - A. Product scope
   - B. Project scope
   - C. Configuration management
   - D. Change log
10. What is a change impact statement?
    - A. A formal request to change the project scope
    - B. A formal response to a request to change the project scope
    - C. A formal response to a request to change the project deliverables
    - D. A formal rejection of a request to change the project scope
11. You are the project manager for your organization. A change has been requested for your project. All of the following are not valid responses to the proposed change request except for which one?
    - A. The proposed change cannot happen, but the project will require an extended timeline due to the time investigating the proposed change.
    - B. The proposed change can happen with the current resources but will require additional funding if the original deadline is to be met.
    - C. The proposed change can happen with the current resources and with the current financial obligations, but a change incorporation processing fee must be applied.
    - D. The proposed change can be completed, but the timeline will need to be extended and an additional project manager will be required.

12. What is a change control board?
    - A. A committee that reviews the project looking for value-added changes to insert into the project scope
    - B. A committee that must approve all changes for a project
    - C. A committee that approves or declines proposed changes to a project
    - D. A committee that studies the change impact a proposed project will have on an organization
13. What component of the project's scope change control system examines all areas of the project to determine the effect of a proposed change?
    - A. Integrated change control
    - B. Configuration management
    - C. Product scope system
    - D. Change control board
14. Of the following, which is a serious problem to a project team but is not a lack of commitment to the project?
    - A. Burnout
    - B. Staff turnover
    - C. Lack of focus
    - D. Reassignment of the project manager
15. When a new team member joins the project, what is the most important thing a project manager can do?
    - A. Get the team member to work immediately.
    - B. Introduce the new team member to the other team members and allow them to assign tasks to the new team member.
    - C. Have the team member research the project to get to know the project plan.
    - D. Spend time with the team member to get them caught up on the implementation and to make them feel welcome.

Answers

1. **C.** A change control system is a formal process to prevent changes to a project without proper justification. Change control allows the project manager to fully understand the proposed change and then determine if the change is necessary.
2. **B.** Martha should submit a Project Change Request form. The Project Change Request form allows Martha to explain the change to the project manager and to offer justification for the proposed change.
3. **A.** As part of the research phase, the project manager should ensure the proposed technology is compatible with the existing OS.

4. **A.** Configuration management is responsible for documenting and controlling changes to the project's product—specifically, the features and functions of the product.
5. **D.** Scope creep consists of small, undocumented changes to the project scope. Scope creep can happen any time a project team member or the project manager makes a small change to the project scope without steering the change request through the project's scope change control system. Scope creep is also known as project poison.
6. **B.** Changes to the project plan can cause the project team's morale to plummet. Team members may view the change as bad news and become disgruntled about the project and the new required work to complete the plan. A project manager must get behind the team members, support them, and encourage them to complete the project even with the new deliverables.
7. **D.** A Project Change Request form allows the IT project manager to determine whether the proposed change is valid. The project manager may have to research the change and determine its impact on the overall project to make a decision.
8. **C.** Once a proposed change proves to have some merit, the project manager has to research the change to determine its impact on the project.
9. **D.** All changes and their information and status are updated in the project's change log. The change log documents both approved and declined changes for the project.
10. **C.** A change impact statement is a formal response from the project manager to the requestor of a change to the product deliverables. It is a summary of the decision to implement or reject the proposed change, and it documents the impact the change may have on the project.
11. **B.** When responding to a Project Change Request form, the project manager can approve the change with one of seven options. In this instance, the project manager can approve the change with additional funding.
12. **C.** A change control board is a group of key project stakeholders that review proposed changes for the project to determine what impact the changes may have on the project, if the changes should be implemented, or if the changes should be declined.
13. **A.** Integrated change control examines all areas of the project to determine the proposed change's effect on the project scope, time, cost, quality, resources, communication, risk, procurement, and stakeholders.
14. **B.** Staff turnover can seriously hamper a project's implementation phase. Each team member is needed to implement the project plan; if a team member leaves the team, then some tasks are delayed, and the project may miss the targeted completion date.
15. **D.** Of course, the project manager needs the new team member to get to work as soon as possible, but some time must be invested in the new team member to ensure that their focus is on their responsibilities. The project manager and the project team must welcome the new team member and help them ramp up on the project implementation.

CHAPTER 11

Enforcing Quality

This chapter covers the following topics:

- Defining quality
- Viewing quality management as a process
- Committing to phases of project management
- Implementing continuous quality management
- Communicating project progress
- Strategizing quality
- Ensuring quality through phases

Picture this: You are having a romantic dinner at a wonderful Italian restaurant. You're seated at an elegant table with white tablecloth, shining utensils, and crystal glasses. The warm glow of the candlelight makes everyone look great. There's Puccini on a distant speaker and the scents of sizzling vegetables, steamy pasta, and crushed garlic drift from the hidden kitchen.

Everything is perfect: The server is attentive, but not overbearing. Warm bread is followed by a crisp salad, and a delicious dinner that looks as good as it tastes. All in all, it's a magnificent evening out. You're feeling swell, so you order two cappuccinos and Italy's classic dessert: tiramisu.

Dessert arrives, you each take a bite, and ... it tastes like wet cardboard! Just awful! The worst taste you've ever had in your life. Now no matter how excellent the evening had been, this one bite has ruined it all.

Hopefully this will never happen to you, at least the dessert part. But what happened? How did such a wonderful experience go from excellent to horrible? Someone, likely the pastry chef, didn't do their job. Now the hard work of the chef, the wait staff, and the proprietors is all ruined, or at least tainted, by a let-down in quality management.

Quality management is the process of ensuring that the entire experience, the entire process for the management and for the customers, is excellent.

VIDEO For a more detailed explanation, watch the *Managing Project Quality* video now.

Defining Quality

Quality. How many times a day do you hear that word? Reports of it come from all around you: upper management, television commercials, salespeople, and the news. At every turn someone is spouting off about their quality carwashes, their process of quality management, the benefits of purchasing their quality products.

But what is quality and how does it relate to project management? Quality, according to *A Guide to the Project Management Body of Knowledge* from the Project Management Institute, is "the degree to which a set of inherent characteristics fulfills requirements." Quality, according to *Webster's New World Dictionary*, is "the degree of excellence of a thing." Hmm.... To a project manager, then, quality could mean many things—and it does. To a project manager, *quality* falls into two areas:

- The quality of the deliverable
- The quality of the processes to produce the deliverable

In this chapter, both areas will be examined, though the focus will be on the quality processes to produce the deliverable. Arguably, to produce a quality product, there must be a controlled, organized process to get to the end result. Not often can a project full of chaos, disorganization, and pandemonium create an excellent deliverable. Both adaptive and predictive projects follow processes in an orderly, anticipated approach to deliver on the project scope.

Quality of the Deliverables

Every project must produce a deliverable to finish. A project to create a new application must, obviously, produce the application. A project to create a new network must result in planning, designing, and producing the expected environment. No project manager would set out to create a new application and end with a print server—it just doesn't make sense. Every project must have clearly stated objectives as to what the project will produce. The objectives help formulate the project requirements. The project requirements evolve, through progressive elaboration, into the project scope baseline. Quality, then, is achieved by satisfying the project scope baseline and satisfying the project objectives.

Producing a Service

Imagine a project that is designed to establish a virtual private network (VPN) for the company's employees. The goal of the project is to allow users from the field to connect to resources on the office network. Resources could include e-mail, printers, file servers, and databases. To remote users, the experience must be just like it is when they are accessing the LAN in the office.

The project manager and the team complete the research, create a plan of action, and implement the new service. Of course, it's a bit more complex than this, but you see the big picture: conceive, plan, and achieve. The users, from home or anywhere in the world, connect to the resources within the LAN through a VPN, as shown in Figure 11-1, which allows users to connect to company resources through an Internet connection.

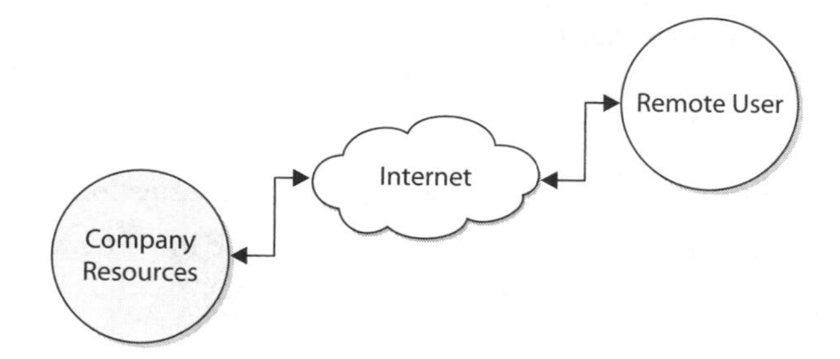

Figure 11-1
A project can deliver a product or a service such as a VPN.

To produce this service, the project manager had to see, and know, what the end results should be. The project manager worked with the project team and the project stakeholders to determine the exact requirements of the project deliverable. Research allowed the project manager to create the vision of the project, leadership allowed the project manager to transfer the vision, and dedication to the project allowed the project team to implement the plan.

The totality of the service is measured in value by several factors:

- **Value of the implementation** What did it cost the company to create the VPN solution, and what are the measurable results? There is a benefit/cost ratio for the project: the cost of the project, let's say $75,000, relative to the benefits of secure, remote access. An ongoing process, through the project's monitoring and controlling, must be used to see the true costs and benefits of the implementation.
- **Value of the service** After the project, there must be a process to measure the value of the service. Metrics are needed to measure the value to the organization before and after the project. The measurement of the service value can be accomplished through tools to log the service's usage. From the logged data, calculations can be used to see the amount of activity over a set period and the cost of each session. For example, the number of users accessing the LAN through the VPN connection over a six-month period will reveal the usability, while an increase in productivity through the VPN connection can show the profitability of the implementation.
- **Value of the experience** How well do the deliverables work? If the project manager has over-promised the deliverables and the speed, reliability, or convenience of the service, then the value of the experience will diminish. For example, if the project manager has stated that the VPN connections will be just as fast as they'd be if the user were on the LAN and this does not prove true, users will consider the service less than excellent. People may use the new ability to connect to the LAN, access resources, and retrieve their e-mail, but their focus will be on the slight delay through the Internet connection. The experience needs to be quantified to measure customer satisfaction and quality.
- **Value of the longevity** How long will the service stay implemented? If the project manager and the project team have failed to research the service adequately and it is replaced within a year by faster, more reliable, and less expensive methods, the value of the service's longevity may be slim. In some instances, the service may

be adequate for the time it is in place; in others, the service may offer little or no ROI. For example, if the project manager had offered the VPN service through analog dial-up connections and offered no support for broadband connectivity, the thrill of the VPN would be diminished by the availability of broadband connections versus outdated analog dial-up connections.

- **Value of the reliability** How reliable is the implemented service? If the project is declared finished but the service consistently fails or is unavailable, the reliability of the deliverable is lacking. The service implemented must be reliable, and the underlying process of the service, whether it is hardware related or it depends on the skill sets of the individuals operating the service, must be reliable and able to fulfill the demands the service requires. For example, if the VPN server is consistently unavailable because the hardware the VPN software is installed on is weak and cannot handle the workload, then the hardware was not addressed properly in the planning phase and must be upgraded. The upgrade in hardware may cause delays in the service availability and incur additional costs to the organization.

Projects that produce services must be planned and implemented toward the end result of the service. A service deliverable must live up to the promises of the project manager, and the project team must have the skill sets and funding available to install the service for reliability, availability, and in proportion to the expected longevity of the service. As Figure 11-2 illustrates, a balance between the reliability and the cost of the implementation must be obtained.

Project managers must work to ensure that the proposed service is not going to be replaced with faster, stronger, and better services within a timeframe that would squelch any ROI on the service. This is derived from the research and planning phases conducted by the project manager rather than the demand from management for an immediate solution.

EXAM TIP The ultimate test of quality is the acceptability and validation of the product by the project customers. Quality is a conformance to requirements and a fitness for use.

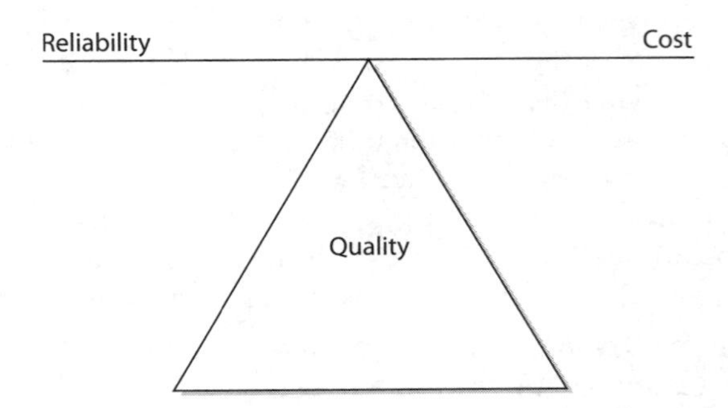

Figure 11-2 Project managers must balance cost and reliability to obtain quality.

Producing Goods

A project that requires deliverables to be a tangible object, such as a network, an application, a database, or an application server, has traits similar to those of a project creating a service. A project that creates a product, however, has different measurements to gauge the quality of the product.

For example, imagine a project that involves creating software to allow customers to design a landscaping scheme. The application will walk users through a wizard that will build an ideal garden based on their climate, the amount of sunlight their yard receives, the amount of color they'd like, the care of the plants, and other factors.

The software will be sold and used online. The interface of the software is not a typical web browser, but it does take advantage of the Internet connection to retrieve plant names, photos, and nursery information in the customer's ZIP code or through a phone app to locate the customer by GPS. This application's quality will be judged differently from that of a service, though it may have similar attributes.

Values used to judge a product are dependent on what the product is. For example, an application will have some characteristics of a service, whereas a laptop, a physical piece of hardware, will have different attributes of quality. For any goods, however, there are measurable values:

- **Value of the product** Is the product worth the cost? A product that must be created, such as an application, has to give the customer some level of satisfaction, enjoyment, or benefit that has a perceived or measurable level of worth. For example, a computer game that sells for $39 must be, to the consumer, worth that money in enjoyment. The $39 investment is measured in the ability of the application to deliver fun in this instance. In other words, does the product deliver on its promises in relation to the cost of the product?
- **Value of the usability** Is the product usable? A product must deliver on its promises to be usable. The usability factor stems from the need for the product to exist. For example, users need a laptop to complete certain duties—in particular, mobile computing. A project manager who manages a project to install and configure laptops for the sales team must know the level of usability the sales team anticipates from the hardware. The product itself is not the deliverable of the project—the satisfaction of the usability is the deliverable.
- **Value of the reliability** Is the product reliable? A product must be reliable, functioning, and usable by the customer. A project manager who implements a device, such as a new drawing tablet, is responsible for the quality of the device implemented. A tablet that has batteries that burn out too quickly, that doesn't synch properly with a workstation's software, or that is difficult to use is not a reliable product. The project failed not because of the hardware—but because either the requirements of the stakeholders were not established or there was inadequate planning to meet the stated requirements.

- **Value of the longevity** What is the product's life cycle? Like the process of delivering a service, a product must also have a life cycle in proportion to its cost. A project that aims to create and sell Software as a Service (SaaS), for example, must have reliable enterprise resource planning (ERP), scalable architecture, documentation, and the ability to grow the SaaS model over time. Today's technology is outdated as soon as it's purchased, or so it seems. A project manager, however, must be able to judge the life cycle of a product in relation to the ROI of the product. The project manager often must calculate the cost of the product and how long the product must be used before it becomes profitable.

Quality vs. Grade

Quality, within a project, is the capability of the project to meet the requirements of the project customer. *Grade,* however, is the ranking or classification of a thing or service. For example, suppose you're managing a project for the art department within your organization. The project requires six new printers for the artists. Two of the printers are inexpensive color inkjet printers, two of the printers are moderately priced color laser printers, and the remaining two printers are high-end image setters that print directly to film.

You have six different printers ranging in price and capability. All of the printers deliver on what they promise, but do they differ in quality? No. The printers differ in grade. Each printer is capable of printing under the specifications its manufacturer states. While the inexpensive inkjet printers are of a lower grade, they can still deliver the quality that they promise.

Consider also the grade of paper you may use with the printers. You can buy slick, photo-ready paper or cheap copy paper. Paper is paper, but the grade of paper can differ.

Low grade may not be a problem, but low quality is always a problem. Say one of your new printers consistently jams the paper, fails to print, produces smoke, or has other defects—that's a quality problem. Low-quality resources affect the overall quality of the project. A low-quality resource, such as a person who's incompetent in their job, will affect the quality of the project. Low quality is poor quality and that will pull the entire project down.

Quality of the Process

Whether you are creating a product or a service, you will follow a *process* to arrive at the deliverables. Predictive and adaptive projects have an approach that the project team and stakeholders must follow. As you've read in all of the earlier chapters, there is a set process, a logical and discrete order of getting any project from start to finish. The project management framework, both for adaptive and predictive projects, is guarded and led by the demand for quality.

No doubt, a project manager who is unorganized, lacks leadership abilities, and fails to motivate the project team will most likely create a project deliverable that is short of excellent. An agile team that doesn't follow the rules, skips the required ceremonies, and

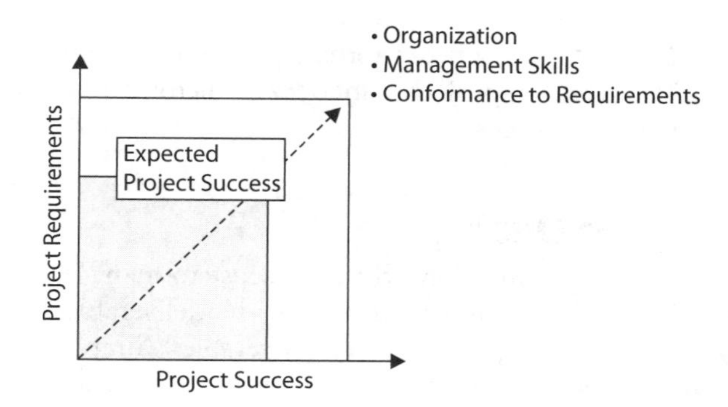

Figure 11-3 The project manager who is organized will generate success.

constantly shifts from requirement to requirement isn't following a quality approach to the project. As Figure 11-3 demonstrates, a project manager who is organized, follows a proven process of getting to the deliverables, has established the project requirements, and inspires the team to success will most likely create a deliverable that is solid, efficient, and valuable to the organization.

The quality of the management process is measured by several factors:

- **Results** The deliverables are a reflection of the ability of the project manager to manage and complete a project. The project team may be doing the actual implementation, but it is the responsibility of the project manager, scrum master, or XP coach to lead the project team throughout the entire process, not just at the beginning and the end. A deliverable that does not meet the expectations of the project's scope represents a project manager who failed to do their job.
- **Experience** The experience of completing the research, the planning, and the implementation of the project should be rewarding and educational for all roles in the project. Not all projects are exciting and thrilling, but the experience of working with an excited project manager who is dedicated to the success of the project is contagious. At the end of the project, all parties involved should possess a sense of pride and satisfaction with the experience of being a part of and contributing to a successful project. The quarterback of the team, the project manager, has to call plays from the line, analyze defense, and discipline the team when it's necessary. Agile teams, which are self-led and self-organizing, can take pride in delivering results in each iteration of the project. Organization, communication, and a desire to achieve are all factors in the sense of accomplishment.
- **Project team** The project team members will measure you by your ability to lead them to finish. They will look to you from day one to inspire, lead, and encourage them. They need you to be decisive, fair, and responsive to their needs. How you work with, talk to, and interact with the individuals on the team will determine their opinion of you. They won't keep their opinion of you a secret, either; news of your ability, or lack thereof, will be shared with their peers and their supervisors throughout your organization.

EXAM TIP Don't take shortcuts in the project management processes. This is for the project manager and all roles in the project. Follow and trust the process.

Managing the Quality

An IT project manager must have the keen sense to manage both the expectations of the deliverable and their own process to obtain the deliverables. The quality of the process is directly related to the quality of the deliverables. Simply put, the greater the project manager's ability to lead the process, the greater the quality of the project deliverables (see Figure 11-4).

A project manager can use numerous tactics to ensure that the project management process is excellent and superior to projects that may be anchored with delays and cost overruns. There are several key managerial skills a project manager needs to have to manage a project successfully:

- **Finance and accounting skills** While the project manager doesn't have to be a certified public accountant, they should have some fundamental accounting experience or training.
- **Planning skills** The project manager must know how to plan for the project implementation. A clear understanding of the project requirements is a fundamental precursor to project planning.
- **Leadership skills** Leadership is the ability to establish direction, align people, motivate, and inspire.
- **Management skills** A project manager must have the management skills to produce the results the project stakeholders are expecting from the project team.
- **Communication skills** Ninety percent of a project manager's time is spent communicating. It's a fundamental skill for a quality project manager.
- **Problem-solving skills** It's key that a project manager has the ability to "figure stuff out." They recognize the problem, find a way to solve it, and then make the decisions necessary to implement the solution.

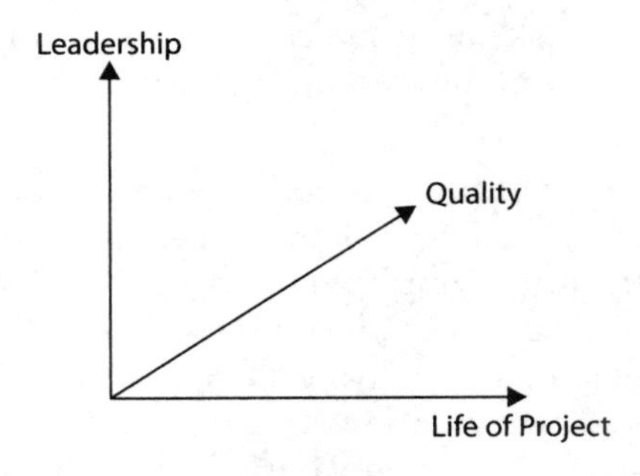

Figure 11-4 A project manager's ability to lead impacts the project quality.

- **Negotiating skills** A project manager must balance stakeholders' interests, keep peace and harmony on the project team, and use the appropriate give-and-take when it's needed.
- **Achievement orientation** A good project manager has to have a drive to get things done.
- **Agility** A project manager must be able to see the big picture, coordinate all of the moving parts of the project, and decompose the project end results into manageable components. Agility isn't just for adaptive projects. Agility also means that you'll timely respond to changes and opportunities.
- **Service-oriented** A project manager works for their manager and the project customers. For projects to be successful, the project manager must serve the project—this includes serving customers, stakeholders, management, and even the project team. This is the concept of servant leadership that is fundamental to adaptive projects.
- **Emotional intelligence** Emotional intelligence is the ability to understand and influence the emotions of others, and the ability to control self-awareness and self-emotions. A project manager won't successfully manage projects if they can't manage themselves. This includes control, temper, flexibility, time management, and so on. A project manager must be personally well organized and forward thinking.
- **Organization** This trait is probably the talent all successful IT project managers have in common. If you are not an organized person, learn how to become organized. Not only will your ability to manage projects increase, but your quality of life will improve.

Project Information Center

One approach of project organization is to create a project information center (also called "the war room"). The size of your project and the available real estate within your office building will determine your ability to create a project information center. This centralized room is a collection of all materials related to the project. Co-located teams are ideal for adaptive and predictive projects, but it's no surprise that more and more teams are not centrally located. Virtual chats, web conferences, and virtual meetings are the new norm for projects.

The project manager, project team, vendors, consultants, and whoever else is involved in the project can drop by the center, virtually or physically, to retrieve information, learn the project status, and review work related to the project. In addition, you can, and should, create a map of the entire PND on a wall to gauge where the project is at any time. If you're using an agile approach, this is where you'll also display your project information radiators to share information quickly about the project work.

Project resources and access to data must conform to organizational requirements for security. This means physical security, such as physical access to servers and network components, but also digital security. Digital security begins with access permissions, multifactor authentication tools, and security policies. Because so much of today's business relies on technology, you'll need to always consider the data security, intellectual property, trade secrets, and even national and government regulations for securing data.

Resources needed by the team can be centrally stored in the project information center, along with books, videos, and magazines related to the technology being implemented. Tools and equipment connected to the implementation are stored here. Finally, the project information center is an excellent location to hold team meetings, as resources are a footstep away.

EXAM TIP Project information centers, or war rooms, are becoming less and less common as employees work remotely. Web solutions, such as web conferencing and chat, are more common in today's work environment.

Web Solutions

Another excellent resource, especially for long-term projects and geographically dispersed teams, is the creation of an intranet solution for the project team. A central web page should be secured for the project team, the project manager, and relevant management. The web solution should offer the same features as the project information center and can be designed to allow for milestone completion, project updates, and a method to communicate with other team members. There are lots of web collaboration tools to move your project information center online. Through these tools you can quickly create links to project resources, status reports, activity lists, budgets, charts, calendars, and just about anything project-related you need to organize and share with the project team and stakeholders.

A web solution for your project may be applicable to the entire organization. Some organizations have a central project management office that coordinates all activities of projects through a web solution. In other words, the process is uniform, with some flexibility, across all projects through the web solution. Projects are kept separated, but costs are streamlined as resources may be used across projects. A web solution allows projects that are dependent on each other to interact, and it allows project managers to see the status of a successor or dependent project to judge the completion of tasks.

Software Solutions

There are many different project management information systems (PMISs) available to assist a project manager. With a PMIS, a project manager can organize, track, complete estimates, and schedule events to happen. A project manager can use traditional project management techniques with the PMIS as the catalyst for reaching the project's end. PMISs all have features that can be designed to track tasks; project flow; and the surge of e-mails, documents, and information. Whichever method you choose, a solid foundation on how to use the application is required to gain the full benefits.

Quality Management as a Process

Quality project management is an activity you need to perform from the concept of the deliverable to the release of the deliverable to ensure quality in all your activities. It is a belief that the process a project manager follows to ensure quality from the start of a project will propagate to the activities of the project team throughout the life span of the project.

Several concepts claim to be the "secret potion" for guaranteed successful projects every time. However, the one weakness and common theme in all project management processes is the reliability and willingness of the project manager and the project team to participate. This situation is comparable to joining a gym to get in shape—you actually have to go to the gym and work out to get the desired results. The same holds true with these concepts: you have to use and follow their principles for them to work.

Quality Phases of Project Management

There are five process groups within a predictive project, as Figure 11-5 demonstrates. Each process group keeps an eye toward the quality of the deliverable or ensures that quality exists within the creation of the deliverable. These are not phases of a project, but groups of processes that bring about a result. Adaptive projects utilize these process groups in each iteration, while predictive projects follow these processes throughout the entire project or in each phase of a longer project.

Within each of the following areas, a project manager must work to implement quality and quality management checkpoints:

- **Initiating** The origin of the project results from a reaction to a need or an opportunity. This realization of the need or opportunity is the concept of the project. The business needs of the organization are addressed to ensure that the project will satisfy these needs first. Once the project charter has been written to authorize the project manager, the project can move into planning. Quality is affected from the start. If the expectations of the quality aren't set, aren't planned for, or aren't quantified, the project's success is doomed.
- **Planning** The cornerstone of a successful project is the planning. The project manager and project team must identify the required activities and estimate the time necessary to complete the activities in order to reach the project goal. Through the research, project managers can identify the necessary resources, funding, and skills required to achieve success. Armed with this information, the project manager can create the project plan. Quality doesn't happen by accident.

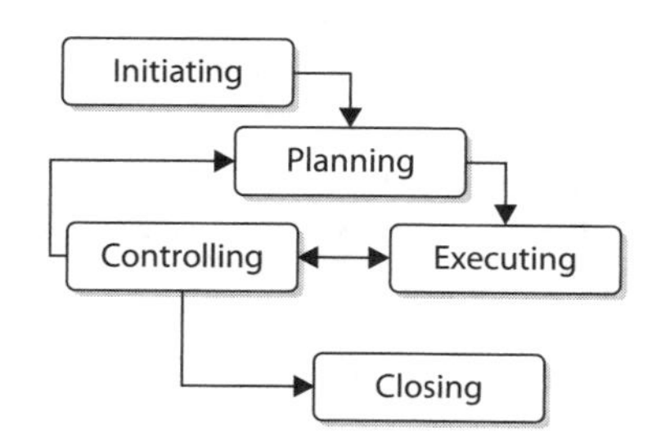

Figure 11-5 Quality is an objective in each of the five process groups of project management.

During the planning processes, the project management and project team must plan how the quality demands will be met. The product owner is responsible for identifying and prioritizing the requirements based on business value, which is ultimately what constitutes quality in the finished product. Agile teams use the Definition of Done to plan the quality required in each of the product's components.

- **Executing** Once the project plan has been approved, the project work can begin. The project manager will rely on the work authorization system to record task completion and allow new project assignments to begin. Adaptive project teams take chunks of requirements from the product backlog and work on those items in each iteration of the project. In either case, quality must be executed as part of the work. The project team must follow the specifics to meet quality demands as defined in the project plan.
- **Controlling** This phase of project management is a continuous cycle to oversee the project. In this phase, more than any other, the project manager ensures quality through quality control. Scope verification is also done here to ensure that the project is delivering what was promised. Project managers also control projects through cost control, schedule control, and risk management. The process of managing the project must be of quality as well. Quality control is inspection driven. Quality control in an adaptive project happens through the testing and demonstrations the development team provides in each iteration.
- **Closing** This portion of project management is the sigh of relief. It requires proof of the project deliverables, approval from management, and satisfaction from the customers or end users. This final stage moves the project from a work in progress to a component of the business. The final reports are submitted and archived, and the lessons learned documentation is completed. Quality also happens in the closing phase. A complete and final review of the project, including whether it met the quality objectives, and the quality of the project management experience, is required.

These five phases of project management contribute not only to the success or failure of the project, but also to the quality of the deliverables. A dedication to doing the required activities properly and with confidence in each phase leads to quality. Any one phase that is lacking a commitment to the success of the project can cause the entire project to be off balance and ultimately fail.

EXAM TIP The CompTIA Project+ exam content outline refers to the five process groups as phases. These aren't phases of the project in most project management circles, but are groups of processes. However, for purposes of achieving the exam passing score, just acknowledge that CompTIA calls them phases, pass the exam, and move forward.

Ensuring Quality Throughout the Project

As your project moves along through each process group, over hurdles and through barriers, you'll need a proven system to check the quality of your progress. You may subscribe to any one of multiple theories in the world of project management to test the quality of your project. All of these theories, however, have one common thread: work completed must be proven to be in alignment with the project deliverables. This is scope verification—the process of ensuring that the project is creating what the customer has asked for.

For example, a project to create a new application for an organization will have several milestones in its path to completion. The desired deliverable of this project is that the application will allow users to submit HR forms through a company website. The project manager can check the work in progress to verify that it is in alignment with the project deliverable. Should the work be out of alignment, the project manager must take immediate corrective actions to nudge the work back on track. Demonstrations in an adaptive project provide opportunities for the customer to see the work results and give immediate feedback to the team for improvements, approval, or changes.

Planning for Quality

Quality planning is a process to determine which quality standards are relevant to the project and how they can be implemented. Planning for quality is a fundamental exercise in the planning phase—each deliverable must have metrics that prove its quality. In IT, this can be bandwidth, latency, database accuracy, the speed of an application, and more. Every predictive project needs a *quality management plan* that defines the project's approach for quality. This plan may be organization-wide, and it usually addresses the quality metrics, control limits, and how often you'll need to measure the results of the project to ensure that quality exists.

The quality management plan also defines how you'll adhere to the quality assurance requirements. Quality assurance in your organization may be known as a quality policy that dictates the expectations of a project in regard to quality, how the expectations are measured, and what the outcomes of those measurements should be. This quality policy is considered and applied to the project scope, which is important because the project scope contains all of the work your project will undertake. What good is a quality policy if it's not implemented with the project work?

Depending on your organization, you may also have relevant standards and regulations that will serve as input to your quality planning. A *regulation* is a law or practice that is not optional in your industry. For example, the health care industry must abide by the Health Insurance Portability and Accountability Act (HIPAA) regulations as well as other regulations. A *standard*, on the other hand, is a rule or generally accepted practice within an industry. For example, most software application windows close using some button in the upper-right corner. While there's no law that says this is a must, it's a generally accepted standard regardless of the application or operating system.

The quality management plan will also define how you'll implement quality control to prove the existence of quality within the project work. When you're planning for quality, there are five major approaches you can rely on:

- **Benefit-cost analysis** Within every project, there will be a demand for quality—and a cost to reach that demand. A benefit-cost analysis considers the cost to reach the level of quality in relation to the benefits of obtaining the quality. For example, a customer may demand that a series of databases provide 100 percent accuracy 24/7. While this seems good, the synchronization of multiple databases after each change may result in a very costly solution. Instead of the expensive 100 percent solution, a better solution, for example, may be a less costly approach that ensures 98 percent accuracy. There will be a balance of cost and acceptable quality in every solution. Sometimes fast and good is better than slow and perfect.
- **Benchmarking** This approach uses other projects as a measure of performance on your current project. It examines the deliverables, the project management processes, and the successes and failures within each project to measure how the current project is performing. The problem with the approach, however, especially in IT projects, is that unless the nature of the IT projects is the same, it's difficult to use. You can't measure the performance metrics of a project to develop an application against the metrics of a project to create a new network. Additionally, because technology is changing so rapidly, benchmarks that were applicable 18 months ago are very likely outdated and inappropriate.
- **Flowcharting** Flowcharting shows how the components within a system are related, as shown in Figure 11-6. This is an ideal approach within any project. Consider an application that follows a client-server model. The front-end and back-end applications must communicate over a network or series of networks. A flowchart can illustrate the various components, how they interact, and their effect on quality. Another example of a flowchart is a cause-and-effect diagram to illustrate the causes that are contributing to quality defects within the project. These diagrams are also called Ishikawa, or fishbone, diagrams.
- **Design of experiments** This approach relies on statistical what-if scenarios to determine which variables within a project will result in the best outcome. The design of experiments approach is most often used on the product of the project, rather than the process of the project itself. For example, a project team creating a new network may experiment with the capacity of the network cable, the network switches, the routers, and the number of the network cards on the servers to determine which is the best combination for the best price. Design of experiments is also used as a method to identify which variables within a project, or product, are causing failures or unacceptable results.

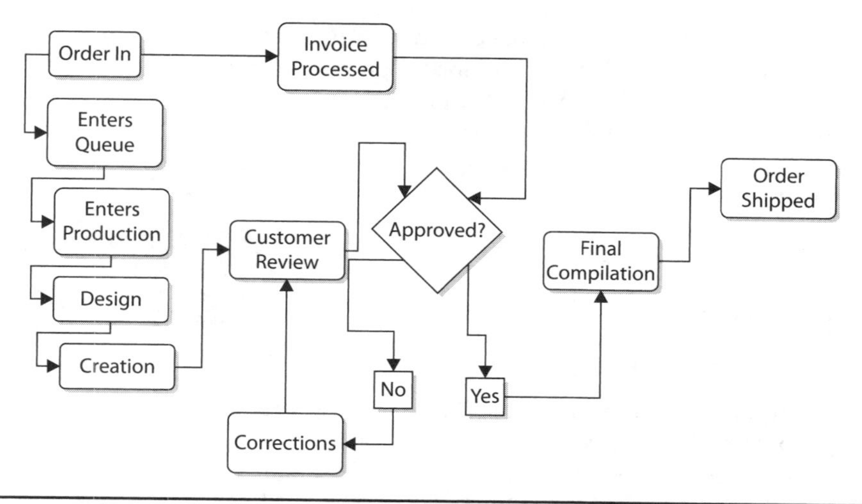

Figure 11-6 Flowcharting shows how the components within a system are related.

- **Cost of quality analysis** Cost of quality is the sum of all the costs to achieve the expected quality the customer demands in the project deliverables. This includes all of the work to conform to the quality requirements and the expense incurred from nonconformance to the quality requirements. The cost of nonconformance is most evident when work needs to be redone and materials are wasted. Technically, the cost of quality has three costs: prevention costs, appraisal costs, and failure costs. Failure costs are associated with what could happen if the project fails: the lack of sales, loss of customers, product returns, fines and fees, downtime, and so on.

Traditional Quality Assurance

Quality assurance (QA) is a series of actions and requirements to assure the organization that each project will meet the relevant quality standards. QA is typically mandated on an organization-wide or departmental-wide program or quality system. For example, if your company uses Six Sigma or is ISO 9000 certified, your project will have quality standards that it will have to map to these guidelines and regulations. QA is concerned with the systematic activities that are applied to each individual project to ensure that quality exists. QA ensures that quality is planned into the project work. It's always more cost-effective, more time-effective, and better for the project stakeholders to complete the project work correctly the first time.

To implement quality assurance, you'll first follow organizational procedures that may be established for your projects. These organizational procedures comprise a form of governance. Governance ensures that all projects operate the same way, with the same rules and adherence, in order to improve the probability of achieving quality within the project for the organization. Governance can include reports, forms, audits, and other quality-assuring activities. Quality assurance is prevention driven, in that the goal is to prevent a lack of quality. The same tools a project manager uses for quality planning can also be applied as part of the QA process:

- Benefit-cost analysis
- Benchmarking
- Flowcharting
- Design of experiments
- Cost of quality analysis

Depending on the nature of your project and organization, your project may also undergo quality audits. A *quality audit* is a formal review of the quality management activities you have in place within your project. Quality audits can be completed in-house by QA professionals your organization employs or by third-party experts.

Traditional Quality Control

As you now know, quality is measured by the end result of a project. Obviously, you cannot wait until the end of a project to determine if quality exists. *Quality control (QC)* is concerned with the quality of the actions and deliverables within a project. QC is inspection driven; QC reviews the deliverables to establish that the quality expected by the project stakeholders is present.

QC is also concerned with the root cause of results that are below the quality standards and with eliminating the issues that are causing quality to slip so that quality issues are not repetitive. It focuses not only on the product of a project, but also on the project management process itself. For example, QC is used to determine why cost and schedule variances have occurred and what corrective actions can be enforced to ensure the same mistakes don't happen again.

QC requires the project manager to have some understanding of statistical analysis, sampling, and probability to track trends, predict quality results, and determine root causes in quality issues. Trend analysis is especially useful in IT projects, as most work within an organization is cyclic. For example, suppose your network servers take a processor hit every morning as users log on to the network, check their e-mail, and open files. In the afternoon the proxy servers may have an increase in Internet traffic as users check the news, the weather, or the traffic for their commutes home. In an IT project, trend analysis can allow the project team to make educated decisions on how to react to conditions within the project.

QC must be managed throughout the project. It's unacceptable to wait until the project has ended to see if the deliverables are of quality. The project manager must get out, look, listen, and inspect. Throughout the project, there are four fundamental facts about quality control to be aware of:

- *Prevention* keeps quality errors out of the project. *Inspection* keeps quality errors away from the customer.

- *Attribute sampling* means the results meet the expected quality standards or they don't. *Variables sampling* tracks the level of acceptability of the results over time.
- Within a project, you have special causes where quality excels or diminishes due to anomalies within the project. Otherwise, you expect the results to vary as part of the project; this typical variance is simply called *random causes.*
- A *tolerance* is an acceptable range of quality for the project or deliverable. *Control limits* are the outer and upper limits that the quality results must fall within. If results are within the limits, the project is in control. If the results are out of the limits, it's considered to be out of control.

Implementing Quality Control

Know this: quality is planned into a project, never inspected in. A goal for any project is to achieve quality by planning for quality—and then following the plan. But how will you know if quality exists on a project unless there is accountability? Sure, you could wait until your project is complete and then test the deliverables, but that's a little late. Quality control must happen throughout the project to ensure that quality exists.

The most accessible method to ensure quality is inspection. Once you inspect the work, you can measure and react to the evidence you and your project team have found. There are many different approaches to inspecting the project deliverables. Following are the most common.

Peer Review

One approach to QC throughout an IT project is to use *peer review.* Peer review, as its name implies, is the process of allowing team members to review each other's work. It is an excellent method to ensure that each team member is completing their work and doing an excellent job. Peer review provides for many things, including

- Ensuring that each task is checked for quality
- Allowing a team member to show others their work
- Allowing a team member to learn about other areas of the project
- Allowing the project manager to ensure the work is being completed
- Holding the team responsible for the quality of the work completed

The risk involved with peer review QA is that not all team members are up to the challenge of reviewing another's work or having their work reviewed by an equal. If you use this approach, your team members must have confidence in each other's ability to review other members' work fairly, and confidence in their own abilities to complete the assigned tasks. Adaptive projects often use *pair programming,* where a pair of programmers works together to develop the code. One person writes the code while the other programmer checks the code as it is being written. The goal is to catch mistakes immediately and save time in finding bugs, refactoring, and providing quality code from the start.

Statistical Sampling

Statistical sampling is the process of choosing a percentage of results at random. For example, a project creating a database and website to sell concert tickets may require a measurement of database accuracy, the speed of the website, and the functionality of the overall program. This testing must be completed on a consistent basis throughout the project, rather than on a hit-and-miss basis.

Statistical sampling can reduce the costs of QC, but mixed results can follow if an adequate testing plan and schedule are not followed. The science of statistical sampling, and its requirements to be effective, is an involved process. There are many books, seminars, and professionals devoted to the process.

Management by Walking Around

One of the most successful methods for managing quality is to allow yourself to be seen. Get out of your office and get into the working environment. You don't have to hover around your team, but let them know you are available, present, and interested in their work. Again, this isn't always a reality in today's workforce with so many of us working remotely.

So many IT project managers have a fear of being disliked or being seen as typical management, or they consider themselves too important to speak with their team. These less-than-successful project managers alienate themselves by hiding in their offices, ignoring the opportunity to work with the project team to ensure quality from the get-go. Don't let this happen to you! Get involved with the project team members and make yourself visible.

Reviews by Outside Experts

Hire an outside expert to review the project as it progresses. This approach allows the project manager, who may not be as skilled as their team on the project's technology, to ensure that the team is completing the assigned work with care and precision. A consultant can be brought into the project at key milestones to make an unbiased review of the work done to date. The consultant can accomplish many things for a project's success. This practice

- Ensures quality and accuracy
- Allows for an unbiased review by a third party
- Creates accountability for the team completing the work
- Allows the project manager to know the true status of the work
- Allows the project manager to make any needed adjustments

Test Plans and Testing Cycles

When leading a software development project through a predictive or adaptive life cycle, the team will need to create a test plan and follow predetermined testing cycles. The CompTIA Project+ exam will likely have a few questions on the most common testing approaches, but the goal is the same: don't allow defects to reach the customer.

Defects that are released into production are called *escaped defects* and are poor quality. The following are common testing approaches to recognize:

- **Unit testing** The smallest testable chunk of code is tested for accuracy and is usually completed by the development team as they create the code.
- **Smoke testing** Where there's smoke, there's fire. Smoke testing is completed after a software build and before the actual release of the software to production.
- **Regression testing** The team does this type of testing after each code change to ensure that the code doesn't break other parts of the software.
- **Stress testing** As its name implies, stress testing aims to confirm the software is stable, reliable, and operates as expected for the end users.
- **Performance testing** This approach, similar to stress testing, tests the performance of the software in a prescribed workload. It can also be used to confirm the scalability, computer resource utilization, and effect on the operating system.
- **User acceptance testing** Commonly called UAT, user acceptance testing is the final test before the software goes live in production. It's a real-world test to confirm the software works as expected according to the design specifications, contractual obligations, regulatory requirements, and user expectations.

EXAM TIP The project team has the largest effect on quality, as they are the people doing the work. Quality assurance is to do the work correctly the first time. Quality control is to inspect the work to confirm that quality exists in the project. You can't inspect quality into a project.

Analyzing Quality

Once you've completed the inspection of the project and the product deliverables, then what? Of course, you'll be doing QC inspections on a regular basis, so you'll need to track and analyze the results. You'll want to complete root cause analysis to determine why quality issues may be random or repetitive. Since quality is the fulfillment of requirements, you can use the scope baseline, the cost baseline, and the schedule baseline together as the quality baseline. In other words, you must satisfy the project scope while balancing the time and cost to complete the project scope objectives. You can't achieve quality in the project if you don't deliver the project scope. There are several approaches to tracking and analyzing quality, as discussed next.

Using Control Charts

A *control chart* displays the results of your inspections over time. The results of inspections are plotted out against a mean, an upper control limit, and a lower control limit. As you can see in Figure 11-7, the results of inspections are measured and then added to the control chart. When results are over the control limit, they're out of control; otherwise, the project is acceptable. However, this approach can be a little tricky in many IT projects. Control charts are best when you have projects that are extremely repetitive, such as manufacturing and construction projects. That's not to say that you can't use

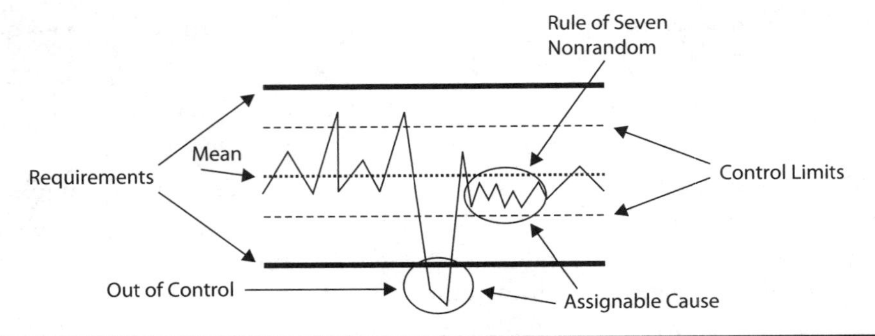

Figure 11-7 Control charts measure results over time.

these charts within IT projects—just be aware that the results of your measurements may fluctuate as the nature of the work within the project changes. You can use control charts to track server usage, update network throughput, and more.

When results of a measurement fall out of control, this is called an *assignable cause.* An assignable cause means there is some reason for this event to occur. It could be a hardware error, a different developer, or some other reason. It's a signal that root cause analysis is needed. In addition, whenever seven results of your testing all fall on one side of the control chart's mean, it's called the "Rule of Seven" and is also an assignable cause. There will always be some reason why the quality has stymied on one side of the mean or the other, which means it's time for root cause analysis.

Using Pareto Diagrams

A *Pareto diagram* is somewhat related to Pareto's Law: 80 percent of the problems come from 20 percent of the issues. This is also known as the 80/20 principle. A Pareto diagram illustrates the problems by assigned cause from highest frequency to lowest frequency, as Figure 11-8 shows. The project team should first work on the most frequent problems and then move on to the smaller problems.

Figure 11-8 A Pareto diagram is a histogram ranking the issues from highest frequency to lowest frequency.

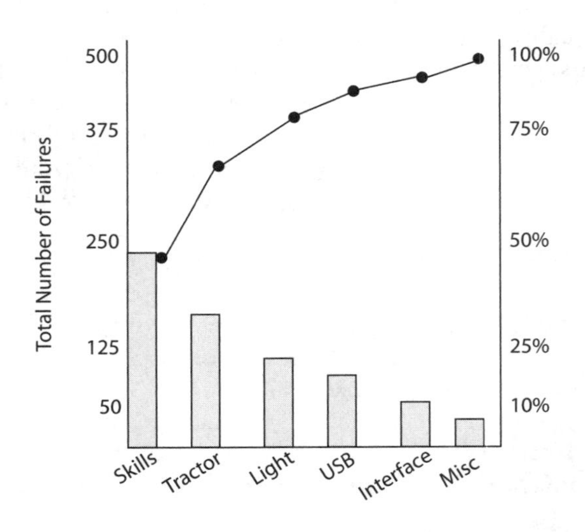

Revisiting Flowcharting

Remember flowcharts? Flowcharting is a method of charting how the different parts of a system operate. Flowcharting is valuable in QC because the process can be evaluated and tested to determine where in the process quality begins to break down. Corrective actions can then be applied to the system to ensure quality continues as planned—and expected.

Applying Trend Analysis

Trend analysis is the science of considering past results to predict future performance. Sports announcers use trend analysis all the time: "The Cubs have never won in Saint Louis, on a Tuesday night, in the month of July, when the temperature at the top of the third inning was above 80 degrees."

The results of trend analysis allow the project manager to apply corrective action to intervene and prevent unacceptable outcomes. Trend analysis on a project requires adequate records to predict results and set current expectations. Trend analysis can monitor two things:

- **Technical performance** Trend analysis can ask, how many errors have been experienced by this point in the project schedule, and how many additional errors were encountered?
- **Cost and schedule performance** Trend analysis can ask, if we are $4,000 over budget now, what is our final cost likely to be?

Relying on Process Diagrams

A *process diagram* is a type of flowchart that shows the flow of a process. The process can be the flow of information through a system, but it may include how people, called actors, interact with the system. Consider a new network and all of the processes that happen by the users: the user logs in; reads and responds to e-mail; has a chat session with a colleague; browses a website; accesses a report from a server; edits, saves, and prints the report; and so on. All of these interactions could be mapped with a process diagram to set and measure expectations for quality.

Creating a Histogram

A *histogram* is just a bar chart. It's nothing too fancy—each bar of the graph represents a different factor. You could make a histogram to show how well each requirement in your project is performing. Or a histogram could show how each vendor is delivering on time or costs. Or it could compare before and after project solutions. A histogram is a great chart for showing multiple requirements and how they're stacking up against what was planned and what was actually experienced. It's a quick way to show how well you're achieving quality among the requirements.

Analyzing a Run Chart

A *run chart* is a chart that shows how a process is running—and by "running," I mean how well the process is operating. You'll often create a run chart near the launch of the project to measure fluctuations in a process that you're aiming to improve. A *run* is a series of increasing or decreasing values in a process. You could create a run chart, as

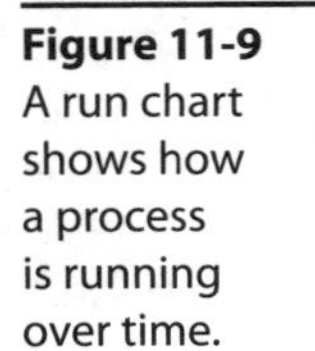

Figure 11-9 A run chart shows how a process is running over time.

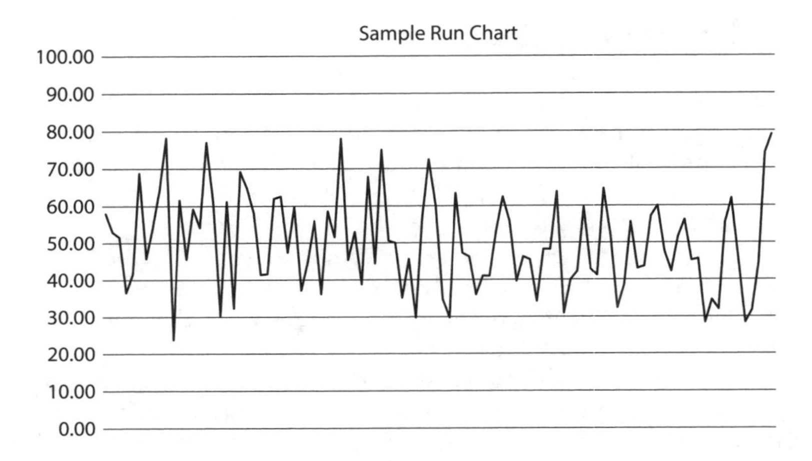

shown in Figure 11-9, to show network bandwidth throughout the project day. Early in the morning, there's not much traffic, but as more and more users log in and use network resources, traffic increases throughout the day. The run chart doesn't show stability of the process, but it shows peaks and valleys of the process.

Analyzing Scatter Diagrams

A *scatter diagram* measures two different variables to see if there's a correlation between the two. For example, in Figure 11-10, on the Y axis is the variable of login timeouts. On the X axis is the number of users logging into the system. As more and more people log in, it's believed that there are more and more timeout errors on the network. A scatter diagram would test this hypothesis to see if these issues are related. If there is a relationship between these two variables, the tighter the points will be to the line in the scatter diagram. Basically, if the points are close together, there's a relationship; if the points are scattered about in the chart, a relationship is unlikely.

Figure 11-10 A scatter diagram measures variables to determine if there is a relationship between them.

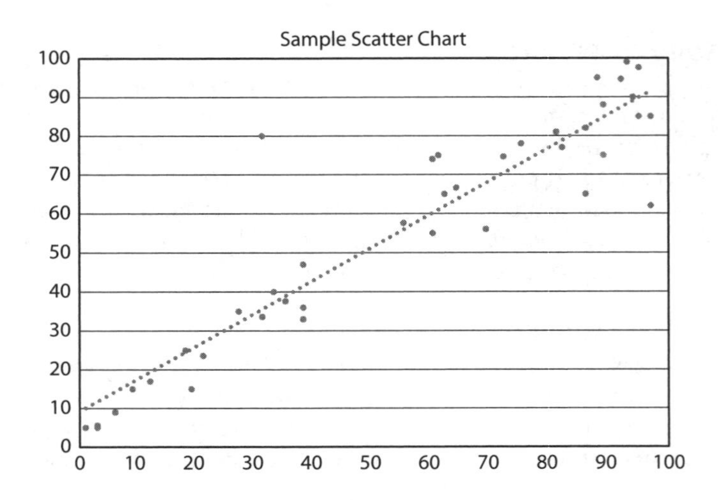

Total Quality Management

No book on project management would be complete without at least a nod to *total quality management (TQM)*. Total quality management is a process that involves all employees within an organization working to fulfill their customers' needs while also working to increase productivity. TQM stems from Dr. W. Edwards Deming and his management principles, which the Japanese adopted after WWII. In the United States, these principles were readily adopted in the 1980s after proof of their success in Japan.

The leading drive of TQM is a theory called *continuous quality improvement*. According to this theory, all practices within an organization are processes, and these processes can be infinitely improved, which results in better productivity and, ultimately, higher profitability.

Here's how this relates to IT project management: the processes a project manager uses to communicate, schedule, and assign resources can be streamlined, improved, and modernized to make the project easier to implement and more profitable as a whole. Examples include Microsoft Project Server and other web solutions for project teams.

In project management, the customer is the end user of the deliverable, and the concept of streamlining processes is dependent on the project manager and the project team. Scores of books have been written about TQM, though one of the best books, *Out of the Crisis* (The MIT Press), was written by Deming himself. Many project managers adapt Deming's "quality circle" of plan, do, check, act (PDCA) as a cycle for quality improvement. The cyclic nature of an adaptive project is not unlike the PDCA cycle.

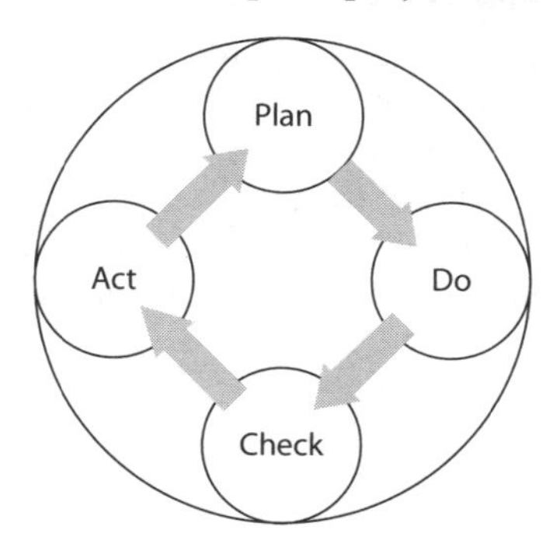

TQM as a process is not a magical formula, an equation you can map out in Excel, or a dissertation from a business professor at Harvard. It is a simple thing to describe, but fairly difficult to implement. TQM comes from the dedication of the project manager and the project team to completing, with gusto, the required activities in each phase to produce an excellent deliverable. Anything less should be unacceptable.

Creating a Strategy for Quality

As with any area of project management, you won't be successful without a plan. Quality control requires a plan, a process, and a strategy to implement and enforce it. You can attack quality enforcement in many ways; the best, however, is to lead by example.

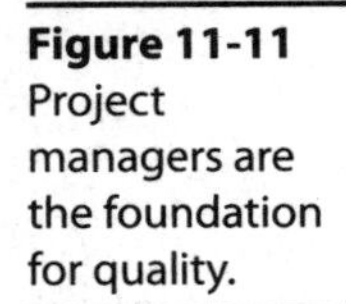
Figure 11-11 Project managers are the foundation for quality.

You should be the focal point of quality for your team, as Figure 11-11 depicts, in all that you do. Leading by example shows your team your own level of dedication to the project and that you expect your team to follow.

Revisiting the Iron Triangle

The second best method of implementing quality, regardless of the project, is a balance of time, cost, and scope. As you can see in Figure 11-12, the quality of the project is dependent on your management of the allotted time, the assigned budget, and the expected scope. Of course, there's leadership, managerial skills, and more—but without balance, the project will fail. Adaptive projects use an inverted triangle, where time and cost are fixed and the scope of the project is expected to change throughout the project.

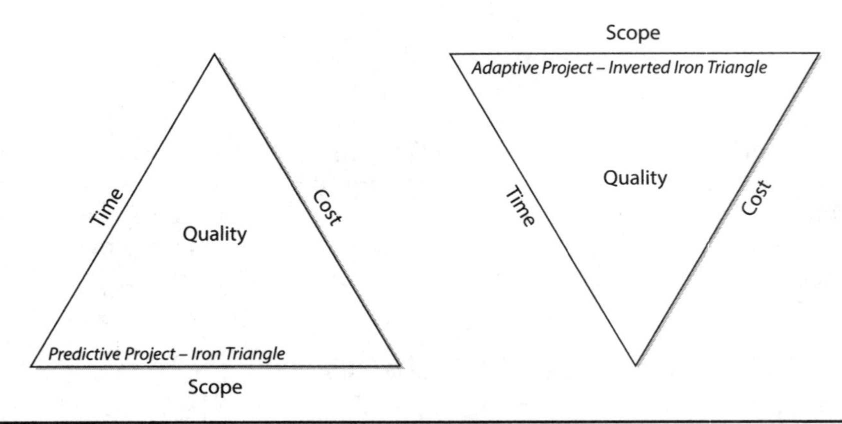

Figure 11-12 Quality can be achieved with a realistic balance of time, cost, and scope.

The one element that you should already have a strong handle on is time. Some projects will have more freedom with time than others. During the planning phase of your predictive project, you should be able to predict what the required time is to complete the project and meet the assigned objectives. Projects that are constrained on time will, no doubt, require you and the team to work diligently and quickly to achieve the objectives. When time becomes an issue, the quality of the project will reflect that.

The budget for the project is approved by management and will be yours to manage—most likely under management's watchful eye. Your planning and implementation of the plan will help determine the budget of the project. Should your plan be full of holes, underdeveloped, and not comprehensive enough to counter foreseeable problems, your budget will be blown and the quality will suffer.

Finally, the scope of the project must be protected from unnecessary change. A change control system must be in place, backed by management, and used. When the scope begins to creep, the project's time and budget must match the changes to the scope. Most often, however, when unapproved changes come into the project, quality begins to diminish because time and funds that should be allotted to complete approved project activities are spent on unapproved activities.

Communicating Project Progress

One method to implement quality is to use *progress reports*. A progress report is simply a formal, informative method of summarizing the status of work completed. Typically, on longer projects, progress reports are essential for keeping a record of the work completed, and they make for handy references in the end phase of the project.

You'll generally use progress reports, or status reports, in a predictive project. Adaptive projects demonstrate progress in the iteration's demo or sprint review. The daily standup meeting, or daily scrum, is a ceremony for the development team to share progress each day of the iteration—another ongoing progress reporting activity. At the end of the iteration, after the sprint review, the sprint team moves into the sprint retrospective. Recall that this is a type of lessons learned meeting to share what has and hasn't worked in the project and to create a plan of how the team can perform better in the next iteration of the project. While adaptive projects don't use a traditional progress report, adaptive projects do share progress through burndown and burnup charts with the project stakeholders through the information radiator.

In regard to quality, the process of creating progress reports allows the project manager and the project team to ascertain where the team is on the project and the amount of work yet to do. It's a great way to visualize the progress the team has made so far and determine if the project is on track with the project vision.

Project sponsors and your project team's functional managers will typically want to see the progress reports, as this allows them to keep in tune with your ability to lead the team and manage the project. Upper management may not want to see these reports, as their time may be limited. These reports can be based on templates that allow you and the project team to complete the progress report quickly and accurately. There are four types of progress reports you'll use as a predictive project manager.

Current Status Reports

Current status reports are quick news on the work completed, or not, since the last status report. For example, you may determine that status reports should be completed every two weeks. Within each two-week window are tasks that must be completed. This report will focus on the scheduled tasks and their status over the last two weeks.

If scheduled tasks were not completed, the report should clearly state why the work has lagged behind and what solution is offered to get the work back on schedule. Distribute this report to the project team and the project sponsor, and keep a hard copy in the project binder. These reports are excellent for record keeping and nudging the project team back on track.

Cumulative Reports

As their name implies, *cumulative* reports focus on the work from the beginning of the project until the current date. Cumulative reports are excellent in long-term projects and should be created based on management's requests, at milestones within the project, or on a regular schedule such as every three months. Use these reports for looking back on the progress accomplished so far on the project. Information in this report should include

- Work completed
- Lagging tasks and plans for recovering lost time
- Significant accomplishments
- Variances
- Budget information

Management Summary Reports

Management summary reports detail the overall status of the project, changes from the original plan, changes in execution, or cost variances within the budget. These reports are created on an as-needed basis and are ideal for upper management, as upper management does not have the time to read detailed reports to discover that everything on the project is going as planned. These reports are quick and to the point—effective when sharing bad news. The purpose of these reports is a fast, honest way to summarize the project status so that management may keep the project in check.

Variance Reports

Variance reports are summaries of any variances within the project, mainly time and cost, but they can also be reports on scope variances. They require the project manager to evaluate the cumulative work against the original project implementation plan. The comparison of the two should indicate where the project is and where it is heading. These detailed, number-oriented reports are an ideal way of enforcing quality and keeping the project on track.

CompTIA Project+ Exam Highlight: Quality

You'll encounter a few questions about quality on the CompTIA Project+ exam. You'll need to be familiar with quality assurance: planning quality into the project work. Know that you achieve quality by satisfying the project scope and balancing the Iron Triangle. Finally, be familiar with quality control: the inspection of the project work to ensure that quality exists. Quality control aims to catch mistakes before the customer completes scope verification.

1.7 Compare and contrast quality management concepts and performance management concepts This is the primary objective of this chapter for both adaptive and predictive project management approaches. In a predictive project, the quality management plan defines how you'll adhere to quality assurance. Recall that QA ensures that the work is done correctly the first time. QA in an organization can be a formal quality program, such as Six Sigma, or it can be a quality policy that all projects must adhere to. Governance of quality ensures a uniform approach to quality management in an organization. The quality management plan also defines the quality control exercises you'll complete in the project. QC is the inspection of the project work to determine the existence of quality in the project deliverables. Your quality management plan should also define how often you'll measure for quality, what the quality baseline metrics are, and what types of charts and measurements you'll be recording.

While it's true that you plan quality into a project, you must still inspect the project deliverables to prove that the project work has been completed as planned. This is quality control, and its goal is to prevent mistakes from reaching the customer. When you inspect the project work and there are errors, corrective actions must be implemented to fix the errors and to get the project work back into alignment with the project scope and the quality baseline. Errors to project deliverables are variances from the project scope, and they'll often create cost and schedule variances too, because of the time, materials, and cost of labor to correct the defects. Adaptive software development projects pass through audits, inspections, and testing cycles to confirm the software is of quality and meets the project requirements.

The quality baseline is achieved by adhering to the time, cost, and scope requirements of the project. You can use a Pareto diagram to categorize areas of defects and rank them from highest to lowest. A Pareto diagram is actually a type of histogram that shows categories of defects. A histogram is a bar chart that shows categories of a measurement. The difference between a simple histogram and a Pareto diagram is the latter ranks the categories of defects and directs the project team to attack the highest categories of defects first, and then to move on to the lower categories of defects. It is a type of chart that tracks measurements of project performance over time. A similar chart is a run chart, which shows the time elapsed between measurements in the project. You might also use an Ishikawa diagram, also known as a fishbone diagram or cause-and-effect chart, to facilitate a conversation on quality defects within the project.

2.3 Given a scenario, perform activities during the project planning phase Quality is planned into a project, not inspected in. It's during the project planning that the quality management plan is created. This plan defines quality assurance for the project and how you and the team will work to ensure that quality is built into the approach and the results of the project. Because quality is a conformance to requirements, the quality management plan directly addresses the customers' expectations of the features and functionality of the requirements and links to the stakeholders' expectations for the finished product. Quality assurance also addresses how the finished product is fit for use when it enters production, so references to how the product will be tested and approved are common in the QA plan.

2.5 Explain the importance of activities performed during the closing phase While the majority of quality activities are performed during project planning and project execution, you'll need to address quality during the closing of the project. Specifically, you'll need to perform a validation of deliverables and acceptance of the deliverables by the project customer. In a formal project, there will likely be a sign-off process that helps lead the project to closure. The validation of deliverables means the product has passed through the defined testing cycles and the customer approves the work's quality, stability, and completeness.

3.3 Given a scenario, analyze quality and performance charts to inform project decisions Throughout the project you'll likely need to create different types of charts to show project performance, progress, and quality in the project. When it comes to quality, you'll most likely create Pareto charts and histograms to show the distribution of defects. Run charts and control charts show the stability of the process and the results of quality control measurements. Agile projects utilize burnup and burndown charts to show the velocity of the project work, the balance and accumulation of the user stories completed or remaining, and how effective the team is in each iteration.

4.2 Explain relevant information security concepts impacting project management concepts Your overall project management approach must address the security needed for your project and its deliverables. This is directly attached to the quality of the project, as security concepts are requirements. You'll need to address the physical security of the hardware, access to software and data, operational security, and any operational security practices of the organization. This can include trade secrets, intellectual property, and national security information, depending on the type of project work you're managing.

4.3 Explain relevant compliance and privacy considerations impacting project management Like the security of the data and physical assets, your project may also need to address compliance and regulatory requirements. When you're dealing with healthcare projects, financial data, or collecting personal customer information, for example, you'll have regulatory compliance concerns that must be planned for and implemented in the project. No one wants a data breach, loose security, or unauthorized access to your systems—these are all poor quality and will affect the acceptability of the project by the customer. In addition, there can be fines and penalties for being out of compliance. You may need to consult with your organization's legal counsel to address the requirements for compliance in your country, state, or province for your industry and type of software solution.

Chapter Review

Quality, quality, quality. Everyone talks about quality, but what is it? Quality is the capability of the project to meet the expected requirements of the project customer. Quality is the good, the worth, the profitability gained from and during the implementation. Quality is also the level of excellence within the project process. An IT project that produces quality results will have quality at its core—which is accomplished through planning, guidance, and leadership. Quality is a conformance to requirements and creating a solution that is fit for use.

The trick to ensuring quality deliverables is to make sure that quality is designed into the project itself. Quality management is a process that ensures quality is a central point of each work unit of a project. As the project moves through its different phases, you must sample and readjust the work to be in alignment with the project deliverables to ensure quality. Quality assurance is prevention driven and aims to ensure that quality is built into the project. Quality control is inspection driven and confirms that quality exists within the deliverables.

Progress reports are used in predictive projects and allow the project team, the project manager, and management to be aware of the status of the project at any given time. The reports can be simple one-page summaries or lengthy, detailed accounts of problems encountered and solutions discovered. Reports designed for management typically are quick and to the point—good news or bad. Management doesn't have the time, or desire, to read a lengthy report only to discover everything is great.

Adaptive projects report on progress through burnup and burndown charts to show how many user stories are remaining in the project and the velocity of the development team in each iteration. An information radiator is an excellent method to publicly share the progress, performance, and quality of the project. Adaptive projects use testing cycles, such as unit testing, smoke testing, and user acceptance testing, to confirm the software is of quality and meets the requirements of the project.

As the project progresses, you must implement a process to ensure quality through each phase of the work. You can use leverage with your team members to ensure quality within their work by implementing peer review, sampling the project, and hiring an outside consultant to review the work.

Ultimately, the quality of a project is measured not by the project manager, the project team, or management, but by the end users of the product. Adaptive projects demonstrate the quality of the work completed in sprint reviews or demos for the stakeholders. Before completed work is released to production, there is validation of the completed work to confirm its acceptability and quality. Stakeholders' experience and the productivity gained by the technology will be the true measure of the worth of the project.

Exercises

These exercises allow you to apply the knowledge you have learned in this chapter and are followed by possible solutions. In these exercises, you will evaluate three different IT projects that are experiencing problems with quality. For each scenario, you will offer a solution to ensure quality.

Exercise 11-1: Addressing Quality in Execution

Hanako is the IT project manager for an implementation of seven new servers. The scope of the project is to install and configure a new network operating system for her organization. Three of the servers are to be located in the company headquarters in Atlanta. Two are to be installed in the Tampa office, and the remaining two are to be installed in the Nashville office.

The servers in Atlanta and Nashville have been installed successfully, but Jerry, the network administrator, has yet to install the servers in Tampa. Jerry reports that he is leery of moving from his current, stable environment to the new operating system the servers provide. Hanako, the project manager, insists that Jerry install the servers as planned to complete the project. A week passes and Jerry still has not completed the installation. What steps can Hanako take to ensure the installation is completed?

Exercise 11-2: Managing Quality in Project Scope

You are the IT project manager for a software development project. Customers of your company, RWE Architects, should be able to use the Internet to check on the status of the building plans, review the work that draftsmen and architects have completed, and communicate with their account managers.

The plan to complete the project calls for a central database that can be queried by customers and the internal staff. The architects will also use a web portal to report the progress of each building plan they are working on so that customers will see the progress of the work.

The timeline for the project development and implementation is four months. At the end of the second month, you want to sample the work to date. You first evaluate the internal web portal that the architects will use to report the status of customers' work. You discover that this application has been designed to track hours worked on the project for billing purposes and not to report the overall progression of the building plans. This is not what the project scope called for. What steps should you take to correct the problem?

Exercise 11-3: Managing Quality in Performance

You are the IT project manager for the installation of four SQL servers. The servers are all clustered for fault tolerance and are installed on multiple-processor machines, with 64TB of RAM for optimal performance. A server will be located in each network within the company.

The scope of the project requires a persistent, fast connection between each of the sites, as the databases will synchronize on a regular, frequent schedule. Currently, the connection between Tempe and Phoenix is restricted to a broadband line, well below optimal speed for fast updates. What are two recommendations you can make to ensure quality for the users in the Tempe office?

Exercise Solutions

The following offer possible solutions for the chapter exercises.

Exercise 11-1: Addressing Quality in Execution

Your proposed solution may be something to this effect: As the project manager, Hanako should first reason with Jerry that this is the plan approved by management and the servers are to be replaced with the new server operating systems. Hanako should then let Jerry know that the old servers in Nashville and Atlanta have been replaced with the new servers. If there were any difficulties in installing the servers, Hanako should pass along that information to Jerry.

Another approach may be to interview Jerry as to why he's hesitant to move on to the new network servers. It may be that Jerry does not understand the technology and feels threatened by it. A solution could be to work with Jerry through the installation and then provide training for Jerry through a training provider. Tasks that need to be completed on Jerry's server can be done remotely until Jerry is comfortable with the solution.

Exercise 11-2: Managing Quality in Project Scope

Your proposed solution may be something to this effect: You should immediately speak with the application designer to discuss the issue. It may be that the designer is aware of the application status, and the hours worked on the project is a portion of the overall formula to communicate the customer's plans and billing information.

Of course, it may also be that the application developer does not have a clear understanding of the project's deliverables. A meeting with the designer is required to ensure that the project is on track or to move the project back in alignment with the expected deliverables.

Exercise 11-3: Managing Quality in Performance

Your proposed solution may be something to this effect: Depending on the available funds, a faster connection, such as a fiber line, should be implemented between Phoenix and Tempe. This would allow the database to replicate its information to all of the sites through the WAN connections. If funds are lacking, you may elect to edit the frequency of updates from Tempe to Phoenix. This second solution, while cost effective, may not be the best, however, because the working data in Tempe may be out of sync with data throughout the rest of the organization.

Questions

1. In order to achieve quality in the project, you must balance the Iron Triangle of Project Management. Which one of the following is *not* a component of the Iron Triangle of Project Management?
   - **A.** Scope
   - **B.** Schedule
   - **C.** Risks
   - **D.** Costs
2. You are the project manager for your organization and you're working to ensure that quality is maintained throughout the project execution. What must every project have to ensure the work in the project sticks to a standard of quality?
   - **A.** A commitment from management
   - **B.** A project manager experienced with the technology
   - **C.** Clearly defined requirements
   - **D.** A budget with plenty of cash reserve
3. Complete the sentence: Research allows the project manager to create the vision of the project. ________________ allows the project manager to transfer the vision.
   - **A.** Dedication
   - **B.** Inspiration
   - **C.** Commitment
   - **D.** Leadership
4. You are coaching a junior project manager on quality and grade. What is the difference between quality and grade?
   - **A.** Quality is the conformance to requirements, while grade is the ranking of the quality.
   - **B.** Quality is the conformance to requirements, while grade is a ranking assigned to a material or service.
   - **C.** There is essentially no difference between quality and grade when it comes to project management.
   - **D.** Quality is the end result of the project, while grade is the ranking of the quality as the project moves toward completion.
5. How are the value of a project deliverable measured?
   - **A.** A service is measured by the initial usage.
   - **B.** A product is measured by the initial usage, and then its worth declines with each usage.

C. A service is measured with each usage; goods are measured only on the first usage.

D. Services and goods are measured with each usage. The more often each is used for productivity, the more worthy the deliverable.

6. Management has asked that you include the details of quality assurance in your project management plan. What is quality assurance?

A. It is the measured value of the goods or service over a set period of time.

B. It is the measured value of the goods or service for the duration of its usage.

C. It is an organization-wide approach to preventing quality defects.

D. It is an organization-wide approach to ensure that the project managers are applying corrective actions on a regular basis.

7. Every predictive project must plan for quality as part of the project's planning processes. What is the purpose of quality planning?

A. It determines which quality standards are relevant to the project.

B. It is not needed on every project, because smaller projects are easier to manage.

C. It ensures that the project manager and the project team are completing the work.

D. It is needed only if the organization is using Six Sigma, TQM, or ISO-certified programs.

8. Of the following, which is a factor that measures the quality of the management process?

A. Project plan

B. Results

C. Project team

D. Budget management

9. All of the following do not describe QC except for which one?

A. Prevention

B. Assurance

C. Inspection

D. Quality standards

10. What is the purpose of a project information center?

A. To centrally organize the resources, planning, and research phases of a project

B. To centrally organize the resources, planning, research, and implementation phases of a project

C. To centrally organize the project team, resolve disputes, and provide additional resources for all projects within an organization

D. To centrally organize all projects within an organization

11. What value determines that when seven consecutive results of testing are on one side of the mean this is an assignable cause?
    A. Six Sigma
    B. Control Limits
    C. Rule of Seven
    D. Pareto's Law
12. What is the purpose of a Pareto chart?
    A. It tracks trends over time.
    B. It plots the results of sampling to determine the root cause of each problem.
    C. It ranks the quality of each component within a project.
    D. It ranks the quality problems within a project from highest frequency to lowest frequency.
13. Which software testing approach is the final type of testing before the product goes live?
    A. Smoke testing
    B. Unit testing
    C. Regression testing
    D. User acceptance testing
14. What is the definition of total quality management?
    A. It is a process that all employees within an organization work to fulfill customers' needs while also working to improve profitability.
    B. It is a process that all employees within an organization work to fulfill customers' needs while also working to improve productivity.
    C. It is a process that upper management leads by enforcing quality in all of their work. The quality implementation will trickle down through the organization.
    D. It is a process that a project manager must implement quality by leadership and a series of risk/reward principles.
15. What is continuous quality improvement?
    A. It is the theory that all practices within an organization are processes and that processes can be infinitely improved.
    B. It is the theory that all practices within an organization can be infinitely improved.
    C. It is the theory that all practices within an organization are projects and that all projects can be infinitely improved throughout the life span of the implementation.
    D. It is the theory that all organizations provide services and that these services can be continuously improved.

Answers

1. **C.** Risks are not part of the Iron Triangle of Project Management. While risks can have an adverse effect on the project, the Iron Triangle of Project Management is actually composed of time, cost, and scope.
2. **C.** Clearly defined requirements are necessary for the project manager to check the status of the project's work. Should the project be moving away from the project's objective, the project manager must take corrective actions to nudge the project back on track.
3. **D.** Leadership allows the project manager to transfer the vision of the project from a personal concept to a goal for the project team.
4. **B.** Quality is the capability of a project to conform to the requirements as expected by the customer. Quality is planned into a project, not inspected into it. Grade is the ranking of a material or service, such as paper, metal, or first-class versus coach.
5. **D.** Goods and services are measured with each usage. The more often the goods and services are used productively, the greater their value. An expensive project that creates a deliverable that is rarely used has a small ROI.
6. **C.** QA is an organization-wide approach to ensuring the prevention of defects within the project. QA is prevention driven, while QC is inspection driven.
7. **A.** Quality planning is the process of determining which quality standards are relevant to a project and then determining how the project work will achieve these quality requirements.
8. **B.** The results of a project are always the ultimate barometer of quality for a project. The planning, the project team, and budget management, among other attributes, must all work in harmony to create quality results.
9. **C.** Inspection is a key activity within QC.
10. **B.** A project information center is useful for organizing all facets of a project. It is typically a room dedicated to the project where resources, testing, and documentation can take place.
11. **C.** The Rule of Seven is a guideline that states when seven consecutive results of testing fall on one side of the mean in a control chart, there is some purpose for the event. This is an assignable cause.
12. **D.** A Pareto chart is a type of histogram, or bar chart, that categorizes quality problems from highest frequency to lowest frequency. The chart helps the project team determine which problems to attack first through corrective actions.
13. **D.** User acceptance testing is the final round of testing to confirm quality and acceptability before the product is released into production.

14. **B.** TQM is the belief that all employees within an organization are consistently working to fulfill customers' needs and at the same time working toward improving productivity by refining processes.
15. **A.** Continuous quality improvement is the theory that all practices within an organization are processes. These processes then can be infinitely improved to streamline the business and better the quality of the organization regardless of its deliverable.

CHAPTER 12

Completing the Project

This chapter covers the following topics:

- Completing the project
- Conducting the project postmortem
- Assessing the project deliverables
- Documenting the project
- Evaluating the team's performance
- Obtaining final sign-off
- Declaring victory or failure

Congratulations! You've made it to the end of your project. The critical path is nearly completed, your team is happy but exhausted, and management is all smiles. You've been sampling the project as you move along through production for quality and scope verification, and you can reminisce about events that tried to shove your project off schedule and how you and the team recovered and pushed it back on track.

Already there's a buzz of excitement among the end users of the technology. As you walk through the halls of the company, you feel a couple of inches taller, and it's hard not to smile like a big, goofy kid.

But hold off on putting that cocktail umbrella in your favorite frosty drink. There's still plenty of work that must be completed to get the project out the door and to finalize the entire process. In this final chapter, you'll learn not only how to finish the project, but also how to formalize the project closure. It's okay to smile and feel a sense of pride at this point of the project—just don't let it get in the way of finalizing a job well done.

Whether you are working on a short-term project that lasted just a few months or a large, complex implementation that has taken over a year, the principles behind completing the project are the same: to complete any project, you and your team have to continue the momentum and supply the final surge to get the project to the finish line.

VIDEO For a more detailed explanation, watch the *Lessons Learned* video now.

Completing the Final Tasks

When you begin to see the final tasks coming into focus, it is not a signal to ease off of the project team and the project. Some project managers make the costly mistake of allowing the project to finish under the guidance of a team leader or having too much faith in the project team to complete the tasks. These project managers allow themselves to relax, begin looking for new projects to lead, or begin their efforts to prove that Microsoft's FreeCell game 11982 can be won. Agile project managers must continue to follow the rules, ceremonies, and servant leadership to ensure the development team continues to work as planned and finish the remaining items in the project.

The problem with relaxing as the project is nearly completed is that the project team will follow your lead and relax as well. Project managers have ownership of the project that sometimes leads them to believe they are superior to the project team and have permission to put their feet up. As the project team sees the project manager ease out of meetings, out of sight, and out of focus, they'll follow suit and do the same, as Figure 12-1 demonstrates.

In the final stages of a project, a project manager must actually do more to motivate and communicate with the project team. The project manager must attend every meeting as has been done throughout the project. They need to get into the trenches and work with the team members to help them complete the work and keep them moving to complete the project on time. A project manager needs to discuss any final issues with the team, with the client, and with management. A project manager's presence is obviously required throughout the project, but even more than usual during the final chunk of the implementation.

In an adaptive project, the project manager meets with the product owner and the development team as planned. The agile project manager continues to ensure that everyone is following the project guidelines, work is being completed according to requirements, and the final features of the project are created, tested, corrected as needed, and demonstrated in the sprint review. Nothing changes just because you're almost done.

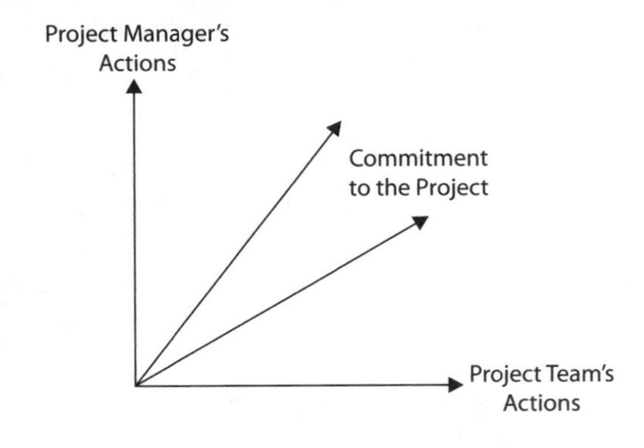

Figure 12-1 The project team will follow the project manager's actions.

Going the Distance

What is, unfortunately, more typical of project managers than easing off of an on-track project is working in a frenzy to complete a project that's gone awry. For example, consider a predictive project that has had six months to complete the implementation of a new e-mail client, develop workflow forms, and convert the existing e-mail servers. In this scenario, the final tasks are the most critical in the entire project. All of the prep work, research, and design has led the project team to this moment. The switch from old to new is when the curtain comes up, and everyone in the organization will see your work, your design, and your implementation. As Figure 12-2 demonstrates, these tasks, close to the finish, reveal your ability as a project manager.

That fact becomes quite evident when the project team and the project manager realize they are not prepared to complete the project on schedule. Now the project manager looks for ways to speed up the process to complete the job on time. This usually means additional hours, nights, and weekends. Be prepared to work the hardest in the final days of a project's implementation if the project is off schedule, even by just a few days.

The secret is to control your emotions, the project team, and any other parties who have volunteered to help with the final tasks. If the folks looking to you to complete the work see you losing control, getting angry, and cutting corners, they'll do the same. Cool heads always prevail.

EXAM TIP Emotional intelligence is the ability to recognize and control your emotions and to recognize the emotional behavior you see in others. Self-control and self-awareness are essential for healthy emotional intelligence as a project manager.

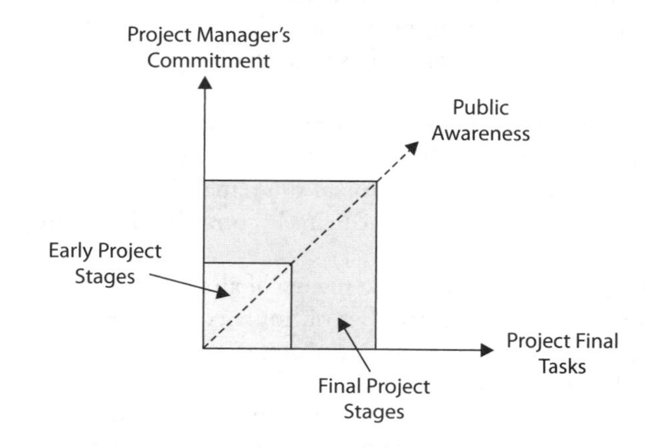

Figure 12-2 The final tasks in a project require the project manager's full attention.

When you find yourself with a huge amount of work to complete in just a few hours, here are a few guidelines for being successful:

- *Remain cool, calm, and collected.* Set the example for your team; think clearly, but quickly.
- *Get organized and treat the final work as a mini-project.* Analyze the work to be completed, break down the tasks, and implement the plan. Create a map of the final implementation in a central war room and color-code the completed tasks.
- *Communicate with the team members, but don't get in the way of their completing tasks.*
- *If you're visiting multiple workstations, organize a method to represent the completed work visually.* For example, if a workstation has been prepped for a new installation, put a red sticky note on the monitor. Once the workstation has been completed, put a green sticky note on the monitor. At a glance, anyone can see the status.
- *Check for quality.* Periodically take a sampling of the work to confirm that what you are attempting to deliver on time is the quality you and the end user will expect.
- *Work in shifts.* If your team must work around the clock, which is not unheard of, break up the team in shifts so that the team can get some rest and be refreshed. It's tempting to have the entire team working on the final phases of an implementation, but as the entire team wears down, the quality of the work suffers.

Examining the Critical Path

As the project begins to wind down, take a close look at the critical path to determine that the tasks to completion are in order, and confirm that the team members who are assigned to the tasks are still motivated. In your team meetings, review the upcoming final tasks to reinforce the importance of their completion. Team members will likely be as excited to complete the project as you, so offer a little urging to continue the momentum to finish.

Within the network diagram, trace the history of the successor paths and determine if there is a history of tasks being late or lagging. If there is, address this issue to the project team and challenge them to complete the remaining tasks on schedule. You must do all you can to ensure that the project team is committed, moving, and excited to complete the project.

A source of motivation can be a review of all the work the project team has completed. You can show the team the number of hours committed to the project and how healthy the project is. Also, remind the team members of the rewards awaiting them once the project has been completed.

At the end of the critical path should be the management reserve. Recall that management reserve is a percentage of the total time allotted for all work within a project, usually 10 to 15 percent. Examine the management reserve to see how much time is still left for slippage. For example, a six-month project would likely have 14 to 20 days in management reserve for tasks that are lagging. In the final stages of the project, examine

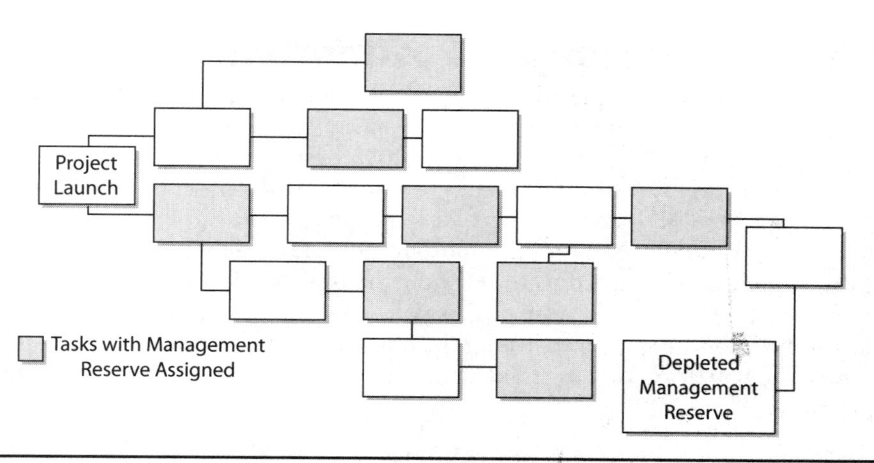

Figure 12-3 A depleted management reserve can impact final tasks.

the balance of time in the management reserve; if things have gone well, you should have a few days left in reserve. If there have been serious mistakes in the project, then all of the available days in the reserve have been used. Figure 12-3 demonstrates the application of management reserve to early tasks in the critical path, which can impact any delays in the final task of a project.

The point is, an examination of the management reserve will give you an idea of the overall health of the project and will predict how much time you truly have to complete the work. Hopefully, you've got a sliver of time left over to complete the project and allotted for any unforeseen troubles. In some instances, management reserve doesn't matter—for example, when a network switch must take place over one weekend. A management reserve with two extra days won't help at all if the network has to be switched when the production is not in place.

98 Percent Done Is Not Complete

When dealing with third parties that will complete the implementation phase of your project, you really need to convey to them that 98 percent done is not complete. Far too often, integrators begin a project with gusto but lag in the final implementation. Vendors sometimes need to be reminded that a project is not complete until it is 100 percent done. You can motivate these folks with these techniques:

- Hold the final payment until the project is completed.
- Specify the project deliverables as 105 percent to get them to aim for 100 percent completion.
- Commit them to specific times and dates to work on the project in the final phases of the implementation.
- Assign a task in the critical path that is a walk-through and sampling of the work they have completed. The critical path is not complete until you approve their work.

Conducting the Project Postmortem

The final task has been completed, and there's a collective sigh of relief from all of the parties. But, sorry, you've still got a touch of work left to go on the project. Once all of the implementation tasks in the critical path have been completed, a project manager and the project team must do a few chores to inspect their own work. This time should be worked into the PND and shouldn't take very long at all, maybe 1 to 3 percent of the total project time. This analysis is important not just because you need to review the project work, but also you need to review how the project deliverables affect your organization's local and global environment. As you move the project into operations, there should also be a conversation confirming the understanding of the effect the project will have on the organization, the customers, and how the project will conform with laws or regulations.

Reviewing for Quality

The primary task that you personally must be involved with is to inspect the quality of the project. Of course, throughout the implementation, you will be sampling the project and confirming the quality, but at completion, you need to experience or test the project deliverables and confirm that they are the required deliverables to complete the project. You want to perform a final inspection before the project customer sees the deliverables. The point of this inspection is to correct any mistakes that may be present before the customer sees the deliverables during the scope verification process.

EXAM TIP Adaptive projects perform quality control in each iteration by providing a demonstration for the project customers.

This final quality control activity should review the deliverables and benefits from the viewpoint of the project customer. To do this, re-create the experience that a typical user would have when using the deliverables. If your project produced an application, use it. If the project was to implement a network, log in to a workstation and test connections, print to a few printers, and access some network resources. Evaluate the product from the end user's point of view and review the results to determine if the project deliverables are acceptable, as shown in Figure 12-4.

If you encounter problems, address them immediately so that the responsible parties can react to them and find a solution. At this point of the project, if you've done your job, there shouldn't be any major surprises. You may encounter a quirk that can be quickly addressed and solved, but overall, things should be smooth and the customer should be happy.

Assessing the Project Deliverables

Once you've completed the final inspection and the quality of the work is acceptable and in alignment with the expected project deliverables, you can enjoy the sense of satisfaction that comes from the success of completing a project. There is a wonderful feeling that comes with taking a project from start to finish. The project is now part of the organization's life, and you helped get it there.

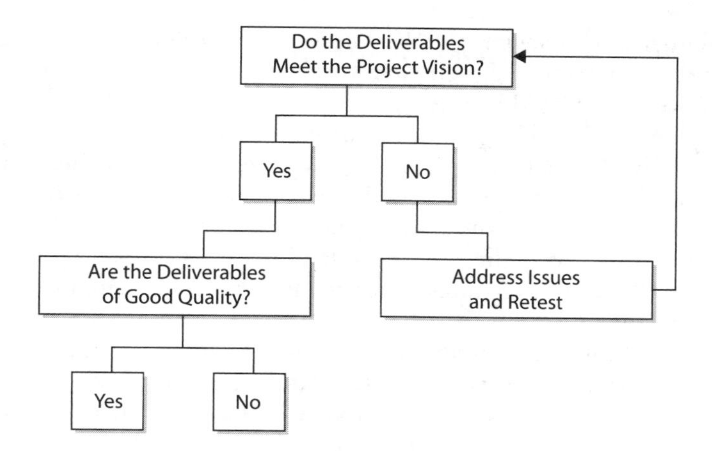

Figure 12-4 The process must be in place to test the quality of the project deliverables.

Examining the Project's Worth

Now that the project is complete, you may want to calculate the worth of the project. This activity involves a bit of math magic, but it allows you to predict the overall usefulness and profitability of your implementation. You calculate the time saved, the new sales earned, and the productivity gained to create a gross value of the project. The expense of the project—the total cost of the implementation—is subtracted from the gross value of the project to learn the project's net value. From here you can create formulas to predict the value of the project over the next six months, the next year, or beyond.

Don't get too excited by the math, however, because eventually, the infrastructure processes of the organization will absorb your deliverables as a matter of doing business. What happens is that your project's deliverables, the wonderful things that they are, will fall victim to the law of diminishing returns. In other words, the 20 minutes you take out of a process will be consumed by some other activity.

The *law of diminishing returns,* sometimes called the *law of variable proportions,* is a rule of economics that grew from Thomas Malthus's *An Essay on the Principle of Population,* published in 1798. The law states that if one factor of production is increased while other factors remain constant, the overall returns will eventually decrease after a certain point, as demonstrated in Figure 12-5. When that point is reached is difficult to say without serious analysis given to the process of a company.

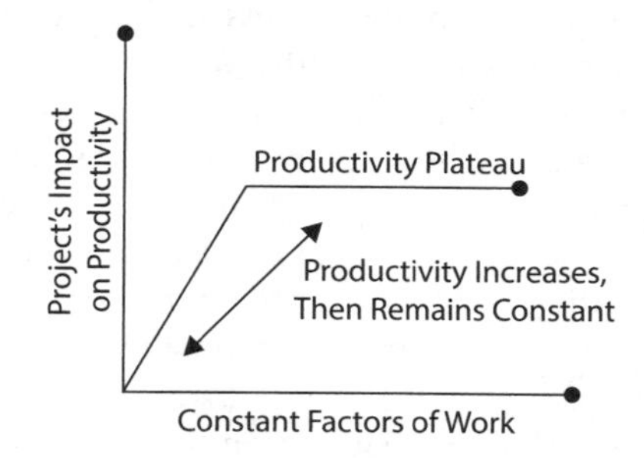

Figure 12-5 The law of diminishing returns prevents exponential productivity.

Another way of viewing the law of diminishing returns is to imagine a cornfield that needs to be harvested. If you were to continue to add workers to the field, each new worker you add would have less to do than the worker added previously because there is less and less corn to harvest as additional labor is added.

To apply this law to a technical implementation, imagine a new application that allows workers to be more productive when entering human resource forms and typical paperwork. Before the application, the workers had to manually enter the forms into Microsoft Excel, save the file, and e-mail it to the human resources department. The wonderful folks in human resources would open the e-mail, open the file, and merge it into some master file.

Your application streamlines the process through a web page and pumps the information into a database. Now when users within the company need to complete insurance forms, request days off work, request new ID badges, or deal with any other HR-related issue, they can complete the process through your company's intranet.

The productivity of this application allows the processes to be faster, better, and easier to complete. However, this level of productivity will not affect other areas of the workers' roles in the organization exponentially. It will allow for additional time to complete other work, but the additional time gained does not continue to grow on, and on, and on. Eventually, productivity reaches a plateau, and the law of diminishing returns reigns.

Third-Party Review

You can also have a third party analyze the before-and-after processes of an organization. For example, imagine an implementation of a special mobile phone app in a manufacturing environment. The goal of this project is to shorten the process a forklift operator must use to deliver a pallet of product to a delivery truck.

In this example, workers pick up pallets of the goods the company manufactures and then move the pallets to the trucks that will deliver the product to the stores and merchandisers. The problem this project resolves is that workers, after dropping off their pallets in the delivery trucks, have to drive the forklift back to a central base to get their next assignment of the product to be loaded on the trucks.

The project created an app that sends a message to the forklift operator on the floor with instructions on the next product to pick up and deliver to a specific truck. The process has been improved; the worker does not have to return to the central base. Additionally, the pallets, which are wrapped in plastic, have a quick response (QR) code that the worker can scan from the forklift to log the goods that are actually placed onto the delivery truck. All steps within the process are logged and can be analyzed from the start of the process to the end of a work shift.

A third party could evaluate the productivity before and after the implementation. The process analysis would allow the consultants to track the amount of product moved per work shift to predict the average amount of productivity before and after the implementation. That information can then be analyzed and tweaked, and the original project deliverables can be adjusted in a new project to streamline the process again.

To complete the project, the information gathered by the completed process would be analyzed and reviewed internally or externally. The review would allow the company to see a true ROI and productivity gained on the implementation.

EXAM TIP Third-party reviews may be required by regulations in your industry. For example, banking and healthcare projects often are subject to review by regulators to ensure the accuracy and safety of the project work.

Far too often, organizations don't invest in the time to validate the promised benefits of the project. The verification of the benefits is needed not only to show the ROI for the current project, but to afford success in future projects.

Obtaining Final Sign-Off

Once you're satisfied with the project deliverables, you will need to move on to the transfer of ownership of the project. You, the owner of the project, will release it to the organization so that the deliverables may go into production.

Using a Project Transition Plan

Some organizations rely on a project transition plan to help ease the transfer of the project deliverables from the project manager to the operations of the organization. This plan defines several things for both the project manager and the organization:

- **Transition dates** There must be either a defined date for the deliverables to be transferred to the organization or a description of the conditions for the deliverables to be moved from the management of the project to the management of operations.
- **Ownership** Projects are temporary endeavors that must eventually come to a close. When the project is completed, and a set of deliverables has been created, someone must now be responsible for the management of the deliverables. The transition plan defines who owns the project deliverables and who is responsible for the maintenance and upkeep on the deliverables.
- **Training** For IT projects, it's typical for there to be some training between the project team and the support team of the deliverable. The project team should train the recipients of the deliverable how to use and maintain the product. This can be done through training manuals, train-the-trainer sessions, hands-on exercises, or a combination of knowledge transfer events. Training can also include training the end users through traditional instructor-led training, web-based training, or one-on-one coaching.
- **Service-level agreement** A service-level agreement (SLA) is a contract for support and service of the solution a vendor provides to the customer. Depending on the type of project and deliverable, the vendor may offer an SLA. The extended support usually requires that the project team and the operations team work together as the new technology is implemented so that the operations team can learn from the project team about the project deliverables and implementation. The terms of the SLA can be negotiated as part of the project contract or through a separate contract and negotiation activity.

- **Warranties** If the project was completed by a vendor for a client, there may be some warranty information about the project deliverables. The warranty should be detailed in the project contract and discussed during the operation transfer. It's ideal to discuss what the warranty will and will not provide early in the project so that there are no surprises when the client needs to enact the warranty.

Your organization may have other details of the transition plan, such as what's to happen to the project team members or other resources on the project—such as equipment and software. The project transition plan can be updated based on changes within the project, but unlike other project management plans, this plan requires both the project manager and the recipients of the deliverables to be in agreement with what the plan promises.

Obtaining Client Approval

The client, whether that be the end user on a workstation or an administrator in a behind-the-scenes release such as a new server installation, will accept the project using one of two methods: informal or formal. You need validation of the deliverables, ideally through a formal process, but there needs to be a signal of acceptance of what's been created for the project customers to allow the project to formally close.

Informal Acceptance

The informal acceptance does not include a sign-off of the completion or even the acknowledgment of the deliverables. An example of the informal acceptance is a project that ends on deadline whether or not the deliverables are finished—for example, a project to organize and build an application for a tradeshow demo. The tradeshow will happen regardless of the completion of the project.

Another example is a project that creates a deliverable that doesn't require additional testing to prove the implementation. Imagine a short-term project to replace all of the printers in the organization with newer models. The implementation of the new printers is obvious. You've configured a script to install the printer driver on all of the workstations in the network to automate the end-user installation. All of the print jobs are controlled through a central printer server, so the end users experience little impact from the implementation other than their print jobs come out of a new print device.

Formal Acceptance

The formal acceptance of a project's deliverables is a process completed by the client of the project and the appropriate members of the project team. This is the preferred method of client acceptance. These acceptances are contingent on a project acceptance agreement. The project acceptance agreement is typically written very early in the project timeline and in alignment with the defined project deliverables. The document clearly explains what qualifies for an acceptance of the deliverables. Project acceptance agreements are typical of application development projects and often consist of a checklist of the required features of the project.

The client and the team members will test the deliverables against the acceptance agreement to confirm that the deliverables exist. The sidebar "Sample Application Development Acceptance Agreement" shows an example.

Sample Application Development Acceptance Agreement

The purpose of the project "Learning Management System" is to create a web-based reporting system to log the hours an employee completes in the corporate continuing education department. The application must provide for the items in the following checklist to be accepted:

- A web-based dashboard triggers a certificate of completion when an employee finishes a class and records the class as completed in the employee's record.
- A web-based dashboard allows human resources, trainers, and management to view the classes completed by each employee. Managers should be allowed to see the classes completed only by the employees within their supervision.
- A web-based dashboard allows employees to view completed classes.
- A web-based dashboard allows employees to view the available classes and to register for upcoming classes.
- Registration requests must be automatically forwarded to the registering employee's managers for approval and to the continuing education registrant.
- Upon approval of the class by the employee's manager, the class enrollment should increase by one to reflect the new class participant.
- Should the class become full during the approval process, the student and the manager should receive an offering of the next available class, for which the student will have a priority over new attendees.
- The registration feature should also allow management and participants to cancel their participation or trade enrollment with another employee.
- When a student enrolls for a class, the corporate calendar should indicate that the student is busy during class hours. Should the student cancel their attendance, their calendar should be updated to indicate the time is available.
- The application should send an e-mail to class participants reminding them of their enrollment and the class hours one day before the class is to start.

The application created in this project meets the preceding criteria and the project is accepted as whole and complete.

________________ ________________

Project Manager Project Client

____

Date

Shutting It Down

When the project work is done and the customer has accepted the project deliverables, it's time to shut it all down. This means you'll need to decommission any project hardware, software, and other tools the project utilized. You'll need to confirm that security permissions and access to resources have been revoked as needed, and you'll release the team from the project according to the project management plan or the resource management plan.

Shutting it all down, especially in a predictive project, likely includes getting feedback from stakeholders, archiving project documentation for future usage, and generating a final project report. Agile projects may have some documentation to archive, but recall that agile projects don't document as extensively as predictive projects do. Agile projects need to release the project team and secure resources.

EXAM TIP Always archive project information and documentation, even if you don't have to. Archived project information can be useful for supporting the product you've created and for future, similar projects. Archived project documentation is also known as historical information.

Post-Project Audit

At the conclusion of the project and before the final project report is submitted, a project manager should complete an audit of the success of the project. The purpose of the audit is to analyze the completed project, the effectiveness of the project team, the success of the project, the value of the deliverables, and the overall approval from the clients. The audit can become part of your lessons learned documentation. This audit can be financial and provide budget reconciliation, and it may lead to creation of a final project report on the project's overall success.

This audit must answer the following questions:

- *Was the project vision achieved?* Remember when you first created the vision of the project? That vision may have changed as the project evolved. The first question should answer if the project accomplished what its original intent was. If the project did not, explain why. Projects have a tendency to change and develop from the concept to the creation—sometimes for the better.
- *Was the project on track from start to finish?* Hopefully, the project was able to stay on plan, on time, and within the allotted budget. If the project wasn't able to stay within the bounds of any of these areas, explain why. Sometimes the scope changed, the resources flexed, or the expenses of the project were not predicted as accurately as they should have been. This should be an honest reflection of each side of the project triangle (scope, time, and budget).
- *Did the project create a recognizable business value?* The deliverable of the project should be to make an organization more profitable in its streamlined process, attract more sales, or gain productivity. This business value needs to be identified and proven to show the ROI of the project.

- *Can you share the knowledge?* Some organizations have a project management system in place that requires project managers to report on their methodology and how it worked for them, or what they may have done during the project to improve the process. These adjustments that you make during the implementation need to be shared so that other project managers within the organization may benefit from your insights.

The post-project audit is an activity that far too many project managers skip. Don't ever skip it. It is an extremely valuable process that will help you become a better project manager. In addition, it is an excellent method for reporting on the work you've completed and the value you've added to an organization. You may choose to use this report as leverage in negotiations for future projects.

Creating the Project Closeout Report

As with every other phase of the predictive project, documentation is required. The good news is that the final documentation of the project does not have to be an in-depth novel of all of the work completed. If you have completed cumulative progress reports throughout the project, consider the final report one last cumulative record with a few extra ingredients. The collection of all of the cumulative reports may serve as a final record of each phase's work with a few extra parts. You will need the following:

- The project vision statement that introduced the project
- The project proposal that you may have used to sell management on the idea of the technical implementation—or the supporting information for the project that was assigned to you
- The scope statement
- The statement of work
- The project schedule
- The WBS and the PND
- Any Project Change Request forms that were approved (some project managers may choose to include the denied Project Change Request forms to verify why the request was not included in the deliverables)
- Variance reports
- All communication relevant to the project deliverables (some project managers include all memos, letters, and e-mail in the report)
- Total cost of the project and the calculated value of the implementation
- Scope verification agreement
- Post-project audit report

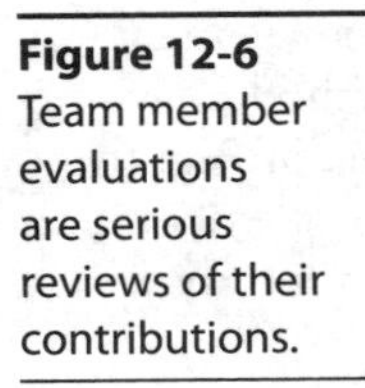

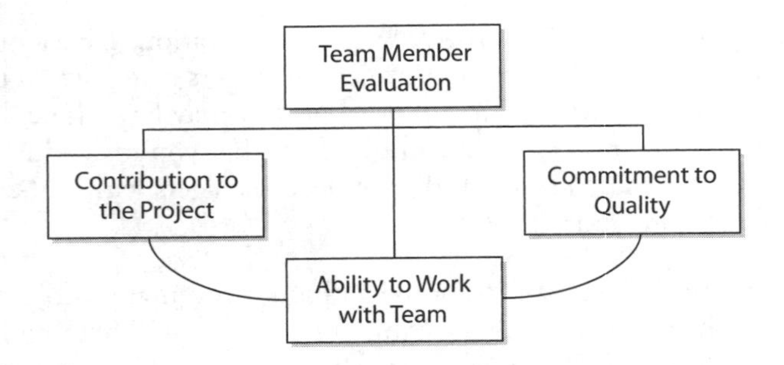

Figure 12-6 Team member evaluations are serious reviews of their contributions.

Evaluating Team Members' Performance

In many organizations, a project manager is called upon to review the work of the team members who were involved with the project. This evaluation is a serious process that may impact their salaries, their jobs, or their opportunities to advance within the organization. Your organization may have you complete a form on your own, conduct an interview with each team member and the immediate manager, or hold a private meeting with the team member's immediate manager to discuss the work.

Whatever method is invoked, use caution to be accurate, fair, and professional. This is another reason why the project manager requires the amount of documentation they do throughout the project. The evaluation process, formal or informal, accomplishes three goals, as shown in Figure 12-6. The evaluation includes the team member's contribution to the project, their ability to work with the team, and their commitment to quality.

Declaring Victory

At the end of the project, you'll have a fairly good idea of whether the project is a victory or a failure. A project is a victory if you have completed the project on time, on scope, and within budget and the project has resulted in the quality deliverable that was expected. A project manager and the project team should take measures throughout the project to ensure that the vision of the project is being worked toward and achieved. Think of any skyscraper. It didn't start out as a two-story building and then get switched midway through construction to a high-rise instead. The same is true in your project management skills: by knowing and recognizing the vision of the project, you constantly check that the work being completed is moving closer to the desired results.

Declaring Failure

No one likes to fail. Unfortunately, some projects can make the project manager, the project team, and all associated parties feel like failures. How many projects are started, stopped, and rearranged, only to complete the cycle again and again without ever getting anywhere near the projected deliverables?

For a project to be a success, it must include these things:

- A vision of the project's deliverables
- Adequate skills by the project manager

- Adequate skills by the project team
- Enough finances to provide for the resources to complete the implementation
- Time to complete the work to produce the deliverables
- Change management to protect the project scope
- Commitment from the project manager, project sponsor, the project team, and management

Without these elements, a project will have a very tough, if not impossible, time succeeding. At some point in a doomed project, the team, the project manager, and management may become so disgusted by the lack of progress that the project needs to be written off and put out of its misery.

In these instances, the project ceases to exist. The attempted implementation is a failure, and it's a general unhappiness for everyone. If you find yourself in this situation, and hopefully you will not, look for an understanding of why the project failed. Evaluate the project requirements, the finances, the talents of the team, and the available time. Evaluate your own performance and the performance of your team. Learn from the mistake and become a better person because of it. As Thomas Edison said, "I have not failed. I've just found 10,000 ways that won't work."

In other instances, the project may be cancelled not because of a lack of leadership, finances, or time, but because of a new influence on the project. Examples include the following:

- A better, cheaper technology is released.
- A better solution is discovered within the current project.
- The need for the deliverables has been eliminated by the client.
- The organization has changed its focus.
- The organization is experiencing financial strains.
- The organization has been absorbed by another organization.

Regardless of the project status as a success or failure, the project manager must create a final project report so that management may review the work of the project. Scope verification and an audit of the project up to the point of the project's cancellation are needed.

Cheers! Celebrating Victory

At the end of the project, give congratulations and kudos to the project team, the project sponsor, and anyone else who helped make the project a success. It is necessary to celebrate the victory of a project and reward the team members for their commitment and hard work. Hopefully, especially if you want future projects to be a success, your organization will spring for something elaborate and in proportion to the project you've completed and the success of the victory.

The team members that you've coached along have formed friendships and trust among themselves and hopefully with you. This group of individuals has worked hard day and night for you to make you look good and make the project a success. Reward them! Offer tickets to an event, take the team to a fancy dinner, go out dancing, or offer individual rewards that the team members can enjoy on their own such as gift certificates or cold, hard cash. The point is, reward them, appreciate them, and they'll come back to work with you again and again.

CompTIA Project+ Exam Highlight: Closing the Project

You can expect plenty of questions about closing projects on the CompTIA Project+ exam. Project closure, while often short in duration, is one of the most important process groups on your project. Also, remember that while the focus of these closing processes is often on the final stage of a predictive project's life cycle, you can also close each phase of a project. Project closure is about the customer accepting the project deliverables and benefits, finalizing the closure of the project, and then archiving the project's lessons learned, documentation, and supporting details as part of the organizational process assets.

2.5 Explain the importance of activities performed during the closing phase The CompTIA Project+ exam outline has one specific objective for closing out the project. During project closure, you'll evaluate the project, confirm validation of deliverables, and generate a final project report. Closing out the project also includes getting feedback from stakeholders, budget reconciliation, and archiving the project documentation for use in future projects.

The *transition plan,* sometimes called the integration plan, defines how the benefits and deliverables of the project will be transitioned from the project team and integrated into the operations of the performing organization. The plan defines who owns the project deliverables and who will be responsible for the maintenance and support of the deliverables. The plan also defines when the transition from the project to operations will happen and how the recipients of the deliverables will be trained. The transition plan should also define any terms for extended support and warranties of the project.

Agile projects also follow some rules when it comes to closure. Agile projects finalize the product for operational transfer and product release with the product owner, and they complete any final tasks to shut down the project work. These final tasks can be decommissioning hardware and software, providing insight to what items may not have been completed in the product backlog for future projects, and perhaps providing some ongoing support to the product that's been created for the customer.

Formal project closure is important, because it defines what the project has completed, and the customer signs off on the acceptance of the project deliverables. It confirms that the project deliverables, objectives, and requirements were met—and in some cases, such as project cancellation, that the project deliverables, objectives, and requirements may not have been achieved. Formal project closure also allows the project manager to formally close project contracts, release the project resources, and complete a post-project review of what went well or poorly in the project.

The closing processes don't happen just at the end of the project. You can use the closing processes at the end of project phases or stages, at milestone completion, and should the project get canceled. Most often, however, the closing processes are completed at the end of each project phase in the project life cycle and at the end of the project. On smaller projects, the project manager may elect to do the closing processes only at the end of the project. On larger projects, it's often better to close each project phase formally to ensure with the project customer that all of the requirements of the phase were met before moving forward. Should a project get canceled, the project manager should formally close out the project and document why the project was cancelled, what was created up to the point of the cancellation, and how the project performed.

4.1 Summarize basic environmental, social, and governance (ESG) factors related to project management activities Throughout project planning and execution, you'll need to address the environmental, social, and governance constraints of the project. In project closing, you'll need to confirm that the project adhered to these requirements. Failure to meet the ESG requirements can result in fines, penalties, or loss of business value and can adversely affect the environment where the project took place. As part of project closing, it is essential to confirm that the deliverables the project created conform to the ESG requirements of the project before the deliverables go into production.

Chapter Review

A successful project manager has to complete many tasks to finish any technical project. As the project is winding down, you must conduct an analysis of the critical path and the resources assigned to the path to ensure that the work is completed and resources are committed. You have to be seen, be available, and be in the trenches with the project team to keep the project momentum moving toward the deliverables.

Once the project work has been completed, the project manager, the project sponsor, and the project team need to review the work for quality, snags, or technical issues that may have popped up in the final tasks. Any issues need to be immediately addressed and resolved. You can evaluate the project's worth to predict the ROI of the product; of course, you'll need to consider the law of diminishing returns as the calculations are made.

Once the work has met the satisfaction of the project manager and client, the client accepts the project either formally or informally, depending on the type of work completed. Some projects may require you and members of your team to work with the client to ensure the client acceptance agreement criteria are met as defined in the early stages of the project.

Once the client has approved the work, you must create a final project report that includes all documentation and facts from the project lifespan. An audit of the work and the success of the project need to be included in the plan. Ultimately, the final project report must answer whether the project vision was met and if the overall health of the project is excellent or otherwise.

At the project's finale, you may determine from the evidence of the goods produced if the project was a success or a failure. Regardless of the project evaluation, you should create a lessons learned document for your own future success, but also for the success of others.

You should review the project together with the project sponsor and focus on the quality of the work, a review of the budget at completion, and the success of the effort. The project sponsor may want to work with you on how you can manage your next project better, or what you did successfully on this implementation.

Finally, once the project is officially finished, the team is relieved, the client is satisfied, and recognition is given. A celebration is in order for the project team, the support of management, and the hard work and commitment poured into the project vision by you, the successful project manager.

Exercises

These exercises allow you to apply the knowledge you have learned in this chapter and are followed by possible solutions. In these exercises, you will be presented with a scenario of a project for a fictitious company. You will then be asked a series of questions about the project.

Exercise 12-1: Reviewing the Project for Completeness

Marcy is the project manager for All Boots Company. Her project is to lead a team to develop a web-based application that allows customers to visit the corporate website and search for boots based on several factors such as size, style, color, heel, and toe of the boot. Throughout the project, Marcy has periodically tested the cumulative work to see how the project is progressing. With just a few days to spare, the application developers report that the project is complete and the website is ready for users.

Of course, Marcy wants to experience the application before management and others see the results. Marcy opens the website and successfully completes a few searches. She then discovers that she cannot search for a combination of factors such as a red boot in a size 10.

Please answer the following questions to complete the exercise:

1. What should Marcy do first to rectify the problem?
2. What could Marcy have done prior in the project to avoid this problem?
3. What should Marcy do if the developers report that the work can be done but it may take two weeks to complete her request?
4. What can Marcy tell management about the project?

Exercise 12-2: Managing Unexpected Changes in a Project

Kerem is the project manager for Farmstead Jams and Jellies. The project he has been managing is a rollout of new laptops and operating systems to 720 employees at seven sites throughout North America. His plan calls for the laptops to be shipped to his central office in Tampa, Florida, where his team would configure the PCs and then his team will ship them to each of the employees throughout the country.

The project itself is designed for tiered delivery of the workstations and addresses just one of the seven North American sites at a time. Unfortunately, the manufacturer of the laptops has been very late in shipping them to Kerem. In fact, the rollout has been successfully deployed at only one site in the company, and the project is to end in one week.

Suddenly, a huge delivery from the manufacturer arrives at Kerem's office and he now has 480 laptops; his team needs to verify the hardware, push the approved disk image to the laptops, pack the laptops, and ship them to each of the remaining cities.

1. What should Kerem do first?
2. What should Kerem tell management?
3. Is there anything that Kerem could have done earlier in the project to avoid this situation?
4. What are the steps you would take to complete this project?

Exercise Solutions

The following offer possible solutions for the chapter exercises.

Exercise 12-1: Reviewing the Project for Completeness

Your answers may be slightly different than the ones presented here, but generally should be similar.

1. Marcy should verify that the flaw exists within the scope of the deliverables of the project, and then she may contact the developers immediately to address the problem.
2. There are many things that Marcy should have done earlier in the project. She should have made certain that the multiple-factor search capability was part of the project requirements. She should have reviewed the work throughout the project to confirm the existence of the search capability. She should have asked her developers for their input and advice on what could make the application better.
3. In this situation, as the work is due in a very short amount of time, Marcy should speak with management about the flaw and assure them it is being addressed. As the website is currently functional to an extent, it should be released in a tiered manner. The current website should go live and then be replaced with the updated application as soon as it is finished and approved.

Marcy may also elect to assign additional developers to the work to shorten the time frame to include the multiple-factor search capability. It may be that the developers are able to complete the request quickly, but they are currently bogged down with other assignments.

4. Marcy can report to management that overall the project is very healthy. She should also emphasize that with a few tweaks the application will be excellent.

Exercise 12-2: Managing Unexpected Changes in a Project

Your answers may be slightly different than the ones presented here, but generally should be similar.

1. Obviously, the shock of 480 laptops being delivered at once was not in the project plan, so Kerem should first calm himself and his staff down. The volume of computers will make completing the project on time impossible.
2. Kerem should report to management the good news and the bad news about all of the laptops having arrived. The good news is the hardware is finally in place; the bad news is they are late and there are 480 workstations to configure and drop-ship.
3. Kerem should have created a closer relationship with the manufacturer to ensure the laptops were delivered earlier in the project or not fully paid for the laptops until they were delivered. Kerem could also have considered ordering laptops from other vendors.
4. First, consider if it is possible to complete the project on time, given all of the hardware is in place. If not, then a plan to place an image on the hard drives and ship the laptops to the cities must be drafted. Hardware security needs to be addressed so that none of the laptops is compromised.

Questions

1. Of the following, which is the greatest danger to a project as it nears completion?
   A. The project scope changes.
   B. The project manager eases off of the project.
   C. The budget has a surplus and the project manager doesn't know what to spend it on.
   D. The project sponsor eases off of the project.
2. What method should a project manager use to ensure that a project maintains momentum in the final stages?
   A. Host a celebration for the project team.
   B. Discipline any team members who stray from the project.

C. Become available to the project team.

D. Get in the trenches and work with the project team.

3. Of the following, which is a reflection of the project manager's ability to lead?

A. The project manager attends regular meetings.

B. The project manager speaks with team members weekly.

C. The project manager leads by example.

D. The project manager speaks with management on a regular basis.

4. Of the following, which method is a likely solution to successfully finish a project that is near completion but may be lagging?

A. Ask for additional time to produce the deliverables.

B. Ask for additional funds to hire outside help.

C. Ask the project team to work additional hours.

D. Remove parts of the project deliverable for delivery at a later date.

5. Of the following, which is *not* a suggestion to quickly and accurately complete a project on schedule?

A. Think clearly, but quickly.

B. Create a visual marker to represent completed work on the workstations.

C. Check for quality as applicable within the final phases.

D. Ask for volunteers to work through the night and into the next day on an implementation.

6. What information can an IT project manager learn from completed tasks in the project?

A. If the tasks have been lagging, it is a sign that the future tasks may lag as well.

B. If the same group of team members has completed the tasks, they may be bored with the work.

C. If the same group of team members has completed the tasks, they may need to be replaced by other team members to keep the team excited.

D. If the critical path has had tasks that can begin in unison (SS), the remaining work units may have SS tasks as well.

7. Why should management reserve be examined near the end of a project?

A. To determine the cost of the errors on the project

B. To determine the additional revenues that are left in place

C. To determine the amount of time that may actually be used, if necessary, for the implementation

D. To determine the amount of time that may be used in addition to the allotted time for the remaining work units

8. You are the project manager for your organization and your project is coming to a close. What project management component will guide you through the process of transferring the project deliverables to operations?
   A. Transition plan
   B. Transfer plan
   C. Project management plan
   D. Operational plan
9. What must the project manager do before the customer accepts the project deliverables?
   A. Inspect the project deliverables for quality.
   B. Inspect the project deliverables for missing components.
   C. Present the invoice for the project completion.
   D. Assign the project team to new projects.
10. What method can an IT project manager invoke to ensure that third parties will reach 100 percent completion on a project?
    A. Retain payment until the integrator reports the project is finished.
    B. Retain payment until the project manager approves the deliverables.
    C. Ask the vendor to add a few extra tasks to complete the project.
    D. Assign the final tasks in the critical path to an internal project to ensure the project is finished as planned.
11. When should a project manager inspect the project for quality?
    A. Throughout the implementation phase
    B. During the final task
    C. In the last 1 to 3 percent of the implementation phase
    D. When the project is complete
12. What should a project manager do if they encounter a flaw when evaluating the finished deliverables of a project?
    A. Note the flaw and add it to a to-do list.
    B. Reassign the project to another team member.
    C. Address the flaw immediately to have it resolved.
    D. Complete a Project Change Request form.
13. What is the purpose of calculating a project's worth?
    A. To determine if the project was worth doing
    B. To determine if the project manager did their job properly
    C. To determine if the project was an expense or an investment
    D. To determine the ROI of the project

14. What is the law of diminishing returns?
    - A. It is a law of economics that states all process cannot be improved infinitely.
    - B. It is a law of economics that states if one area of profitability increases, the other areas of an organization must increase also to produce higher profits.
    - C. It is a law of economics that states if one area of production is increased while other factors remain constant, the overall returns will eventually increase as well.
    - D. It is a law of economics that states if one area of production is increased while other factors remain constant, the overall returns will eventually diminish.
15. What is the purpose of having a third party review the completed project?
    - A. To ensure the deliverables were worthy of the budget and time required to produce them
    - B. To ensure the deliverables are of quality and to second-guess the project team
    - C. To ensure the deliverables are of quality and in alignment with the project scope
    - D. To ensure the project manager did their job as defined in the project plan

Answers

1. **B.** Some project managers make the mistake of easing off a project as it concludes. The inverse is what is really needed, because project team members have a tendency to ease off of the project as well.
2. **D.** A project manager should always be available to the project team, but getting in the trenches and helping the project team as much as possible to complete the project inspires a team to work harder and commit to the conclusion of a project.
3. **C.** A project manager who sits behind a desk and manages the team through memos is not as effective as a project manager who gets involved with the team members and helps them succeed. Leading by example is never a bad idea.
4. **C.** When a project is lagging behind on tasks in the final stages of a project, a project manager may need to ask the project team to work extra hours to complete the project on time.
5. **D.** Asking team members to work an excessive number of hours without rest can be wasteful. Team members need a break from the project to be productive and useful.
6. **A.** History of performance is a good indicator of what future tasks may bring. If the project manager can see that prior tasks have been lagging, they can predict that the remaining tasks may also lag. This can allow the project manager to stress the urgency that these final tasks must be completed on time. The project manager may also elect to reassign resources, add resources, or join the effort himself.

7. **C.** A project manager should evaluate the remaining time in management reserve near the end of the project to determine the whole amount of time available for the project. Just because there is additional time in reserve does not mean that the project manager should assign the time to remaining tasks; it means the project manager can calculate the room for error.
8. **A.** The project's transition plan defines how the project will be transitioned from the project into operations. It defines the ownership of the project deliverables, when the transition will happen, how the recipients will be trained, and what type of extended support is offered. It also offers a review of the project's warranty information.
9. **A.** Before the customers accept the deliverables of the project, the project manager must inspect the final product for quality. This inspection keeps mistakes out of the hands of the customers.
10. **B.** A project manager can withhold payment until they have approved all of the work that the vendor was assigned to complete. The project manager should, however, take time to review the work as soon as the vendor reports that the work is done. It is unprofessional to use this leverage and then take weeks to review the submitted work.
11. **A.** A project manager should work throughout the project to ensure quality in each phase of the implementation.
12. **C.** When a project manager discovers a flaw in the implementation during the review, the project manager must address that flaw immediately to rectify the problem.
13. **D.** The project manager and management should calculate the ROI of the project to determine if the budgeted cost and the actual cost outweigh the benefits of the project's deliverables. Hopefully, if everyone has done the proper research and preproject calculations, the ROI will be excellent.
14. **D.** The law of diminishing returns predicts if one area of production increases, but the other areas remain the same, the overall returns will eventually diminish. For an IT project manager, that means the productivity and the profitability gained by the implementation will eventually reach a plateau because the remaining factors in production have not been improved.
15. **C.** In some instances an organization should hire a third-party consultant to review the work of the project to determine whether the deliverables are of quality and are in alignment with the project scope.

APPENDIX

Exam Objectives Map

Exam PT0-005

Official Exam Domain/Objective	Chapter Coverage
1.0 Project Management Concepts	
1.1 Explain the basic characteristics of a project and various methodologies and frameworks used in IT projects	1, 3
1.2 Compare and contrast Agile vs. Waterfall concepts	1, 2, 5, 7
1.3 Given a scenario, apply the change control process throughout the project life cycle	6, 10
1.4 Given a scenario, perform risk management activities	6
1.5 Given a scenario, perform issue management activities	6, 10
1.6 Given a scenario, apply schedule development and management activities and techniques	6, 9
1.7 Compare and contrast quality management concepts and performance management concepts	6, 9, 11
1.8 Compare and contrast communication management concepts	6, 8, 9
1.9 Given a scenario, apply effective meeting management techniques	3, 8
1.10 Given a scenario, perform basic activities related to team and resource management	3, 7, 8, 9
1.11 Explain important project procurement and vendor selection concepts	5, 6, 7
2.0 Project Life Cycle Phases	
2.1 Explain the value of artifacts in the discovery/concept preparation phase for a project	2, 4, 5
2.2 Given a scenario, perform activities during the project initiation phase	1, 4
2.3 Given a scenario, perform activities during the project planning phase	2, 4, 5, 6, 11
2.4 Given a scenario, perform activities during the project execution phase	3, 4, 5, 7, 8, 9, 11
2.5 Explain the importance of activities performed during the closing phase	11, 12

Official Exam Domain/Objective	Chapter Coverage
3.0 Tools and Documentation	
3.1 Given a scenario, use the appropriate tools throughout the project life cycle	6, 7, 10
3.2 Compare and contrast various project management productivity tools	2, 8
3.3 Given a scenario, analyze quality and performance charts to inform project decisions	2, 4, 11
4.0 Basics of IT and Governance	
4.1 Summarize basic environmental, social, and governance (ESG) factors related to project management activities	3, 6, 11, 12
4.2 Explain relevant information security concepts impacting project management concepts	2, 6, 11
4.3 Explain relevant compliance and privacy considerations impacting project management	2, 3, 6, 9, 11, 12
4.4 Summarize basic IT concepts relevant to IT project management	2, 3, 4, 5, 6, 9, 11, 12
4.5 Explain operational change-control processes during an IT project	1, 2, 10

APPENDIX B

Critical Exam Information

If you're reading this book you are prepping to pass your CompTIA Project+ examination. Good for you! The intent of this book is to explain a practical approach to IT project management and to serve as a study guide to help you pass, not just take, the CompTIA Project+ exam. If you're looking for exam-specific information as a last-minute review for your exam, this is the place for you. You will not pass the exam if you're unfamiliar with the specifics in this appendix.

Test-Passing Tips

For starters, don't think of this process as preparing to take an exam; think of it as preparing to *pass* an exam. Anyone can prepare to take an exam: just show up. Preparing to pass any exam, especially the CompTIA Project+ exam, requires project management experience, diligence, and a commitment to study. I highly recommend that you read the exam objectives from Appendix A as part of your study efforts; after all, this is what you'll be tested on.

CompTIA will present scenarios based in IT, so think through how you'll move the project forward while also conforming to industry standards, governance, and government compliance in technology. While the principles of project management span all disciplines, this exam will focus on project management in technology. Pay attention to the project management aspect of the question and answer it accordingly. You do not need to know anything about construction, health care, manufacturing, but you should have 6-12 months of hands-on experience managing IT projects in an IT environment for this test—all of the focus is on project management.

Days Before the Exam

In the days leading up to your scheduled exam, you should do these basics to prepare yourself for success:

- *Get some moderate exercise.* Find time to go for a jog, lift weights, take a swim, or do whatever workout routine works best for you.
- *Eat healthfully and wisely.* If you eat healthful food, you'll feel good—and feel better about yourself. Be certain to drink plenty of water, and don't overdo the caffeine.

- *Get your sleep.* A well-rested brain is a sharp brain. You don't want to sit for your exam feeling tired, sluggish, and worn out.
- *Time your study sessions.* Don't overdo your study sessions—long, crash study sessions aren't that profitable. In addition, try to study at the same time every day for the same amount of time your exam will last.

Practice the Testing Process

If you could take one page of notes into the exam, what information would you like on this one-page document? Of course, you absolutely cannot take any notes or reference materials into the exam area. However, if you can create and memorize one sheet of notes, you're allowed to re-create it once you're seated in the exam area.

You'll be supplied with several sheets of blank paper and a couple of pencils. Once your exam process begins, immediately re-create your reference sheet. The following are key pieces of information you'd be wise to include on this sheet (you'll find all of this key information in this appendix):

- Activities within each process group (which CompTIA calls phases)
- Estimating formulas
- Communication formula
- Agile project terms
- Organizational structures and their characteristics
- Earned value management (EVM) formulas
- Project management theories

Testing Tips

The multiple-choice questions on the CompTIA Project+ exam are fairly direct and not too verbose, but they may include a few red herrings. For example, you may face questions that state, "All of the following are correct options expect for which one?" The question wants you to find the incorrect option, or the option that is not appropriate for the question, and then find a suitable answer. Carefully read the question to understand what type of answer fits.

A tip that can work with many of the questions is to identify which answer matches the question and then look for an option that doesn't fit with the other possible choices. In other words, find the answer that doesn't fit with the other three options. Here's an example: EVM is used during the _______________.

A. Controlling

B. Executing

C. Closing

D. Entire project

Notice how options A, B, and C are now exclusive? If you choose A, Controlling, it eliminates EVM from being used anywhere else in the project. The answer that doesn't fit here is D, Entire project; it's considered the odd choice because it, by itself, is not an actual process group. Of course, this tip won't work with every question—but it's handy to know. In this exam, you'll also encounter performance-based questions. These are question types that require you to complete a task to show you understand the skill.

Some answer choices may appear to have two of the four options as possibly correct answers. However, because you may choose only one answer, you must discern which one is the best choice. Within the question, there will usually be a hint describing the progress of the project, the requirements of the stakeholders, or some other clue that can help you determine which answer is the best for the question.

Answer Every Question—Once

As of this writing, the CompTIA Project+ exam has 90 questions, and you need to score at least 710 to pass. The exam is scored between 100 and 900, not 1000—which means you have to get roughly 72 of the 90 questions correct to pass. You'll have 90 minutes to complete the exam. Do not leave any questions blank—even if you don't know the answers. A blank question is the same as a wrong answer. As you move through the exam and find questions that stump you, use the "mark question" option in the exam software: choose an answer you suspect may be correct and then move on. When you have answered all the questions, you are given the option to review your marked answers.

Some questions in the exam may reveal or prompt your memory of answers to questions you have marked for review. Resist the temptation to review questions you've already answered with confidence, however. More often than not, your first instinct is correct. When you completed the exams at the end of each chapter, did you change correct answers to wrong answers? If you do it in practice, you'll do it on the actual exam.

Use the Process of Elimination

When you're stumped on a question, use the process of elimination. For each multiple-choice question, there'll be four choices. On your scratch paper, write down "ABCD." If you can safely rule out A, mark it out of the ABCD you've written on your paper. Then focus on which other answer won't work. If you determine C won't work, cross it off your list. You've then got a 50/50 chance of selecting the correct choice.

If you cannot determine which answer is best, B or D in this instance, here's the best approach:

1. Choose an answer in the exam. (No blank answers, remember?)
2. Mark the question in the exam software for later review.
3. Circle ABCD on your scratch paper, jot down any relevant notes, and then record the question number next to the notes.
4. When you do your review, or answer questions further on in the exam, you may realize which choice is the better of the two. Return to the question and select the best answer.

Everything You Must Know

As promised, this section now covers all of the information you must know going into the exam. It's highly recommended that you create a method to recall this information. Here goes.

The 49 Project Management Processes

Project management has 49 generally accepted processes I've covered throughout this book. You'll need to be familiar with the project management processes and what each process accomplishes in the project. Processes are what were discussed throughout this book for both waterfall and predictive projects. Here's a quick rundown of each process group and its processes.

Initiating the Project

There are just two processes to know for project initiation:

- Create the project charter.
- Identify the project stakeholders.

Planning the Project

There are 24 processes to know for project planning:

- Develop the project management plan.
- Plan scope management.
- Gather project requirements.
- Define the project scope.
- Build the work breakdown structure.
- Plan schedule management.
- Define the project activities.
- Sequence the project activities.
- Estimate activity durations.
- Develop the project schedule.
- Plan cost management.
- Estimate the project costs.
- Establish the project budget.
- Create the quality management plan.
- Write the resource management plan.
- Estimate activity resources.
- Create the project communications management plan.

- Create the project risk management plan.
- Identify the project risks.
- Complete qualitative risk analysis.
- Complete quantitative risk analysis.
- Create risk responses.
- Create the procurement management plan.
- Plan stakeholder engagement.

Executing the Project

There are ten executing processes:

- Direct and manage the project work.
- Manage project knowledge.
- Manage quality.
- Acquire resources.
- Perform team development.
- Manage your project team.
- Manage communications.
- Implement risk responses.
- Conduct procurement.
- Manage stakeholder engagement.

Monitoring and Controlling the Project

There are 12 monitoring and controlling processes:

- Monitor and control the project.
- Administer integrated change control.
- Validate scope.
- Control the project scope.
- Perform schedule control.
- Perform cost control.
- Administer quality control activities.
- Control resources.
- Monitor communications.
- Control the project risks according to the risk management plan.
- Control and monitor procurement activities.
- Monitor stakeholder engagement.

Closing the Project

There is just one closing processes:

- Close out the project or the project phase.

EVM Formulas

Earned value management (EVM) shows project performance. You may have a few questions about earned value management and the EVM formulas on your CompTIA Project+ exam. Here's a summary of the EVM formulas and sample mnemonic devices you can use to memorize the formulas.

Name	Formula	Sample Mnemonic Device
Planned value	PV = percent complete of where the project should be	Please
Earned value	EV = percent complete × budget at completion	Eat
Cost variance	CV = EV – AC	Carl's
Schedule variance	SV = EV – PV	Sugar
Cost performance index	CPI = EV / AC	Candy
Schedule performance index	SPI = EV / PV	S (this and the following two spell "SEE")
Estimate at completion	EAC = BAC / CPI	E
Estimate to complete	ETC = EAC – AC	E
To-complete performance index (BAC)	(BAC – EV) / (BAC – AC)	The
To-complete performance index (EAC)	(BAC – EV) / (EAC – AC)	Taffy
Variance at completion	VAC = BAC – EAC	Violin

Quick Exam Facts

This section has some quick facts you should know at a glance. Hold on—this moves pretty fast.

Organizational Structures

Organizational structures are relevant to the project manager's authority. A project manager's authority will vary depending on the organizational structure they are operating within. The structures that offer the least amount of power to highest amount of power are listed in order in the following:

- Functional
- Weak matrix

- Balanced matrix
- Strong matrix
- Projectized

WBS Facts

The work breakdown structure (WBS) is the big picture of the project deliverables. It is not the activities that occur to create the project, but the components the project will create. The WBS helps the project team and the project manager create accurate cost and time estimates. The WBS also helps the project team and the project manager create an accurate activity list. The WBS is an input to five planning processes:

- Cost estimating
- Cost budgeting
- Resource planning
- Risk management planning
- Activity definition

Project Scope Facts

Projects are temporary endeavors to create a unique product. Projects are selected by one of two methods:

- **Benefit measurement methods** These include scoring models, cost-benefit ratios, and economic models.
- **Constrained optimization** These involve mathematical models based on linear, integer, and dynamic programming models. (You probably won't see this one on the CompTIA Project+ exam as a viable answer.)

The project scope defines all of the required work, and only the required work, to complete the project. Scope management is the process of ensuring the project work is within scope and protects the project from scope creep. The scope statement is the baseline for all future project decisions, as it justifies the business need of the project. There are two types of scope:

- **Product scope** Defines the attributes of the product or service the project is creating
- **Project scope** Defines the required work of the project to create the product

Scope verification is the process completed at the end of each phase and project to confirm that the project has met the requirements. It leads to the formal acceptance of the project deliverable.

Project Time Facts

Time can be a project constraint. Effective time management is the scheduling and sequencing of activities in the best order to ensure the project completes successfully—and in a reasonable amount of time. Here are some key terms for time management:

- **Lag** The wait time between activities.
- **Lead** Activities come closer together and even overlap.
- **Free float** The amount of time an activity can be delayed without delaying the next scheduled activity's start date.
- **Total float** The amount of time an activity can be delayed without delaying the project finish date.
- **Slack and float** These are synonymous.
- **Duration** How long an activity lasts. This may be abbreviated as "du."

There are three types of dependencies between activities:

- **Mandatory** This hard logic requires a specific sequence between activities.
- **Discretionary** This soft logic prefers a sequence between activities.
- **External** Due to reasons outside of the project, such as vendors, the sequence must happen in a given order.

Project Cost Facts

There are several methods for providing project estimates:

- **Bottom-up** Project costs start at zero; each component in the WBS is estimated for costs, and then the "grand total" is calculated. This is the longest method to complete, but it provides the most accurate estimate.
- **Analogous** Project costs are based on a similar project. This is a form of expert judgment, but it is also a top-down estimating approach, so it is less accurate than a bottom-up estimate.
- **Parametric modeling** Price is based on cost per unit; examples include cost per metric ton, cost per yard, and cost per hour.

There are four types of costs attributed to a project:

- **Variable costs** The costs depend on other variables. For example, the cost of a food-catered event depends on how many people register to attend the event.
- **Fixed costs** The cost remains constant throughout the project. For example, a rented piece of equipment is the same fee each month even if it is used more in some months than others.
- **Direct costs** The cost is directly attributed to an individual project and cannot be shared with other projects; examples include airfare to attend project meetings, hotel expenses, and leased equipment that is used only on the current project.

- **Indirect costs** These are the costs of doing business; examples include rent, phone, and utilities.

Quality Management Facts

The cost of conformance to quality is the money spent investing in training, requirements for safety, laws and regulations, and steps added to ensure quality acceptance. The cost of nonconformance to quality is the cost associated with rework, downtime, lost sales, and waste of materials.

Some common quality management charts and methods include:

- **Ishikawa diagrams** These are also called fishbone diagrams. They are used to find causes and effects that contribute to a problem.
- **Flow charts** These charts show the relationship between components and the flow of a process through a system.
- **Pareto diagrams** These diagrams identify project problems and their frequencies. These are based on the 80/20 rule: 80 percent of project problems stem from 20 percent of the root causes.
- **Control charts** These charts plot out the result of samplings to determine if projects are "in control" or "out of control."
- **Kaizen techniques** These techniques make small improvements in an effort to reduce costs and consistency.
- **Just-in-time ordering** This method reduces the cost of inventory but requires additional quality because materials are not readily available should mistakes occur.

Human Resource Facts

There are several human resource theories you should be familiar with for the CompTIA Project+ exam:

- **Maslow's Hierarchy of Needs** There are five layers of needs for all humans: physiological, safety, social, esteem, and—the crowning jewel—self-actualization.
- **McClelland's Theory of Needs** David McClelland states that needs are acquired over time and shaped by experiences. There are three needs in this theory: need for achievement, need for affiliation, and need for power.
- **Herzberg's Theory of Motivation** There are two catalysts for workers: hygiene agents and motivating agents.
  - **Hygiene agents** These do nothing to motivate workers, but their absence demotivates them. Hygiene agents are the expectations all workers have: job security, a paycheck, clean and safe working conditions, a sense of belonging, civil working relationships, and other basic attributes associated with employment.
  - **Motivating agents** These are the elements that motivate people to excel. They include responsibility, appreciation of work, recognition, the opportunity to excel, education, and other opportunities associated with work other than financial rewards.

- **McGregor's Theory X and Theory Y** This theory states that X people are lazy, don't want to work, and need to be micromanaged. Y people are self-led, motivated, and can accomplish tasks.
- **Ouchi's Theory Z** This theory states that workers are motivated by a sense of commitment, opportunity, and advancement. Workers will work if they are challenged and motivated. Think participative management.
- **Vroom's expectancy theory** People will behave based on what they expect as a result of their behavior. In other words, people will work in relation to the expected reward of the work.

Communication Facts

Communicating is the most important skill for the project manager. With that in mind, here are some key facts on communications:

- The communication channels formula is $N(N-1)/2$. N represents the number of stakeholders. For example, if you have ten stakeholders, the formula would read $10(10-1)/2$, for 45 communication channels. Pay special attention to questions asking how many additional communication channels you have based on added stakeholders. For example, you have 25 stakeholders on your project and have recently added 5 team members. How many additional communication channels do you now have? You'll have to calculate the original number of communication channels, $25(25-1)/2 = 300$; and then calculate the new number with the added team members, $30(30-1)/2 = 435$; and, finally, subtract the difference between the two: $435 - 300 = 135$ additional communication channels.
- Fifty-five percent of communication is nonverbal. Effective listening is the ability to watch the speaker's body language, interpret paralingual clues, and decipher facial expressions for insight. The next step is to follow these messages with questions for clarity and to offer feedback. Active listening requires the receiver of the message to offer clues, such as nodding the head to indicate they are listening. It also requires the receiver to repeat the message, ask questions, and continue the discussion if clarification is needed.
- Communication can be hindered by trendy phrases, jargon, and extremely pessimistic comments. In addition, communication can be blocked by noise, hostility, cultural differences, and static, among other communication barriers.

Risk Management Facts

Risks are uncertain events that can affect the project for good or bad. Risks should be identified as early as possible in the planning process. A person's willingness to accept risk is the utility function (also called the utility theory). The Delphi Technique can be used to build consensus on project risks.

The two outputs of planning are the risk management plan and the risk register. There are two broad types of risks:

- **Business risk** The loss of time and finances. A downside and an upside exist.
- **Pure risk** The loss of life, injury, and theft. Only a downside exists.

Risks can be responded to with one of seven methods:

- **Avoidance** Avoid the risk by planning a different technique to remove the risk from the project. This is for negative risks.
- **Mitigation** Reduce the probability or impact of a negative risk.
- **Transference** The negative risk is not eliminated, but the responsibility and ownership of the risk is transferred to another party; for example, it's transferred to insurance.
- **Enhancing** Conditions are created to encourage a positive risk to occur.
- **Exploiting** The benefits of a positive risk are amplified so that the project can take advantage of the positive risk.
- **Sharing** The positive risk is shared with another project, organization, or company through a teaming agreement.
- **Acceptance** The positive or negative risk's probability or impact may be small enough that it can be accepted.

Procurement Facts

A statement of work (SOW) is provided to the potential sellers so that they can create accurate bids, quotes, and proposals for the buyer. A bidders' conference may be held so that sellers can query the buyer on the product or service to be procured.

A contract is a formal agreement, preferably written, between a buyer and a seller. On the CompTIA Project+ exam, procurement questions are usually from the buyer's point of view. All requirements the seller is to complete should be clearly written in the contract. Requirements of both parties must be met, or legal proceedings may follow. Contract types are as follows:

- Cost-reimbursable contracts require the buyer to assume the risk of cost overruns.
- Fixed-price contracts require the seller to assume the risk of cost overruns.
- Time-and-material contracts are good for smaller assignments but can impose cost overrun risks to the buyer if the time by the seller is not monitored.
- A purchase order is a unilateral form of contract.
- A letter of intent is not a contract, but it shows the intent of the buyer to purchase from a specific seller.

Adaptive Project Management Facts

Adaptive project management doesn't try to predict the entire project like a waterfall project management approach does. Agile begins with a product backlog of all the requirements, then the team selects a chunk of work to accomplish in an iteration, which lasts up to four weeks. At the end of the iteration, the team demonstrates the work accomplished to the customers. Next, the team goes into a type of lessons learned meeting, called a retrospective, where the team discusses how to improve the project in the next iteration. And then the entire cycle starts over.

Every day in an adaptive project there is a standup meeting, which can last up to 15 minutes. This is a meeting just for the team members and the project manager role. The team answers three questions:

- What did you accomplish since our last meeting?
- What will you accomplish today?
- Are there any impediments that need to be addressed?

These three questions are asked of each team member, and a daily touchpoint is to discuss accomplishments, what people are working on, and to be transparent about any issues or blockers to project progress.

There are several nuances of adaptive project management, but the most common is scrum. Scrum has three roles in the scrum team you should recognize:

- **Scrum master** Ensures that everyone is following the rules of the project, provides coaching, gets the team members the things they need in the project, and prevents interruptions in the project.
- **Product owner** Works with the project customer to gather requirements, write user stories, and prioritize the product backlog. When a change enters the project, the product owner assesses the change and will add it to the product backlog and prioritize where it should be in the product backlog.
- **Development team** The people that do the work to create the product. These roles are considered generalizing specialists, meaning the people can do more than one activity in the project.

In each iteration, the total number of story points assigned to the project work that the team completes is called velocity. Velocity can be graphed in a burnup chart or burndown chart. A burnup chart shows the accumulation of story points accomplished in each iteration and the cumulative total of all story points done in the project. A burndown chart shows the diminishing number of story points left to accomplish in the project and the velocity of the development team in each iteration. Both charts can help predict when the project will likely finish based on the number of story points remaining in the product backlog and the stabilized velocity of the development team.

The CompTIA Project+ exam will have a few questions on agile. If you're not too familiar with agile project management, you may want to spend some extra time brushing up on the approach and the key terms associated with the methodology. The key thing to remember in agile project management is that you don't do upfront planning in an agile project, but rather plan in shorter segments as you go into each iteration of the project.

A Letter to You

My goal for you is to pass the CompTIA Project+ exam. As I teach project management seminars online and in-person for organizations around the globe, I'm struck by one similarity among the most excited course participants: these people want to pass their exam. Sure, project management is not the most exciting topic, but the individuals are excited about passing their exam. I hope you feel the same way. I believe that your odds of passing the CompTIA Project+ exam are like most things in life: you're going to get out of it only what you put into it. I challenge you to become excited, happy, and eager to pass the exam.

Here are ten final tips for passing your CompTIA Project+ exam:

- Prepare to pass the exam, not just take it.
- If you haven't done so already, schedule your exam. Having a deadline makes the exam even more of a reality.
- If you haven't done so already, create a clutter-free area for studying.
- Study in regular intervals right up to the day before your exam.
- Repetition is the mother of learning. If you don't know the formula, repeat and repeat. And then repeat it again.
- Create your own flashcards from the glossary and terms used in this book.
- Practice creating one page of notes that you'll re-create at the start of your exam.
- Create a significant reward for yourself as an incentive to pass the exam.
- Make a commitment to pass.

If you're stumped on something I've written in this book, or if you'd like to share your CompTIA Project+ success story, drop me a line: cs@instructing.com. You can drop by my project management site whenever you'd like to: https://instructing.com. Finally, I won't wish you good luck on your exam—luck is for the ill-prepared. If you follow the strategies I've outlined in this book and apply yourself, I am certain you'll pass the exam.

Keep moving forward,
Joseph Phillips
PMP, PMI-ACP, PSM, Project+, CTT+

APPENDIX C

About the Online Content

This book comes complete with

- TotalTester Online customizable practice exam software with 190 practice exam questions
- Video training from the author
- Downloadable templates and worksheets

System Requirements

The current and previous major versions of the following desktop browsers are recommended and supported: Chrome, Microsoft Edge, Firefox, and Safari. These browsers update frequently, and sometimes an update may cause compatibility issues with the TotalTester Online or other content hosted on the Training Hub. If you run into a problem using one of these browsers, please try using another until the problem is resolved.

Your Total Seminars Training Hub Account

To get access to the online content you will need to create an account on the Total Seminars Training Hub. Registration is free, and you will be able to track all your online content using your account. You may also opt in if you wish to receive marketing information from McGraw Hill or Total Seminars, but this is not required for you to gain access to the online content.

Privacy Notice

McGraw Hill values your privacy. Please be sure to read the Privacy Notice available during registration to see how the information you have provided will be used. You may view our Corporate Customer Privacy Policy by visiting the McGraw Hill Privacy Center. Visit the **mheducation.com** site and click **Privacy** at the bottom of the page.

Single User License Terms and Conditions

Online access to the digital content included with this book is governed by the McGraw Hill License Agreement outlined next. By using this digital content you agree to the terms of that license.

Access To register and activate your Total Seminars Training Hub account, simply follow these easy steps.

1. Go to this URL: **hub.totalsem.com/mheclaim**
2. To register and create a new Training Hub account, enter your e-mail address, name, and password on the **Register** tab. No further personal information (such as credit card number) is required to create an account.

   If you already have a Total Seminars Training Hub account, enter your e-mail address and password on the **Log in** tab.
3. Enter your Product Key: `6ztn-57jf-vbbj`
4. Click to accept the user license terms.
5. For new users, click the **Register and Claim** button to create your account. For existing users, click the **Log in and Claim** button.

   You will be taken to the Training Hub and have access to the content for this book.

Duration of License Access to your online content through the Total Seminars Training Hub will expire one year from the date the publisher declares the book out of print.

Your purchase of this McGraw Hill product, including its access code, through a retail store is subject to the refund policy of that store.

Restrictions on Transfer The user is receiving only a limited right to use the Content for the user's own internal and personal use, dependent on purchase and continued ownership of this book. The user may not reproduce, forward, modify, create derivative works based upon, transmit, distribute, disseminate, sell, publish, or sublicense the Content or in any way commingle the Content with other third-party content without McGraw Hill's consent.

Limited Warranty The McGraw Hill Content is provided on an "as is" basis. Neither McGraw Hill nor its licensors make any guarantees or warranties of any kind, either express or implied, including, but not limited to, implied warranties of merchantability or fitness for a particular purpose or use as to any McGraw Hill Content or the information therein or any warranties as to the accuracy, completeness, correctness, or results to be obtained from, accessing or using the McGraw Hill Content, or any material referenced in such Content or any information entered into licensee's product by users or other persons and/or any material available on or that can be accessed through the licensee's product (including via any hyperlink or otherwise) or as to non-infringement of third-party rights. Any warranties of any kind, whether express or implied, are disclaimed. Any material or data obtained through use of the McGraw Hill Content is at your own discretion and risk and user understands that it will be solely responsible for any resulting damage to its computer system or loss of data.

Neither McGraw Hill nor its licensors shall be liable to any subscriber or to any user or anyone else for any inaccuracy, delay, interruption in service, error or omission, regardless of cause, or for any damage resulting therefrom.

In no event will McGraw Hill or its licensors be liable for any indirect, special or consequential damages, including but not limited to, lost time, lost money, lost profits or good will, whether in contract, tort, strict liability or otherwise, and whether or not such damages are foreseen or unforeseen with respect to any use of the McGraw Hill Content.

TotalTester Online

TotalTester Online provides you with a simulation of the CompTIA Project+ exam. Exams can be taken in Practice Mode or Exam Mode. Practice Mode provides an assistance window with references to the book, explanations of the correct and incorrect answers, and the option to check your answer as you take the test. Exam Mode provides a simulation of the actual exam. The number of questions, the types of questions, and the time allowed are intended to be an accurate representation of the exam environment. The option to customize your quiz allows you to create custom exams from selected domains or chapters, and you can further customize the number of questions and time allowed.

To take a test, follow the instructions provided in the previous section to register and activate your Total Seminars Training Hub account. When you register, you will be taken to the Total Seminars Training Hub. From the Training Hub Home page, select your certification from the Study drop-down menu at the top of the page to drill down to the TotalTester for your book. You can also scroll to it from the list on the Your Topics tab of the Home page, and then click the TotalTester link to launch the TotalTester. Once you've launched your TotalTester, you can select the option to customize your quiz and begin testing yourself in Practice Mode or Exam Mode. All exams provide an overall grade and a grade broken down by domain.

Other Book Resources

The following sections detail the other resources available with your book. You can access these items by selecting the Resources tab, or by selecting **CompTIA Project+ Certification** from the Study drop-down menu at the top of the page or from the list on the Your Topics tab of the Home page. The menu on the right side of the screen outlines all of the available resources.

Video Training from the Author

Video MP4 clips from the author of this book provide detailed examples of key project management concepts in audio/video format. You can access these videos by navigating to the Resources tab and selecting Videos.

Downloadable Content

The Resources tab also includes a link to download additional content that accompanies this book. Supplemental worksheets and templates are available from the author as Microsoft Word files, PowerPoints, Excel spreadsheets, and a PDF for specified chapter exercises.

Technical Support

For questions regarding the TotalTester or operation of the Training Hub, visit **www.totalsem.com** or e-mail **support@totalsem.com**.

For questions regarding book content, visit **www.mheducation.com/customerservice**.

acceptance A risk response that chooses to accept the risk within the project without creating a strategy to counteract the risk event. Acceptance is often used with small risks.

action items Actions or assignments that team members are to complete, usually determined in a project meeting.

activity-on-the-arrow project network diagram Origin of the arrow is the "begin activity" sign, and the end of the arrow is the "end activity" sign. Also commonly referred to as the Arrow Diagraming Method (ADM).

activity-on-the-node project network diagram A project network diagramming method that allows the project manager to map relationships between activities. With the activity-on-the-node (AON) method, the focus is on the project activities rather than on the start and end of activities. AON illustrates the sequencing of activities from the start of the project through the end of the project.

actual costs (AC) Used in earned value management (EVM) and represents the actual cost of the work performed.

ad hoc reporting Impromptu reports on schedule, cost, scope, or other key performance indicators (KPIs) as requested by stakeholders with enough authority to cause the project manager to create the report.

add/move/change projects Generally smaller projects that, as the name implies, add, move, or change some element within an organization. Approximately 10 percent of the project time is allotted to planning.

adjourning Finality of project calls for the team to disperse, or adjourn, to other projects and join different project teams. The project team, like the project, is not a permanent fixture in the organization.

administrative closure When the customer or project sponsor documents and accepts the project results; also needed if a project is terminated.

agile project management A change-driven approach to project management, where the project team moves through iterations, or sprints, of planning and executing. Requirements are prioritized for each iteration with the project manager, sometimes called the scrum master, and the product owner.

analogous estimating A form of expert judgment that relies on historical information to predict estimates for current projects. Also known as top-down estimating.

architect A role that creates the high-level plan or solution for the project scope. Architects generally create the software, network, database, or communication structure for the project scope solution.

as late as possible (ALAP) constraint When you specify a task as ALAP, Microsoft Project will schedule the task to occur as late as possible without delaying dependent tasks. This is the default for all new tasks when scheduling tasks from the end date. This constraint is flexible.

as soon as possible (ASAP) constraint When you specify a task as ASAP, Microsoft Project will schedule the task to occur as soon as it can. This is the default for all new tasks when assigning tasks from the start date. This constraint is flexible.

assumptions Beliefs considered to be true, real, or certain for the sake of planning. All project assumptions should be evaluated later in planning to determine their risk for the project should the assumptions prove untrue.

assumptions log You often have to make assumptions in planning, and these assumptions are documented in the assumptions log.

asynchronous communication Any communication that doesn't happen in real time, such as e-mail, text, or project management information system (PMIS) solutions.

avoidance One response to a risk event. The risk is avoided by planning a different technique to remove the risk from the project.

backlog The list of things to complete in the project, such as defect resolution, added features, or documentation. The product backlog is a prioritized list of features the development team is to create in a scrum project.

benchmarking The process of using prior projects within, or external to, the performing organization to compare and set quality standards for processes and results.

benefit measurement methods A method used to compare the value of one project against the value, or benefits, of another. It's often used in project selection models.

benefit-cost analysis *See* cost-benefit analysis.

bid A document from the seller to the buyer, used when price is the determining factor in the decision-making process.

bidder conference A meeting with prospective sellers to ensure that all sellers have a clear understanding of the product or service to be procured. Bidder conferences allow sellers to query the buyer on the details of the product to help ensure that the proposal the seller creates is adequate and appropriate for the proposed agreement. Also called a contractor conference or vendor conference.

bottom-up cost estimating The process of creating a detailed estimate for each work component (labor and materials) and accounting for each varying cost burden. These estimates are based on the WBS and the WBS dictionary, as these documents define each element of the project deliverables.

brainstorming An approach that encourages participants to generate ideas about an opportunity or business problem. Brainstorming at the research stage is useful to determine different types of outcomes for the project. The project manager should encourage the participants to come up with as many ideas as possible, and then these ideas can be sorted and researched more in-depth after the session.

branding restrictions Organizational rules regarding logos, colors, trademarks, and other branding materials that the project must adhere to in order to be in organizational compliance.

budget The finances allotted for the completion of a project.

budget at completion (BAC) The sum of the budget for each phase of your project; the estimated grand total of your project.

budget burndown chart Shows the amount of funds the project has spent on the overall project budget to reach a specific point in the project. The funds spent are represented in a declining line from left to right, where left represents the start of the project and right represents the target project end date.

budget estimate This estimate is somewhat broad and is used early in the planning processes and also in top-down estimates. The range of variance for the estimate can be, for example, –10 percent to +25 percent.

burn rate The rate of consumption of resources, budget, and tasks. A burn rate can describe how quickly the project is "burning" through the project funds, but often is associated with the number of tasks to be completed in the project and the rate of how many tasks per day, week, or month the project team is able to complete.

burndown chart Shows the estimated effort remaining needed to complete the project or phase of the project. The burndown chart shows the diminishing number of tasks left to complete the project work. Burn charts are often used in agile project management.

burnup chart Shows the effort spent completing users stories in an agile project. A burnup chart shows the accumulation of tasks toward the project completion.

business analyst A role that is responsible for eliciting, documenting, and supporting the product requirements of the project.

business case A document that helps the organization determine if it can justify the cost of the project in proportion to the return on investment. The business case links the value of the project's solution to the organization.

business cycles A time of the business productivity when activities are very high or low. For example, an accounting firm may experience a busy business cycle during tax season.

business partners The sellers, vendors, and contractors that may be involved in a project through a contractual relationship. Business partners can provide goods and services such as hardware and software, and subject matter experts like developers, technical writers, and software testers that the project manager might need on the project.

business rules analysis If the project outcome will likely affect the way the organization does business, the business rules should be studied. Business rules define the internal processes to make decisions; provide definitions for operations; define organizational boundaries; and afford governance for projects, employees, and operations.

capital expenses (CapEx) Any purchase by an organization to provide future benefits; examples include new computer hardware, buildings, and property. These are usually long-term acquisitions, can be costly, and are seen as an investment in the organizational business value.

capital resources Money or financing to charter new projects, purchase equipment, contract labor, or other physical resources.

cause-and-effect diagrams Diagrams that are used for root cause analyses of what factors are creating the risks within the project. The goal is to identify and treat the root of the problem, not the symptom. Also called Ishikawa diagrams and fishbone diagrams.

centralized contracting All contracts for all projects need to be approved through a central contracting unit within the performing organization.

champion The individual who defends the project, ensures resources for the project work, and has authority over the project resources. The champion works with the project manager and project team to ensure that the project is successful in the environment. This person is often the project sponsor but could also be the project customer.

change control board (CCB) Determines the validity and need for project change requests and approves or denies them.

change control system (CCS) An internal process the project manager can use to block anyone, including management, from changing the deliverables of a project without proper justification. Change control requires the requestor to have an excellent reason to attempt a change, and then it evaluates the proposed change's impact on all facets of the project.

change impact statement A formal response from the project manager to the originator of a Project Change Request form. It is a summary of the project manager's proposed plan to incorporate the changes. Usually this is a listing of the paths and trade-offs the project manager is willing to implement.

change log A document that records all proposed changes in the project, the effect of each change, the change request status, and relevant information about each change request.

chart of accounts A coding system used by the performing organization's accounting system to assign the cost for project work. This is a predefined table of costs for project or organization use for commonly completed activities. For example, a programmer's time is worth $150 per hour regardless of which programmer is assigned to the project.

checklist A list of activities that workers check to ensure the work has been completed consistently. Checklists are used in quality control.

closing The period when a project or phase moves through formal acceptance to bring the project or phase to an orderly conclusion.

cloud-based solutions vs. on-premises solutions Cloud-based solutions are technical solutions that are provided via web technologies, such as Amazon Web Services (AWS) or Google Cloud. On-premises solutions are also technical solutions but are provided via local hardware and solutions on the organization's private network.

code of accounts A numbering system that shows the different levels of WBS components and identifies which components belong to which parts of the WBS.

coercive power The type of power that comes with the authority to discipline the project team members. Also known as penalty power. It is generally used to describe the power structure when the team is afraid of the project manager.

collective bargaining agreements Contractual agreements initiated by employee groups, unions, or other labor organizations. They may act as a constraint on the project.

communication channel formula A formula to predict the number of communication channels within a project; the formula is $N(N - 1)/2$, where N represents the number of stakeholders.

communications management plan A plan that documents and organizes stakeholder needs for communication. This plan covers the communications system, its documentation, the flow of communication, modalities of communication, schedules for communications, information retrieval, and any other stakeholder requirements for communications. The plan may also address communications for virtual teams and consider time zone differences, language differences, and cultural norms in each location where the project exists.

compromising A conflict resolution method that requires both parties to give up something. The decision ultimately made is a blend of both sides of the argument. Because neither party completely wins, it is considered a lose-lose solution.

conferencing platforms Web conferencing software, such as Zoom, that allows organizations to collaborate online without the need to be co-located to have the meeting.

configuration management Activities focusing on controlling the characteristics of a product or service. A documented process of controlling the features, attributes, and technical configuration of any product or service. It is sometimes considered a rigorous change control system.

constrained optimization methods Complex mathematical formulas and algorithms that are used to predict the success of projects, variables within projects, and tendencies to move forward with selected project investments. Examples include linear programming, integer algorithms, and multi-objective programming.

consultative decision-making process Occurs when the project team meets with the project manager, and together they may arrive at several viable solutions. The project manager can then take the proposed solutions and make a decision based on what they think is best for the project.

contingency plan A predetermined decision that will be enacted should the project go awry.

contingency reserve A time or dollar amount allotted as a response to risk events that may occur within a project.

contingency/fallback plans Plans often used in technical projects to fall back or roll back to the last known good status of the environment should something go awry when the solution is implemented into production.

continuous quality improvement The theory that all practices within an organization are processes and that processes can be infinitely improved.

contract A legal, binding agreement, preferably written, between a buyer and seller detailing the requirements and obligations of both parties. It must include an offer, an acceptance, and a consideration.

contract administration The process of ensuring that the buyer and the seller both perform to the specifications within the contract.

contract change control system A system that defines the procedures for how contracts may be changed. Includes the paperwork, tracking, conditions, dispute resolution procedures, and the procedures for getting the changes approved within the performing organization.

contract closeout A process for confirming that the obligations of the contract have been met as expected. The project manager, customer, key stakeholder, and, in some instances, seller complete the product verification together to confirm the contract has been completed.

contract file A complete indexed set of records of the procurement process incorporated into the administrative closure process. These records include financial information as well as information on the performance and acceptance of the procured work.

control account plans A control tool within the project that represents the integration of the project scope, project schedule, and budget. It allows management to measure the progress of a project.

control charts Charts that illustrate the performance of a project over time. They map the results of inspections against a chart. Control charts are typically used in projects or operations that have repetitive activities such as manufacturing, test series, or help desk functions. Upper and lower control limits indicate whether values are within control or out of control.

controlling The project is controlled and managed. The project manager controls the project scope and changes, and monitors changes to the project budget, schedule, and scope by comparing plans to actual results and taking corrective action as necessary.

cost baseline Shows what the project is expected to spend. It's usually shown in an S-curve and allows the project manager and management to predict when the project will be spending monies and over what duration. The purpose of the cost baseline is to measure and predict project performance.

cost budgeting A process of aggregating the assigned cost to arrive at a budget for the entire project. This process shows costs over the execution of the project. The cost budget results in an S-curve that becomes the cost baseline for the project.

cost change control Part of the integrated change control system that documents the procedures to request, approve, and incorporate changes to project costs.

cost control An active process to control causes of cost change, document cost changes, and monitor cost fluctuations within the project. When approved changes occur, the cost baseline must be updated. Unapproved changes are usually seen as a variance from the cost baseline.

cost estimating The process of calculating the costs, by category, of the identified resources to complete the project work.

cost management plan A plan that details how changes to costs within the project will be estimated, planned, and controlled and what procedure will be used to report and document cost changes.

cost of conformance The cost of completing the project work to satisfy the project scope and the expected level of quality. Examples include training, safety measures, and quality management activities.

cost of nonconformance The cost of completing the project work without meeting the quality standards. The biggest issue here is the money lost by having to redo the project work; it's always more cost effective to do the work right the first time. Other nonconformance costs are loss of sales, loss of customers, downtime, and corrective actions to fix problems caused by the incorrect work.

cost performance index (CPI) A ratio of the amount of actual cumulative dollars spent on a project's work and how closely that value is to the predicted budgeted amount. The CPI formula is earned value/actual costs.

cost variance The difference in the amount of budgeted expense and actual expense. A negative variance means that more money was spent on the service or goods than was budgeted for it. The cost variance formula is earned value – actual costs.

cost-benefit analysis The ratio of the number of costs (not just financial) and the number of benefits to help an organization decide on the value of an implementation or solution. These are sometimes called cost-benefit ratios (CBR).

cost-plus contract A contract that represents a set fee for the procured work plus a fee for the actual cost of the work. Some unscrupulous vendors try to use a cost plus a percentage of costs contract where they expect the project manager to pay for the cost of the materials plus a percentage fee for the materials. Cost-plus contracts are risky for buyers, as the vendor can drive the price up by actually wasting materials. There are some cost-plus contracts that include incentives and penalties if the vendor finishes early or late—though the project manager can add these terms to a fixed-fee contract.

crashing The addition of more resources to activities on the critical path in order to complete the project earlier. Crashing results in higher project costs.

critical path Shows the latest finish and the early finish for the project activities. The critical path will reveal the project duration. The critical path is represented in a project network diagram as one or more paths that equate to the longest duration of sequenced activities to reach the completion of all activities in the project.

critical path method (CPM) The most common approach to calculating when a project may finish. It uses "forward" and "backward" paths to reveal which activities are considered critical and which contain float. If activities on the critical path are delayed, the project end date will be delayed.

customers and end-user stakeholders These stakeholders could be internal to the organization or quite literally customers that purchase the deliverable the project creates.

daily standup A daily 15-minute timeboxed meeting, called a standup as the participants should stand, if able, rather than sit during the meeting to keep conversations brief and to the point.

date constraints There are three types of date constraints:

- **No earlier than** This constraint specifies that a task may happen any time after a specific date, but not earlier than the given date.
- **No later than** This constraint is deadline oriented. The task must be completed or must start by this date or else.
- **On this date** This constraint is the most time oriented. There is no margin for adjustment, as the task must be completed or must start on this date, no sooner or later.

decision tree analysis A type of analysis that determines which of two decisions is best. The decision tree assists in calculating the value of the decision and determining which decision costs the least.

decoder A part of the communications model, a decoder is the inverse of the encoder. If a message is encoded, a decoder translates it back to a usable format.

defect log Escaped defects are any defects that leave the project environment and make it into production. Defects are documented in the defect log for resolution and tracked for their priority, ownership, and date of resolution.

definitive estimate This estimate is one of the most accurate. It is used late in the planning process and is associated with bottom-up estimates. The range of variance for the estimate can be –5 percent to +10 percent.

Delphi Technique A method to query experts anonymously on foreseeable risks and other factors within the project, phase, or component of the project. The results of the survey are analyzed and organized and then circulated to the experts. There can be several rounds of anonymous discussions with the Delphi Technique. The goal is to gain consensus on a given topic, and the anonymous nature of the process ensures that no one expert's advice overtly influences the opinion of another participant.

demotivators An element of Herzberg's Motivation-Hygiene Theory that employees are motivated or demotivated by effects within an organization. The hygiene factors are the expected benefits a company has to offer, such as insurance, vacation time, and other benefits. The presence of these elements is expected by the motivation seekers, and the absence of these elements has a negative impact.

design of experiments An approach that relies on statistical "what-if" scenarios to determine which variables within a project will result in the best outcome. It can also be used to eliminate a defect. The design of experiments approach is most often used on the product of the project, rather than on the project itself.

detailed variance report A detailed explanation of any quality, scope, cost, or schedule variance within the project.

DevOps A software development approach that combines the project of software development with the day-to-day operations of supporting existing software.

DevSecOps A software development approach similar to DevOps that combines software development, security, and day-to-day operational duties.

directive decision-making process The project manager makes the decision with little or no input from the project team. Directive decision-making is acceptable, and needed, in some instances, but it isolates the project manager from the project team.

discretionary dependencies The preferred order of activities. Project managers should adhere to the order at their discretion and should document the logic behind the ordering. Discretionary dependencies have activities happen in a preferred order because of best practices, conditions unique to the project work, or external events. Discretionary dependencies are also known as soft logic.

earned value (EV) The value of the work that has been completed and the budget for that work: EV = % Complete × BAC.

earned value management (EVM) Integrates scope, schedule, and cost to give an objective, scalable, point-in-time assessment of the project. EVM calculates the performance of the project and compares current performance against the plan. EVM can also be a harbinger of things to come. Results early in the project can predict the likelihood of the project's success or failure.

effective listening The receiver is involved in the listening experience by paying attention to the speaker's visual clues and paralingual intentions and by asking relevant questions.

encoder As part of the communications model, an encoder is the device or technology that packages the message to travel over the medium.

end-of-life (EOL) software When software is to be entirely replaced by a new version or new software, it is considered to be at its end-of-life stage.

enhance A positive risk response that tries to make the conditions just right for a positive risk to happen. For example, suppose the project manager could save a tremendous amount of time and project costs if the project manager were able to finish a particular milestone by a given date. To reach the milestone, the project manager adds extra resources to help the effort-driven work so that the team can complete the milestone by the specific date.

enterprise environmental factors Factors that describe the rules, policies, and governance the project manager must adhere to within the organization. The rules and policies of the organization may require the project manager to deal with a risk management department, follow particular risk analysis rules, or complete risk assessment forms. Always follow the rules of the organization.

enterprise resource planning (ERP) Software used to manage business activities, such as project management, procurement, accounting, and risk management.

epic A large user story in an agile project that can be broken down into smaller components and is too large to be wholly completed within one defined iteration of an agile project.

estimate at completion (EAC) A hypothesis of what the total cost of the project will be. Before the project begins, the project manager completes an estimate for the project deliverables based on the WBS. As the project progresses, there will likely be some variances between what the cost estimate was and what the actual cost is. The EAC is calculated to predict what the new estimate at completion will be.

estimate to complete (ETC) Represents how much more money is needed to complete the project work: ETC = EAC – AC.

evaluation criteria Criteria used to rate and score proposals from sellers. In some instances, such as a bid or quote, the evaluation criterion is focused just on the price the seller offers. In other instances, such as a proposal, the evaluation criteria can be multiple values: experience, references, certifications, and more.

executing The project plans are carried out, or executed; the project manager coordinates people and other resources to complete the plan.

expectancy theory Theory that people will behave on the basis of what they expect as a result of their behavior. In other words, people will work in relation to the expected reward for the work. This is also known as Vroom's expectancy theory.

expert power A type of power in which the authority of the project manager comes from experience in the area that the project focuses on.

exploit A positive risk response that aims to take advantage of a positive risk. It aims to increase the probability of the risk event to 100 percent. Imagine, for example, that the IT project creates a by-product that could be sold to help offset the project costs. Or perhaps the project manager and the project team could take advantage of a slow business cycle to work on a project uninterrupted.

eXtreme Programming (XP) An agile project management approach that aims for high quality in software development. It has specific rules and ceremonies for its approach.

fast tracking Performing project phases in parallel that are normally done sequentially.

feasibility study A documented expression of what the research has told the project manager. This plan is written to help determine the validity of a proposed project, of a section of a project, or the scope of a given project. The feasibility study is divided into eight sections: executive summary, defined business problem or opportunity, requirements and purpose of the study, description of the options assessed, assumptions used in the study, audience impacted, financial obligations, and recommended action.

final project report The collection of all of the cumulative reports may serve as a final record of each phase's work, with a few additions. It includes the project vision statement, the project proposal, the project plan, the WBS, the PND, meeting minutes, any Project Change Requests forms, all written notices relevant to the project deliverables, the client acceptance agreement, and the post-project audit.

finish no earlier than (FNET) constraint Semiflexible constraint that requires that a task be completed on or after a specified date.

finish no later than (FNLT) constraint Semiflexible constraint that requires that a task be completed on or before a specified date.

finish-to-finish (FF) tasks A finish-to-finish relationship specifies that the successor tasks cannot finish until the predecessor task finishes. An example is rolling out a new software package while users are in software training sessions. The new software should be installed and configured on their workstations by the time the training sessions end.

finish-to-start (FS) tasks These tasks are successors and cannot begin until the predecessor task is completed. An example is installing network cards before connecting PCs to the Internet.

fishbone/Ishikawa diagrams A cause analysis tool that shows the causal factors and the effect to be solved. The diagram looks somewhat like a fishbone, hence the name. Ishikawa, a Japanese quality engineer, is known for popularizing the chart.

fixed-fee contract A contract that provides a set price for the work defined in the statement of work and in the contract. Fixed fee contracts are generally a low-risk solution for the buyer, as any cost overruns go back to the vendor. Also known as a fixed-priced contract.

flexible constraints Constraints that do not have dates assigned to their activities and are bound only by their predecessor and successor activities. Use flexible constraints as much as possible.

flexible deadline A deadline that doesn't assign an exact time for completion.

float The amount of time a task can be delayed without delaying the project completion. Technically, there are three different types of float: *Free float* is the total time a single activity can be delayed without delaying the early start of any successor activities. *Total float* is the total time an activity can be delayed without delaying project completion. *Project float* is the total time the project can be delayed without passing the customer's expected completion date. Float is also known as slack.

flowchart A chart that illustrates how the parts of a system occur in sequence.

focus groups A type of stakeholder analysis that aims to gather requirements for the project. Stakeholders are led through a discussion about the opportunity or problem by an impartial moderator. A scribe or recorder keeps the minutes in the meeting, and then the results are analyzed. An average focus group has six to twelve participants. The participants can be considered homogenous if they all share the same characteristics, such as all salespeople. Or the project manager can use a heterogeneous group, where the participants are stakeholders with different backgrounds, such as users of a software application from different departments within the organization.

force majeure Often a clause in contracts to release the parties from obligations in the case of a powerful and unexpected event, such as a hurricane or other disaster.

forcing A decision-making approach in which the person with the power makes the decision. The decision made may not be the best decision for the project, but it's fast. As expected, this autocratic approach does little for team development and is a win-lose solution. Used when the stakes are high and time is of the essence or if relationships are not important.

forecasting An educated estimate of how long the project will take to complete. It can also refer to how much the project may cost to complete.

formal acceptance The formal acceptance of a project's deliverables is a process that is completed by the client of the project and the appropriate members of the project team. These acceptances are contingent on a project acceptance agreement.

formal power The type of power in which the project manager has been assigned by senior management to be in charge of the project.

forming This stage of team development allows the project team to come together and learn about each other. Project team members feel each other out, find out who's who, and learn about other team members.

fully burdened workload The amount of work, in hours, required by the staff to complete each phase of the project.

functional decomposition This method simply takes a large problem and breaks it down into smaller, manageable components. While it sounds easy, it's tricky. The project manager will break down the problem into as small of subcomponents as possible, so that each "subproblem" can be managed independent of the other problems. The project manager needs to link the components together so that one solution to a subproblem doesn't adversely affect the solutions to other components in the decomposition.

functional structure An organizational structure that groups staff members according to their area of expertise (sales, marketing, construction, and so on). Functional structures require the project team members to report directly to the functional manager. In this type of structure, the project manager's authority and decision-making ability is less than the functional manager's.

future value A formula to calculate the future value of present money. The future value formula is $PV \times (1 + i)^n$, where i is the interest rate and n is the number of time periods.

Gantt chart A chart that allows a project manager to see the intersection of dates until completion and the tasks within a project. Henry Gantt, an engineer and social scientist, invented this method of tracking deliverables in 1917.

Graphical Evaluation and Review Technique (GERT) Conditional advancement, branching, and looping of activities based on probabilistic estimates. Activities within GERT are dependent on the results of other upstream activities.

hard logic The logical relationship between activities based on the type of work. For example, the foundation of a house must be created before the frame of the house can be built. This is also known as mandatory dependency.

hardware decommissioning When hardware has become old, outdated, or no longer serves a purpose in an organization, it is taken out of production and considered to be no longer viable.

Herzberg's Motivation-Hygiene Theory Posits that there are two catalysts for workers: hygiene agents and motivating agents. Hygiene agents do nothing to motivate workers, but their absence demotivates them. Hygiene agents are the expectations all workers have: job security, paychecks, clean and safe working conditions, a sense of belonging, civil working relationships, and other basic attributes associated with employment. Motivating agents are components such as rewards, recognition, promotions, and other values that encourage individuals to succeed.

histogram A bar chart that shows the distribution of data; for example, the number of defects found per month, cost of different contractors, or the number of laptops in each region.

historical information Information the project may use from previous projects.

hygiene seekers An element of Herzberg's Motivation-Hygiene Theory that employees are motivated or demotivated by effects within an organization. A hygiene seeker takes comfort in salary, management, and job security.

implementation tracking As tasks are completed on time or over time, the number of time units used can accurately display the impact on dependent tasks within the project.

indirect costs Costs attributed to the cost of doing business. Examples include utilities, office space, and other overhead costs.

inflexible constraints Constraints that have date values associated with them but are very rigid. Constraints that are inflexible require that activities happen on a specific date. Use these constraints very sparingly.

influence diagram A diagram that charts out a decision situation. It identifies all of the elements, variables, decisions, and objectives and how each factor may influence another.

informal acceptance The informal acceptance does not include a sign-off of the completion or even the acknowledgment of the deliverables. An example is a project that ends on deadline whether or not the deliverables are finished—such as a project to organize and build an application for a tradeshow demo. The tradeshow will happen regardless of the completion of the project.

initiating The process group that begins the project. The business needs are identified, and a product description is created. The project charter is written, and the project manager is selected.

integrated change control A project-wide method of researching and documenting the effect of change on the entire project. It reviews the effect of change in scope, schedule, costs, quality, human resources, communication, and procurement. When a scope change request is being considered, there should be communication between the project manager and the appropriate stakeholders.

integration phase The phase of the project when the project plan is put into action.

intellectual property Ownership of an approach, system, or idea that an individual or organization retains; examples include the approach taken in a software development project, software code, manuscripts, and inventions.

internal audience The stakeholders targeted within an organization that will utilize, see, support, or otherwise interact with a project solution. These individuals can see the solution from the customers' perspective, notice problems or issues before the customers, or otherwise support the solution the project creates.

interview questions, closed Questions that must be answered with specific information. For example, "Have you ever created a batch file?" Or "What's your birthday?"

interview questions, essay Questions that allow the candidate to tell the project manager information, and allow the project manager to listen and observe. For example, "Why are you interested in working on this project?"

interview questions, experience Questions that allow the project manager to see how a candidate has acted in past situations to predict how they may act in future situations that are similar. For example, "How did you react when a teammate did not complete a task on a past project and you had to do their work for them in order to complete your own? How was the situation resolved?"

interview questions, reactionary Questions that evolve from the candidate's answers. When the project manager notices a gap or an inconsistency in an answer, use a follow-up question that focuses on the inconsistency without directly calling it a lie. This gives the candidate the opportunity to provide a better explanation or leaves the candidate floundering for an explanation. Reactionary questions also allow the project manager to learn more information that may be helpful on the project. For example, "You mentioned you had experience with Visual Basic. Do you also have a grasp on VBScript?"

interviewing Interviewing subject matter experts and project stakeholders is an approach to identify project requirements and risks on the current project based on the interviewees' experience.

invitation for bid (IFB) A document from the buyer to the seller that requests the seller to provide a price for the procured product or service.

Ishikawa diagram *See* fishbone/Ishikawa diagram.

ISO 9000 An international standard that helps organizations follow their own quality procedures. ISO 9000 is not a quality system, but a method of following procedures created by an organization.

issue log All issues are documented in the issue log. The issue log requires the issue definition, date discovered, an assigned issue owner, updates, a deadline for resolution, and the actual date the issue was resolved.

issues Decisions that have yet to be made, in which there may be opposing sides and opinions. Issues are recorded in the issue log, where the issue owner is recorded and a date for resolution is determined.

IT project management The ability to balance the love for and implementation of technology while leading and inspiring the team members. This domain of project management focuses on technical solutions and includes the waterfall and agile approaches to completing projects.

job description Details the activities of a team role, the scope of the position, the responsibilities, and the working requirements of the team member who fills the role. A job description should be clear, concise, and easily summed up.

joint application development/joint application review sessions A project management approach that heavily involves the customer or end user in the software development approach through a series of workshops. The customers review the work completed and offer input and suggestions to the application development team.

Kanban Kanban means signboard and shows each task, usually on a sticky note, as it moves through each phase of the project to completion. This approach is often used in agile project management.

key performance indicators (KPIs) Objectives of the project; common KPIs are time, cost, and scope. These elements are measured for project performance and provide an overall picture of how well the project is performing.

lag The scheduled time between project tasks, or the amount of time a project task must wait before it can begin.

law of diminishing returns A law of economics, sometimes called the law of variable proportions, which grew from Thomas Malthus's *An Essay on the Principle of Population,* published in 1798. The law states that if one factor of production is increased while other factors remain constant, the overall returns will eventually decrease after a certain point.

lead The negative time added to a task to bring it closer to the project start date. The lead is calculated by subtracting time between activities.

lessons learned An ongoing documentation of things the project manager and project team have learned throughout the project. Lessons learned are supplied to other project teams and project managers to apply to their ongoing projects. Lessons learned are documented throughout the project, not just at the end of the project.

licensing, per connection A license is required for each workstation-to-server connection. This scheme allows a maximum number of connections to a server.

licensing, per station A license that covers the software application at the workstation where it is installed. Think of Microsoft Office installed on each workstation within an organization.

licensing, per station (server-based) This licensing method allows an unlimited number of connections to a server covered by the licensing plan. Each additional server would require its own licensing to allow connections to that server.

licensing, per usage This licensing plan allows a user to run an application for a preset number of days or a preset number of times, or charges the user a fee for each instance that the application is used.

macro project A project that takes more than 2,000 hours of implementation and/or more than $250,000 to complete. Thirty percent of the project time is allotted to project planning.

management by projects This approach characterizes organizations that manage their operations as projects. These project-centric entities can manage any level of their work as a project. They apply general business skills to each project to determine its value, efficiency, and, ultimately, return on investment.

management by walking around A method to manage quality and to allow yourself to be seen. Get out of the office and get into the working environment. The project manager doesn't have to hover around the team members, but let them know the project manager is available, present, and interested in their work.

management reserve An artificial task that is added at the end of the project. The time allotted to the reserve is typically 10 to 15 percent of the total amount of time to complete all the tasks in a project. When a task runs over its allotted time, the overrun is applied to the management reserve at the end of the critical path rather than to each lagging task.

management summary reports Detail the overall status of the project, changes from the original plan, change in execution, or cost variances within the budget. These reports are created on an as-needed basis and are ideal for upper management.

managerial constraints Dependency relationships imposed because of a decision by management, which includes the project manager.

mandatory dependency The logical relationship between activities based on the type of work. For example, the foundation of a house must be created before the frame of the house can be built. This is also known as hard logic.

Maslow's hierarchy of needs A theory that states that there are five layers of needs for all humans: physiological, safety, social, esteem, and, the crowning jewel, self-actualization.

master service agreement An agreement between two parties that serves as a framework of what the overall terms and conditions will be. This allows the parties to get to work and quickly create contracts for future endeavors under the master service agreement. It is a type of umbrella agreement that provides the terms and conditions for future work.

matrix structures A type of organizational structure. There are three matrix structures: weak, balanced, and strong. The different structures are reflective of the project manager's authority in relation to the functional manager's authority.

McClelland's Theory of Needs David McClelland developed his acquired-needs theory based on his belief that a person's needs are acquired and develop over time. People's needs are shaped by life experiences and circumstances. This theory is also known as the Three Needs Theory, because there are just three needs for each individual and one need is considered the driving motivation behind the actions people take. Depending on the person's experiences, the order and magnitude of each need shifts the need for achievement, the need for affiliation, and the need for power.

McGregor's Theory X and Theory Y This theory states that X people are lazy, don't want to work, and need to be micromanaged, whereas Y people are self-led, motivated, and strive to succeed.

medium As part of the communications model, this is the path the message takes from the sender to the receiver. The modality in which the communication travels typically refers to an electronic model, such as e-mail or a cell phone.

meeting coordinator An individual who runs the business of a meeting to keep the topics on schedule and according to the agenda.

meeting minutes A document that represents a record of a meeting, the problems and situations that were discussed, and documentation of the project as it progresses. Meeting minutes are an excellent method for keeping the team aware of what has already been discussed and settled, resolutions of problems, and proof of the attendees in the meeting.

metrics A standard of project measurement; often applied to cost, schedule, scope, quality, and performance.

micro project A project that takes fewer than 2,000 hours of implementation and/or less than $250,000 to complete. Approximately 25 percent of the project time is allotted to planning the project.

micromanage The negative approach to managing a subordinate's work in a meddlesome, counterproductive manner.

milestones Represent the completion of significant tasks within a project's schedule.

minimally viable product (MVP) A software product that has the smallest amount of features to satisfy the customer requirements, for use in production. MVP answers what is the fastest route to satisfy the requirements of the project for the customers to receive business value from the solution.

mitigation Reducing the probability or impact of a risk.

Monte Carlo analysis Process that predicts how scenarios may work out given any number of variables. It doesn't actually create a specific answer, but offers a range of possible ones. When Monte Carlo is applied to a schedule, it can present, for example, the optimistic completion date, the pessimistic completion date, and the most likely completion date for each activity in the project.

motivation seekers An element of Herzberg's Motivation-Hygiene Theory that states that employees are motivated or demotivated by effects within an organization. The effects a motivation seeker takes comfort in are achievement, recognition, the work, responsibility, and advancement.

Motivation-Hygiene Theory *See* Herzberg's Motivation-Hygiene Theory.

multifactor authentication Requires additional security measures, such as texting or e-mailing the user a one-time use code, facial recognition, or other challenges to prove their authenticity as a legitimate user.

must finish on (MFO) constraint An inflexible constraint that requires that a deadline-oriented task be completed on a specific date.

must start on (MSO) constraint An inflexible constraint that requires that a task begin on a specific date.

national security information As its name implies, this is secretive information governed by the country where the software is developed or released that may not be shared with certain parties or users. For example, U.S. government projects often include requirements for data security, security clearance, and audits of how the data in the software is stored, accessed, and continually secured.

need-to-know basis Not all information in a project is available to all users, such as pay rates, contract fees, human resource records, and other information. There is an ethical requirement of IT users to not access certain data, just because they may have the technical authority to access the data.

nondisclosure agreements (NDAs) Often part of the contractual terms between the buyer and seller, NDAs forbid the seller from discussing the details of the contracted work with others outside of the project or procurement engagement.

nonverbal communication Facial expressions, hand gestures, and body language that contribute to the message. Approximately 55 percent of communication is nonverbal.

norming A stage of team developement. Once control on the project team has been established, the project team's focus shifts toward the project work. This is where people learn to work together.

on-premises solutions *See* cloud-based solutions vs. on-premises solutions.

operational definitions The quantifiable terms and values used to measure a process, activity, or work result. Operational definitions are also known as metrics.

operational expenses (OpEx) The normal day-to-day costs of doing business, such as rent, leases, payroll, utilities, and other business expenses. These are sometimes called indirect project expenses as they can't be assigned directly to the cost of completing a project.

operational handoff At various stages of a project, the completed portions of the project scope leave the project and are entered into production. In some projects, operational handoff may happen throughout the project, while other projects may not have the operational transfer until all project scope components are completed.

operations management Managers of the core business area such as design, manufacturing, and product development. Operations managers oversee and direct the salable goods and services of the organization.

organizational constraints Within the organization, there may be multiple projects that are loosely related. The completion of another project may be a key milestone for your own project to continue.

organizational process assets Historical information, templates, software, and other project management support materials that are provided to the project manager to help plan and manage the current project.

Ouchi's Theory Z Posits that workers are motivated by a sense of commitment, opportunity, and advancement. Workers will work if they are challenged and motivated.

paired programming Used in XP and calls for one person to program and the other person to watch the development for errors. The two programmers change roles periodically.

paralingual Relating to the pitch, tone, and inflections in the sender's voice that affect the message being sent.

parametric modeling A mathematical model based on known parameters used to predict the cost and time required for a project. The parameters in the model can vary based on the type of work being done. A parameter can be the cost per cubic yard, time per unit, and so on.

Pareto diagram A diagram that illustrates problems by assigned cause, from largest to smallest, in a histogram and is related to Pareto's Law: 80 percent of the problems come from 20 percent of the issues (this is also known as the 80/20 principle).

Parkinson's Law Work expands so as to fill the time available for its completion.

participative decision-making process In this ideal model, all team members contribute to the discussion and decision process. Through compromise, experience, and brainstorming, the project team and the project manager can create a buzz of energy, excitement, and synergy to arrive at the best possible solution for a decision.

peer review The process of allowing team members to review each other's work.

performance testing Software development projects are required to test the solution to see how well the code works in a simulated production environment. These tests can evaluate the performance of the network, hardware, software,, and data to predict how well the actual production environment will interact with the solution, which can set expectations for the software performance or show that additional work is needed to meet performance requirements.

performing The project team has settled into their roles and they focus on completing the project work as a team. During this stage of team development, a synergy is developed; this is the stage at which high-performance teams come into play.

personal health information A term commonly interchanged with protected health information. *See* protected health information (PHI).

personally identifiable information (PII) Any data that can be used to personally identify a specific individual. This classification is used in data security to protect what should be personal and private information from unethical hackers, users, or organizations. Examples include Social Security numbers, phone numbers, addresses, or photos.

PERT chart A Program Evaluation and Review Technique chart can graphically illustrate tasks, their durations, and their dependency on other tasks in the work unit.

PERT estimate The Program Evaluation and Review Technique is ideal for time and cost estimates. PERT uses a weighted average to predict how long the activity may take. PERT uses the formula of "pessimistic plus the optimistic, plus four times the most likely, divided by six." It's divided by six because of one count for pessimistic, one count for optimistic, and four counts for most likely.

phase A portion of the project that typically must be completed before the next phase can begin. Each phase has a set deadline.

phase gate estimating Dividing a project into phases and extracting cost estimates for each phase of the project. This cost approach helps the project team get to work on immediate deliverables as they work toward milestones at the end of each phase. The immediate actions of a project should be foreseeable, as opposed to actions that will happen way off in the future.

phase gate review In a waterfall project, a review of the project performance, product performance, and financial performance that is conducted before allowing the project to move forward. If the project passes the defined reviews, it is allowed to move into the next phase of the project. If the project doesn't pass the review, corrections to the project are needed. If project performance doesn't meet a threshold, it's possible the project could be cancelled.

physical resources Equipment, tools, technology, and facilities. Resource management includes human resources and physical resources.

pilot team A collection of users who agree to test the project deliverables before the rest of the organization sees the implementation. Their input to the project allows the project manager to realize if the project deliverables are on target or not.

planned value (PV) The worth of the work that should be completed by a specific time in the project schedule.

planning group This process group is iterative. All planning throughout the project is handled within the planning process group.

PMBOK Guide The book, *A Guide to the Project Management Body of Knowledge*, which includes all knowledge and practices within the endeavor of project management.

points of escalation The project manager may have limited authority over some project decisions that exceed a dollar or schedule variance and must be escalated to management or a project steering committee. Escalation points can be assigned to risk, issues, human resources, and other project factors.

portfolio review board This group of stakeholders is responsible for determining which projects are worthy of the company's capital. The board defines the governance of projects and programs within an organization and oversees the selection of the projects based on a number of factors such as return on investment, project value, risk to reward of proposed projects, and predicted financial outcomes of launching a new project.

post-implementation support Once the project is completed there may be a post-implementation warranty or service period where members of the project will continue to support the solution for the customer.

postmortem Also referred to as post-project audit. *See* post-project audit.

post-project audit The purpose of this audit is to analyze the completed project, the effectiveness of the project team, the success of the project, the value of the deliverables, and the overall approval from the clients.

precedence diagramming method (PDM) This method requires the project manager to evaluate each work unit and determine which tasks are its successors and which tasks are its predecessors to create the PND.

predetermined client When one piece of software must interact with another piece of software, such as a web browser calling for an e-mail application, the target e-mail application is the predetermined client. The web browser needs to know which specific software it should open, such as Microsoft Outlook or Google Gmail. Some solutions, for example Microsoft Explorer, will specifically call the Outlook client.

prequalified vendor Organizations may have a preferred vendors list of prequalified vendors the project managers can choose from for contract work in the project.

problem management meeting A meeting to resolve problems as they arise on a project.

problem solving This conflict resolution approach confronts the problem head-on and is the preferred method of conflict resolution. Problem solving (aka confronting) calls for additional research to find the best solution for the problem. Problem solving is a win-win solution. It can be used if there is time to work through and resolve the issue, and it works to build relationships and trust.

process boundaries Identify where a process begins and where a process stops. By understanding the process boundaries, the project manager can document what things are needed for a particular process to begin and what conditions must be true for that process to stop.

process configuration identification The project manager identifies all of the components within the process. The project manager documents how a process is completed, what the process interfaces are, and what each process in the workflow accomplishes.

process groups The five process groups—initiating, planning, executing, controlling, and closing—make up projects and project phases. These five process groups have sets of actions that move the project forward toward completion.

process metrics The project manager can measure a process's speed, cost, efficiency, throughput, or whatever metric is most appropriate for the type of work the process is participating in.

procurement The process of a buyer soliciting, selecting, and paying for products or services from a seller.

procurement audit The successes and failures within the procurement processes are reviewed from procurement planning through contract administration. The intent of the audit is to learn from what worked and what did not work during the procurement processes.

procurement management plan A subsidiary project plan that documents the decisions made in the procurement planning processes. It specifies how the remaining procurement activities will be managed.

product acceptance criteria The project scope defines either directly or by reference the technical requirements, the expected deliverables, and/or the detailed design documents that constitute the product deliverables. The product acceptance criteria clearly define what the project must create in order for the project to be accepted by the customer and for the project to be considered completed.

product manager An organizational role that is responsible for a product, such as a software product. The product manager is responsible for the selection of software and its requirements and oversees future releases of the product to the organization.

product owner A scrum project management role that oversees the features and functions of a solution, prioritizes the product backlog, and approves or declines changes to the list of features in the product backlog. The product owner is available to work with the development team to answer questions and clarify requirements for the product.

product scope The attributes and characteristics of the deliverables the project is creating.

program A collection of related projects working in alignment to realize benefits that the organization could not realize if the projects were managed independently of one another.

Program Evaluation and Review Technique (PERT) A scheduling tool that uses a weighted average formula to predict the length of activities and the project. Specifically, the PERT formula is $(O + 4ML + P)/6$.

program manager Oversees all of the orchestrated projects in their program. If a project is operating within a program, then the program manager is a stakeholder.

progress reports Reports that provide current information on the project work completed to date.

progressive elaboration The process of providing or discovering greater levels of detail as the project moves toward completion.

project A temporary endeavor undertaken to create a unique product or service.

project acceptance agreement A document that is typically written very early in the project timeline and in alignment with the defined project deliverables. The document is a clearly written explanation of what qualifies for an acceptance of the deliverables. These agreements are common in application development projects and often consist of a checklist of the required features of the project.

project calendar A calendar that defines the working times for the project. For example, a project may require the project team to work nights and weekends so as not to disturb the ongoing operations of the organization during normal working hours. In addition, the project calendar accounts for holidays, work hours, and work shifts the project will cover.

Project Change Request form Formalizes requests from anyone to the project manager. It requires the requestor not only to describe the change but also to supply a reason why this change is appropriate and needed. Once the requestor has completed the form, the project manager can determine whether the change is indeed necessary, should be rejected, or should be delayed until the completion of the current project.

project charter Similar to the project goal, but more official, more detailed, and in line with the organization's vision and goals.

project closeout report Often used in predictive projects to document the successes and failures of the project and to provide final information on the cost, schedule, and other key performance indicators of the project.

project closure meeting A final meeting to discuss the project performance and its accomplishments.

project closure phase A phase of project management that requires proof of the project deliverables, approval from management, and satisfaction from the customers or end users.

project constraints Anything that limits the project manager's options. Predetermined budgets, deadlines, resources, preferred vendors, and required technology are all examples of constraints. Project management always has three constraints: time, cost, and scope—sometimes called the Triple Constraints of Project Management. The project manager should identify and document all of the known constraints.

project control phase This phase of project management is a continuous cycle to oversee the project. It allows the project manager to manage task reporting, team meetings, reassignment of resources, change, and quality through software, communications, and the project team.

project coordinator A role that has limited authority over the project, works with the functional manager, and may serve under a more experienced project manager. The project manager is often called a project coordinator in a functional environment.

project dashboard Like a car's dashboard, a project dashboard visualizes the current project performance. It can automate controls, reports, and communications. Common components of the project dashboard include cost and schedule performance, burndown charts, and status of each project task.

project deliverables The end result of the project. They are what the project produces.

project exclusions The project scope statement must define the boundaries of the project to communicate what will not be included in the project deliverable. It's important to define what's excluded so that there's no confusion when the project manager wants to close the project and the project customers are expecting more deliverables.

project execution Consists of the project team completing the project work according to the project plan.

project genesis The origin of the project; a reaction to a need or an idea to improve operations within an organization. This realization of an opportunity to fulfill a need is the concept of the project.

project goal The clearly stated result a project should meet or deliver.

project information center A centralized room that contains a collection of all materials related to the project. The size of the project and the available real estate within an office building will determine the ability to create a project information center. This could also be cloud-based to host project information.

project kickoff A meeting or an event to introduce the project, the management backing the project, the project manager, and the team members. It should be casual but organized and used as a mechanism to assign ownership of the project to the team.

project life cycle The project life cycle is unique to each project and is composed of the unique phases of a project.

project management information system (PMIS) Typically a computer program that assists with project management activities, recordkeeping, and forecasting.

project management life cycle The project management life cycle is composed of five process groups: initiating, planning, executing, monitoring and controlling, and closing.

project management office (PMO) Centralizes and coordinates the management of projects within an organization, line of business, or department. PMOs can vary by organization, though most offer project management support, guidance, and direction for projects within their business domain. It's not unusual for a PMO to direct the actual project management of a project.

project manager The individual accountable for all aspects of a project. The project manager is a stakeholder for the project and is responsible for developing the project plans, keeping the project on track, monitoring and controlling the project, and communicating the project status and performance.

project network diagram (PND) A fluid mapping of the work to be completed. PNDs allow the project manager and the project team to tinker with the relationships between tasks and create alternative solutions to increase productivity, profitability, and the diligence of a project.

project planning phase The cornerstone of a successful project. The project manager and project team must identify the required activities and estimate the time required to complete the activities in order to reach the project goal.

project resources Employees, contractors, or equipment used on a project.

project scope The defined range of deliverables a project will produce. The project scope is concerned with the work, and only the required work, necessary to complete the project.

project scope statement Defines all of the deliverables the project will create, the boundaries of the project, and the work that the project team will need to complete in order to create the project deliverables. This document is based on the project requirements, the feasibility study, the business goals and objectives, and the business case document.

project sponsor A person in the organization who has the authority to grant the project manager power over the project resources, assign a project budget, and support the presence of the project. This person also signs the project charter to launch the project officially and assign the project manager to the project.

project status report Often a weekly report on how the project is performing on key performance indicators, accomplishments in the project, and goals for the project work over the next week. The report is distributed to key project stakeholders, such as management, sponsors, the project steering committee, and customers.

project steering committee A group of managers, executives, customers, and other stakeholders that oversees project decisions. Some decisions, such as costs and schedule changes, are escalated to the project steering committee.

project team People who work on planning and executing the project plan. Depending on the organization and the staffing management plan, the project team may work full time or part time on the project. Team members may intermittently work on the project work as the plan warrants or be assigned to the project manager for the duration of the project.

project vision In project management terms, the ability to see the project deliverables clearly and recognize the actions required to produce them.

projectized structure An organizational structure in which the project manager has the greatest amount of authority. The project team is assigned to the project on a full-time basis. When the project is complete, the project team members move on to other assignments within the organization.

Projects IN Controlled Environments (PRINCE2) Often used in the UK and Europe, PRINCE2 is a structured project management approach that segments the project work into defined stages. PRINCE2, like the Project Management Professional (PMP), is a certification program.

protected health information (PHI) Addresses privacy concerns for medical patients. The Health Insurance Portability and Accountability Act (HIPAA) Privacy Rule provides patient privacy rights that patient health information cannot be shared without the patient's consent.

purchase order A unilateral form of a contract where the purchasing organization and the vendor have an agreement for the rate and cost of materials and services.

purpose statement A statement indicating why the research was initiated and reflecting the proposed project.

qualitative risk analysis An examination and prioritization of the risks based on their probability of occurring and their impact on the project if they do occur. Qualitative risk analysis guides the risk reaction process.

quality assurance (QA) A process in which the overall project performance is evaluated to ensure the project meets the relevant quality standards.

quality audit A process used to confirm that the quality processes are performing correctly on the current project. The quality audit determines how to make things better for the project and other projects within the organization. Quality audits measure the project's ability to maintain the expected level of quality.

quality control (QC) An inspection-driven process in which the work results are monitored to see if they meet relevant quality standards.

quality management plan Document that describes how the project manager and the project team will fulfill the quality policy. In an ISO 9000 environment, the quality management plan is referred to as the project quality system.

quality policy The formal policy an organization follows to achieve a preset standard of quality. The quality policy of the organization may follow a formal approach, such as ISO 9000, Six Sigma, or total quality management (TQM), or it may have its own direction and approach. The project team should either adapt the quality policy of the organization to guide the project implementation or create its own policy if one does not exist within the performing organization.

quantitative risk analysis A numerical assessment of the probability and impact of the identified risks. Quantitative risk analysis also creates an overall risk score for the project.

quote (or quotation) A document from the seller to the buyer in response to the buyer's request for quote (RFQ). Quotes are used when price is the determining factor in the decision-making process.

RACI chart A chart that uses the legend of Responsible, Accountable, Consult, and Inform: the first letter of each responsibility spells RACI. This is a type of responsibility assignment matrix (RAM).

real-time, multi-authoring editing software A software solution that allows multiple people to edit and view the project work in real-time. The software provides for live collaboration by multiple people as the solution is being created.

receiver As part of the communications model, the receiver is the recipient of the message.

referent power Power that is present when the project team is attracted to, or wants to work on the project with, the project manager. Referent power also exists when the project manager references another, more powerful person, such as the CEO.

regression planning When a change is approved for the project there may also be a regression, aka rollback or fallback, plan to define how to undo the change if the change creates risks or issues within the project.

regression testing After completing a change to software, a regression test confirms the added change didn't break the existing code in the solution.

request for bid (RFB) Like an invitation for bid, this is a document sent from the buyer to the seller with specific requirements on the product or service to be purchased. The buyer only wants a price for the product or services the vendor is selling.

request for information (RFI) A formal request from the buyer to the seller for additional information about goods and services the seller may provide. It's not an official sign of business to come but is often a starting point of research to start the procurement processes.

request for proposal (RFP) A formal request from an organization to a prospective vendor or vendors to create a proposal for the work to be completed and provide a cost estimate. An RFP does not guarantee anyone the job; it simply formalizes the proceedings of the selection process.

request for quote (RFQ) A document from the buyer to the seller asking the seller to provide a price for the procured product or service.

requirements traceability matrix (RTM) A table that tracks each requirement through each phase of the project's stages to completion and implementation.

residual risks Risks that are left over after mitigation, transference, and avoidance. These generally are accepted risks. Management may elect to add contingency costs and time to account for the residual risks within the project.

resource calendar Shows when resources, such as project team members, consultants, and subject matter experts, are available to work on the project. The resource calendar takes into account vacations, other commitments within the organization, restrictions on contracted work, overtime issues, and so on.

resource constraint Describes a situation where a project may need additional resources but doesn't have access to the resources—for example, a particular type of equipment, a particular project team member, or a facility such as a training room.

resource histogram A bar chart reflecting when individual employees, groups, or communities are involved in a project. It is often used by management to see when employees are most or least active in a project.

resource leveling heuristics A method to flatten the schedule when resources are overallocated or allocated unevenly. Resource leveling can be applied in different methods to accomplish different goals. One of the most common methods is to ensure that workers are not overextended on activities.

resource pool An organization's potential team members that can join a project based on their skillsets, availability, and experience.

resources Resources can be both workers and physical objects such as bandwidth, faster computers, or leased equipment.

retrospective The final meeting or ceremony in a scrum iteration that allows the product owner, scrum master, and development team to discuss the success and failures of the last iteration and how the team can work more effectively and efficiently in the next iteration of the project.

return on investment (ROI) The attitude that IT projects are not an expense but an investment that will allow an organization to become more profitable.

reward power The project manager's authority to reward the project team.

risk An unplanned event that can have a positive or negative influence on the project's success.

risk database A database of recognized risks. The planned response and the outcome of the risk should be documented and recorded in an organization-wide risk database. The risk database can serve other project managers as historical information. Over time, the risk database can become a risk lessons learned program.

risk management plan A subsidiary project plan for determining how risks will be identified, how quantitative and qualitative analyses will be completed, how risk response planning will happen, how risks will be monitored, and how ongoing risk management activities will occur throughout the project life cycle.

risk register A document that defines each risk event, its characteristics, signs that the risk may be occurring, any responses for the risk event, and the risk's eventual outcome in the project.

risk report A report that communicates pending risks, risks that have occurred, risk responses, and the effectiveness of the risk response.

roles and responsibilities matrix A method to identify all of the roles within a project and the associated responsibilities to the project work. This matrix is an excellent way to identify the needed roles of the project participants, identify what actions they'll need to be able to perform in the project, and ultimately determine if the project has all of the roles to complete the identified responsibilities.

root cause analysis A study of the effect that's being experienced with a determination of the causal factors of the effect. This is one of the purest forms of analysis, and the results are often graphed in a cause-and-effect diagram. The project manager will also use this approach in quality control.

rough order of magnitude This estimate is "rough" and is used during the initiating processes and in top-down estimates. The range of variance for the estimate can be –25 percent to +75 percent.

run chart A line chart of data plotted over time. Run charts can be used to show funds spent, tasks completed, quality control inspection scores, or performance on other key performance indicators such as time or costs.

runaway project A project that starts out well and then gains speed, momentum, and scope, causing budget overages and increased man-hours, and possibly hurting the project manager's reputation or career. The biggest element of a runaway project is the budget.

Scaled Agile Framework (SAFe) An agile project management approach that has specific activities, workflows, meetings, and approval mechanisms. It is an approach, like scrum or eXtreme Programming.

scales of probability and impact Each risk is assessed according to its likelihood and its impact. There are two approaches to ranking risks: Cardinal scales identify the probability and impact by a numerical value, ranging from 0.01 as very low to 1.0 as certain. Ordinal scales identify and rank the risks from very high to very unlikely.

scatter chart A diagram that tracks the relationship between two variables within a project. The closer the two variables track together on the scatter diagram, the more likely the variables are related and can influence each other (for example, maintenance on a piece of equipment and improved quality in outputs from the maintained equipment).

schedule management plan A subsidiary plan that details how the schedule will be created and how changes to the schedule may be allowed. This plan also defines how the actual changes themselves will be managed and how the changes may affect other areas of the project.

schedule performance index (SPI) Reveals the efficiency of work. The closer the quotient is to 1, the better: SPI = earned value/planned value.

schedule variance Describes the difference between the earned value of the project work and the planned value of the project work. Schedule variances often prompt the project manager to create a schedule variance report explaining the variance in the project. The formula for schedule variance is earned value – planned value.

scheduler A special project management role. This person works with the project team, the project manager, and management to create, monitor, and control the project schedule. The team will meet regularly with the scheduler to provide updates on project work and progress.

scope creep A process that happens when a project manager allows small changes to enter into the project scope without formal approval. Eventually the scope of the project swells to include more deliverables than the project budget or team is able to deal with.

scope management plan A plan that details how the project scope will be created, how the WBS will be defined, how the scope will be protected from change, where changes to the scope may be permitted, and how the management of approved changes will be handled.

scope statement A document that describes the work, and only the required work, to meet the project objectives. The scope statement establishes a common vision among the project stakeholders to establish the point and purpose of the project work. It is used as a baseline against which all future project decisions are made to determine if proposed changes or work results are aligned with expectations.

scope verification The process of the project customer accepting the project deliverables. Scope verification happens at the end of each project phase and at the end of the project. Scope verification is the process of ensuring the deliverables the project creates are in alignment with the project scope.

scrum A type of agile project management. Scrum uses a prioritized product backlog of requirements to determine which requirements will be created during the next sprint of the project work. Scrum has special rules, ceremonies, and resources that traditional, predictive project management does not use.

scrum master An agile project role that gets the team the information and supplies they need in the project, confirms everyone in the project is following the agile rules, protects the team from interruptions, and works to remove obstacles or blockers from the team's progress.

secondary risks Risks that stem from risk responses. For example, the response of transference may call for hiring a third party to manage an identified risk. A secondary risk caused by the solution is the failure of the third party to complete its assignment as scheduled. Secondary risks must be identified, analyzed, and planned for just as any identified risk must be.

semiflexible constraints Constraints that have a date value associated with them but require that the task begin or end by the specified date. Use these constraints sparingly.

sender As part of the communications model, the sender is the person or group delivering the message to the receiver.

sharing A positive risk response that allows the project team to partner or team with another entity to realize an opportunity that the team may not have been able to realize on their own. Sharing examples are teaming agreements, joint ventures, partnerships, and special-purpose companies.

short message service (SMS) Text messages shared over mobile phones, tablets, or other telephone devices. These messages aren't encrypted.

should-cost estimates Estimates created by the performing organization to predict what the cost of the procured product should be. If there is a significant difference between what the organization has predicted and what the sellers have proposed, either the statement of work was inadequate or the sellers have misunderstood the requirements. Should-cost estimates are also known as independent estimates.

simulation An exercise that allows the project team to play "what-if" games without affecting any areas of production.

single source A specific seller that the performing organization prefers to contract with.

smoke testing Preliminary testing to confirm that the major components and functions of a software solution are working. Smoke testing is done after a software build and before the software is released into production.

smoothing A conflict resolution method that smooths out the conflict by minimizing the perceived size of the problem. It is a temporary solution, but it can calm team relations and temper boisterous discussions. Smoothing may be acceptable when time is of the essence or any of the proposed solutions would work.

soft logic The preferred order of activities, aka discretionary dependency. Project managers should use these relationships at their discretion and document the reasoning behind making soft logic decisions. Discretionary dependencies allow activities to happen in a preferred order because of best practices, conditions unique to the project work, or external events.

software development life cycle (SDLC) A waterfall project management approach that defines the stages the software development project will follow. The stages are planning, analysis, design, implementation, and maintenance. This is a predictive project management approach.

sprint A component of agile project management. Sprints are executing phases in the agile project management approach and can last from two to four weeks in duration.

sprint review A demonstration of what the development team has created in the iteration for the project customers. Only completed items of the sprint can be demonstrated. The meeting allows the development team to hear feedback from the project customers.

staffing management plan A subsidiary project plan that determines how project team members will be brought on to and released from the project. This plan is useful if the project does not require the project team members to be on the project for the duration of the project schedule.

stakeholder management strategy The project will likely have negative, neutral, and positive stakeholders. The stakeholder management strategy defines how the project manager will bolster support for the project, fend off the negative stakeholders, alleviate fears and concerns, and promote the support for the project. The project manager will use the stakeholder management strategy along with the project's communications management plan.

stakeholders The individuals, groups, and communities that have a vested interest in the outcome of a project. Examples include the project manager, the project team, the project sponsor, customers, clients, vendors, and communities.

standard costs The budget department may have preassigned standard costs for labor to do certain tasks like programming lines of code, installing hardware, or adding new servers. This preassignment of values helps the project manager estimate labor costs of a project easily and without having to justify each labor expense as a line item.

STAR method One of the best interview methods, especially when dealing with potential integrators. STAR is an acronym for situation, task, action, result.

start no earlier than (SNET) constraint This semiflexible constraint requires that a task begin no earlier than a specific date.

start no later than (SNLT) constraint This semiflexible constraint requires that a task begin by a specific date at the latest.

start-to-finish (SF) tasks These rare tasks require that the predecessor doesn't begin until the successor finishes. This relationship could be used with accounting incidents. For example, the predecessor task is to physically count all of the network jacks that have been installed—once they have been installed.

start-to-start (SS) tasks These tasks are usually closely related in nature and should be started, but not necessarily completed, at the same time. An example is planning for the physical implementation of a network and determining each network's IP addressing configurations. Each task is closely related and should be done in tandem with other tasks.

statement of work (SOW) A document that fully describes the work to be completed, the product to be supplied, or both. The SOW becomes part of the contract between the buyer and the seller. The SOW is typically created as part of the procurement planning process and is used by the seller to determine whether it can meet the project's requirements.

statistical sampling A process of choosing a percentage of results for inspection. Statistical sampling can reduce the costs of quality control.

status reports Reports that provide current information on the project cost, budget, scope, and other relevant information.

status review meetings Regularly scheduled meetings to record the status of the project work. These commonly employed meetings provide a formal avenue for the project manager to query the team on the status of its work, to record delays and slippage, and to forecast what work is about to begin.

storming A stage of team development marked by a struggle for project team control and momentum of who's going to lead the project team. It is during this phase that people figure out the hierarchy of the team and the informal roles of team members.

stress testing A software development test to see how much stress the solution can manage—often beyond the normal performance expectations the solution will experience in production.

subteams Project teams in a geographically diverse project. Each team on the project in each site is a subteam. On a larger project, subteams can be created for different disciplines used within the larger project.

SWOT analysis An approach that examines the project based on its strengths, weaknesses, opportunities, and threats to test where the project could fail and where the project could improve. SWOT is a great technique to use on resources the project managers have not used before or on requirements that the project manager and the project team haven't tried before.

synchronous communication Communication that happens in real-time, such as a phone call.

system or process flowcharts Show the relation between components and how the overall process works. They are useful for identifying risks between system components.

task list A list of the major steps required from the project's origin to its conclusion. A task list is created after the technology has been selected and before the project manager creates the implementation plan. The task list is based on the work packages within the work breakdown structure.

team leaders Each subteam has a team leader who reports directly to the project manager and oversees the activities of the team members on the subteam.

terms of reference (TOR) Provides the terms of how the project, team, or organization will operate and sets the framework and boundaries for the project. Terms of reference can be used for project teams, vendors, and organizations.

testers/quality assurance (QA) Software development role that applies performance and security tests to the software solution to confirm that the software is of quality, meets expectations, and is secure before the software may move forward to operational transfer to production.

three-point estimating An estimating approach that uses worst-case, most likely, and optimistic approaches to create an average cost or time duration for a project component.

tiered structure A structure in which the deliverables are released over multiple dates, allowing the IT project manager to agree to meet new requirements for the project scope.

time and material (T&M) billing method Most technology integrators like to bill time and materials because there may be some additional problem discovered in the midst of the project that can result in the vendor working extra hours to solve it. T&M contracts should have a not-to-exceed (NTE) clause to contain costs.

time-tracking tools Software and hardware solutions that track task completion time. Can be used for billing, payroll, and performance analysis.

to-complete performance index (TCPI) A formula used to forecast the likelihood of a project to achieve its goals based on what's happening in the project right now. There are two different formulas for the TCPI, depending on what the project manager wants to accomplish. To see if the project can meet the budget at completion, use this formula: TCPI = (BAC – EV)/(BAC – AC). To see if the project can meet the newly created estimate at completion, use this version of the formula: TCPI = (BAC – EV)/(EAC – AC).

top-down estimating A technique that bases the current project's estimate on the total cost of a similar project. A percentage of the similar project's total cost may be added to or subtracted from the total, depending on the size of the current project.

total budgeted costs The amount of dollars budgeted for the project prior to the start of the project implementation phase. This is the same as budget at completion (BAC).

total quality management (TQM) A process that has all employees within an organization working to fulfill their customers' needs while also working to increase productivity. TQM stems from Dr. W. Edwards Deming's management principles, which the Japanese adopted after WWII. In the United States, these principles were readily adopted in the 1980s after proof of their success in Japan.

total slack The total time an activity can be delayed without delaying project completion.

trade secrets An organization's private information that is related to a product the organization owns, such as the creation process, a recipe, or underlying code in a software solution.

transference A response to risks in which the responsibility and ownership of the risk is transferred to another party (for example, through insurance).

transition plan/release plan A plan that can help ease the transfer of the project deliverables from the project manager to the operations of the organization. This plan defines several things for both the project manager and the organization: transition dates, ownership, training, extended support, and warranties.

trend analysis Using past results to predict future performance.

triggers Warning signs or symptoms that a risk has occurred or is about to occur. An example is a vendor failing to complete their portion of the project as scheduled.

unit price The price of one unit of a product or service; for example, cost per laptop or cost per hour. This is used in parametric estimating to predict the price of a project.

unit testing Testing the smallest testable part of a software solution for conformance to requirements. Unit testing is usually performed by the software developers and testers.

upper management The chief executive officers of an organization, such as the CEO, CIO, and COO.

usability laboratory A place where the project team can test the technology prior to implementing it in production.

user acceptance testing (UAT) The final testing process before a software solution goes live in production. UAT aims to confirm that the final build and solution performs as expected according to the project requirements and product scope of the project.

utility function A person's willingness to accept risk.

value-added change A change that positively affects the project deliverables. The cost or schedule costs incurred by the change may be less than the value the change will contribute to the project or its deliverables.

variance The difference between what was planned and what was experienced; typically used for costs and schedules.

variance at completion (VAC) The difference between the budget at completion (BAC) and the estimate at completion (EAC); its formula is VAC = BAC – EAC.

velocity chart Shows the amount of user stories the agile project team has completed in each iteration over the project life. The chart can also show how much total work the team has completed over time and how much work remains in the product backlog to be completed.

version control tool A software solution that controls, tracks, and provides a record of changes to software solutions to ensure that all team members are working on the most current build of the code.

virtual team A team composed of geographically dispersed project team members. Communication demands increase when managing virtual teams.

Vroom's expectancy theory *See* expectancy theory.

war room A centralized office or locale for the project manager and the project team to work on the project. It can house information on the project, including documentation and support materials. It allows the project team to work in close proximity.

warranty period After the project has been completed, there is often a predefined period of time where the performing organization is responsible for their solution should there be any bugs, defects, or other performance issues.

waterfall A predictive project management approach where the stages of the project are predefined and the flow of the work moves from the top of the flow through each stage to the final stage of the project.

whiteboard A physical board or software to draw solutions, facilitate a conversation, or illustrate a key idea.

withdrawal A conflict resolution method that is used when the issue is not important or the project manager is outranked. The project manager pushes the issue aside for later resolution. It can also be used as a method for cooling down. The conflict is not resolved, and it is considered a yield-lose solution.

work authorization system A tool that can control the organization, sequence, and official authorization to begin a piece of the project work.

work breakdown structure (WBS) A deliverable-oriented collection of project components. Work that isn't in the WBS isn't in the project. The point of the WBS is to organize and define the project scope.

work breakdown structure (WBS) dictionary A reference tool to explain the WBS components, the nature of the work package, the assigned resources, and the time and billing estimates for each element.

work breakdown structure (WBS) template A master WBS that is used in organizations as a starting point in defining the work for a particular project. This approach is recommended, as most projects in an organization are similar in the project life cycles, and the approach can be adapted to fit a given project.

work unit A chunk of work that must be completed to ensure a phase ends on schedule so that the next phase can begin.

workaround An unplanned response to a risk that has occurred when there is no contingency plan in place.

workshops One-day or multiple-day events, often held offsite, that focus on gathering requirements, writing the project management plan, or writing user stories in an agile project management approach.

zero-based budgeting The budget to be created must always start at the number zero rather than factoring new expenses into a budget from a similar project.

INDEX

References to figures are in italics.

NUMBERS

A

D

E

I

J

K

L

M

N

O

P